Antimicrobial Resistance in Wastewater and Human Health

Antimicrobial Resistance in Wastewater and Human Health

Edited by

Dharm Pal (Associate Professor)
Department of Chemical Engineering, National Institute of Technology Raipur, Raipur, Chhattisgarh, India

Awanish Kumar (Associate Professor)
Department of Biotechnology, National Institute of Technology Raipur, Raipur, Chhattisgarh, India

ELSEVIER

ACADEMIC PRESS
An imprint of Elsevier
elsevier.com/books-and-journals

Academic Press is an imprint of Elsevier

125 London Wall, London EC2Y 5AS, United Kingdom
525 B Street, Suite 1650, San Diego, CA 92101, United States
50 Hampshire Street, 5th Floor, Cambridge, MA 02139, United States
The Boulevard, Langford Lane, Kidlington, Oxford OX5 1GB, United Kingdom

ISBN: 978-0-323-96124-0

For Information on all Academic Press publications visit our website at https://www.elsevier.com/books-and-journals

Publisher: Stacy Masucci
Acquisitions Editor: Elizabeth A. Brown
Editorial Project Manager: Sam Young
Production Project Manager: Sajana Devasi PK
Cover Designer: Christian J. Bilbow

Typeset by Aptara, New Delhi, India

Dedication from Dr. Dharm Pal

To My Grandpa
Thakur Jagdish Singh
The person I admire the most
&
To My Kids,
Simone, Sasa & Sheenu
My love, my life, my light

Dedication from Dr. Awanish Kumar

To
My wife and daughter
This book is affectionately dedicated.

We also dedicate this book to people impacted by the book and readers of the book.

Contents

Contributors

Manisha Agrawal, Rungta College of Engineering & Technology, Bhilai, Chhattisgarh, India

Muhammad Arif, Faculty of Pharmacy, Integral University, Dasauli, Kursi Road, Lucknow, Uttar Pradesh, India

Shahina Bano, Rungta College of Sciences & Technology, Chhattisgarh, India

Sonia Chadha, Amity University Uttar Pradesh, Lucknow Campus, Amity Institute of Biotechnology, Lucknow, India

Radha Chaube, Department of Zoology, Institute of Science, Banaras Hindu University, Varanasi, India

Niketa Chauhan, Amity Institute of Biotechnology, Amity University, Lucknow, Uttar Pradesh, India

Parmesh Kumar Chaudhari, Department of Chemical Engineering, National Institute of Technology Raipur, Raipur, Chhattisgarh, India

Nidhi Dewangan, School of Studies in Life Sciences, Pt. Ravishankar University, Raipur, Chhattisgarh, India

Sandeep Dharmadhikari, Department of Chemical Engineering, School of Studies of Engineering and Technology, Guru Ghasidas Vishwavidyalaya (A Central University), Bilaspur, Chhattisgarh, India

Syed Misbahul Hasan, Faculty of Pharmacy, Integral University, Dasauli, Kursi Road, Lucknow, Uttar Pradesh, India

Amit Jain, Department of Chemical Engineering, School of Studies of Engineering and Technology, Guru Ghasidas Vishwavidyalaya (A Central University), Bilaspur, Chhattisgarh, India

K.V.S.G. Murali Krishna, Civil Engineering Department, Jawaharlal Nehru Technological University, Kakinada, India

Arun Kumar, Faculty of Pharmacy, Integral University, Dasauli, Kursi Road, Lucknow, Uttar Pradesh, India

Awanish Kumar, Associate Professor, Department of Biotechnology, National Institute of Technology Raipur, Raipur, Chhattisgarh, India

Hemant Kumar, Department of Biotechnology, Govt. V.Y.T. P.G. Auto. College, Durg, Chhattisgarh, India

Shom Prakash Kushwaha, Faculty of Pharmacy, Integral University, Dasauli, Kursi Road, Lucknow, Uttar Pradesh, India

Sayali Mukherjee, Amity Institute of Biotechnology, Amity University, Lucknow, Uttar Pradesh, India

Kumud Nigam, Amity Institute of Biotechnology, Amity University, Lucknow, Uttar Pradesh, India

Dharm Pal, Associate Professor, Department of Chemical Engineering, National Institute of Technology Raipur, Raipur, Chhattisgarh, India

Suchitra Kumari Panigrahy, Department of Biotechnology, Guru GhasidasVishwavidyalaya (A Central University), Bilaspur, Chhattisgarh, India

Dhruti Sundar Pattanayak, Department of Chemical Engineering, National Institute of Technology Raipur, Raipur, Chhattisgarh, India

Nitin Pawar, Department of Chemical Engineering, Visvesvaraya National Institute of Technology, Nagpur, Maharashtra, India

A.D. Prasad, Civil Engineering Department, National Institute of Technology Raipur, Raipur, India

M. Narayana Rao, National Institute of Technical Teachers Training and Research, Chennai, India

Somali Sanyal, Amity Institute of Biotechnology, Amity University, Lucknow, Uttar Pradesh, India

Aradhana Sharma, Kalyan P.G. College, Bhilai Nagar, Chhattisgarh, India

Karuna Singh, Department of Zoology, Mahila Maha Vidyalaya, Banaras Hindu University, Varanasi, India

Prerna Tandon, Amity University Uttar Pradesh, Lucknow Campus, Amity Institute of Biotechnology, Lucknow, India

Chandrakant Thakur, Department of Chemical Engineering, National Institute of Technology Raipur, Raipur, Chhattisgarh, India

Rachana Tiwari, Rungta College of Sciences & Technology, Chhattisgarh, India

Munendra Mohan Varshney, Raj Kumar Goel Institute of Technology (Pharmacy), 5 Km Stone, Delhi-Meerut Road, Ghaziabad, Uttar Pradesh, India

Biographies

Dr. Dharm Pal, an Associate Professor of Chemical Engineering at NIT Raipur, has more than 14 years of teaching and research experience. Before joining NIT Raipur, he was an Assistant Professor at SLIET Longowal, Punjab. He is BTech from HBTI Kanpur, MTech from IIT Delhi, and PhD in Chemical Engineering from NIT Raipur.

Dr. Dharm Pal research work is mainly focused on developing smart (active & intelligent) food packaging bionanocomposites and edible films & coatings. His other expertise/interest includes reactive extraction, antimicrobial resistance, functional nanomaterials & nanocomposites, nanophotocatalysis & electrocoagulation. With h-index-18 & i10-index-31, Dr. Dharm Pal has over 100 research communications in various Journals & Conferences, including 51 SCI/Scopus indexed papers with >1000 citations. He is the author of four books & many book chapters, editor of two conference proceedings, and one special issue (Elsevier). He has organized various conferences/workshops at various capacities and delivered many expert lectures. He has received Auropath Global Awards 2019 for the "Excellence in Research." Dr. Dharm Pal is an active member of various professional societies such as IIChE, ISTE, IE (I), ICS, CRSI, OTAI, ASSET & ISRD, SESI.

Awanish Kumar (PhD), is graduated in molecular parasitology from CSIR-Central Drug Research Institute, Lucknow, India and Jawaharlal Nehru University, Delhi, India. He has done postdoctoral study from McGill University, Montreal, Canada. The major research area of Dr. Kumar is antimicrobial resistance, drug discover, and healthcare. Dr. Kumar has served various national and international organizations with different academic/research capacities. With h-index- 28, he has published more than 134 research papers in SCI journals (https://www.scopus.com/authid/detail.uri?authorId=18437170100). He is the author of several monographs/books/chapters. Currently, he is working as an Associate Professor in the Department of Biotechnology, National Institute of Technology,

Raipur (CG), India (http://www.nitrr.ac.in/viewdetails.php?q=bt.akumar) and giving his active/proper contribution in academic and research. Currently, he is serving many national committees, scientific society, and advisory panels. He is the member of many international professional research societies and reviewer/editorial board member of reputed and refereed journals. He has guided many PhD, MS, and BS students and such supervisions are going on.

Foreword

The development of antibiotic resistance is of significant global concern that has a direct impact on human health. The book entitled "*Antimicrobial Resistance in Wastewater and Human Health*" is written in order to highlight this growing concern. I congratulate the editors Drs. Dharm Pal and Awanish Kumar on a very insightful book. The chapters in this book highlight issues related to human health hazards due to the spread of antimicrobial resistance. It also complements previous studies. Importantly, this book identifies current knowledge gaps. The authors have provided cutting-edge information on antimicrobial resistance in wastewaters that interweave biotechnology and chemical engineering. Although the emphasis of this book is on wastewater and antimicrobial resistance the impact of waste treatment, genetic modification in wastewater flora, genetically engineered microorganisms to degrade waste and reuse of treated water on human health, etc., are discussed in detail. The authors have arranged the book into 12 well-conceived chapters that amply illustrate this area of tremendous societal impact. I am sure that the authors can be confident that there will be many grateful readers who will gain a broader perspective into the general area of antimicrobial resistance in wastewater and its significant effects on human health. This book provides fundamental scientific knowledge as well as practical information and suggestions for operators of wastewater treatment facilities.

Dr. Ranil Wickramasinghe
Professor, University of Arkansas, United States
May 30, 2022

Preface

Due to various humans, agriculture and aquaculture activity, animal husbandry work, and pharmaceutical manufacturing, high amount of antibiotics are released into the wastewater. It is mostly evident that due to the activities, antibiotics are ending up in wastewater that increases antibiotic resistivity of microbiota and causing a threat to global public health. This book is written with the intent to provide valuable information on antimicrobial resistance in wastewater and human health risks. Editors are working in the area of antibiotic resistance. The various matter discussed in this book collectively address the important and update material on this topic. This is an attempt to expand the perspective of the reader on the complex problem of antimicrobial resistance in wastewater and their risk on human population globally. Chapters and content of the book target the broad readership of academicians, college students, researchers, and investigators working in this particular area.

Acknowledgments

We are overwhelmed in all humbleness and gratefulness to acknowledge our depth gratitude to all those who have helped us to put these ideas, well above the level of simplicity and into something concrete.

At the outset, we would like to express our special thanks of gratitude to Prof. A.M. Rawani, Director of our organization "National Institute of Technology Raipur," all the colleagues, and research scholars who have helped directly or indirectly in completing this task.

Our heartfelt thanks to all the chapter contributors and publishing team of Elsevier.

Completion of this project could not have been accomplished without the support of our family members. I, Dr. Dharm Pal, especially thanks to my caring, loving, and supportive wife, Pinky: my deepest gratitude. I am eternally grateful to my father Shri C.D. Singh & my world's best mother Mrs. Lalkeswari Devi for their blessings and making me what I am today. Handling my naughty kids during my busy schedule is much appreciated and duly noted.

At last, we bow our head before God Almighty for his blessings in the successful & timely completion of this book.

Dr. Dharm Pal
Dr. Awanish Kumar

Introduction

The spread of bacterial antibiotic resistance through wastewater is an anxious matter and treating them in wastewater treatment plants has been gaining momentum and attracting interest among researchers. Our aqua ecology is the reservoir of resistant bacteria, resistance genes, its multidirectional flow, and amount of bacteria release into the environment is very dangerous for living beings. The reuse of treated wastewater for irrigation is a practical solution for surmounting the scarcity of water, but there are several human health-related and agricultural risks associated with it. Based on the current state-of-the-art in this field, the proposed book aims to provide an updated knowledge on the antimicrobial resistance of wastewater and human health risks with different chapters which target the broad readership of college student, academicians, researchers, and investigators working in this particular field.

As per the best of our knowledge, this is the first dedicated book on antimicrobial resistance in wastewater & related human health risks. The book covers major aspects of the proposed area. Easy to follow the flow between the chapters. Chapters are supplemented with ample illustrations and figures, wherever necessary. In this book, we have gathered the expertise of a wide variety of scientific disciplines, in an attempt to expand the perspective of the reader on the complex problem of antimicrobial resistance in wastewater and their risk on human. It is expected that, the collectively proposed book would be a fundamental & valuable guide that conveys the knowledge needed to understand.

This book mainly targets the readership of college student, scientists, as well as research investigators working in the field of antimicrobial resistance, wastewater treatment, and human health risk. People from biotechnology, chemical engineering, chemistry, waste treatment, and environmental science would be surely benefited with the content of this book. The book would be collective wholesome and up-to-date information on antimicrobial resistance in wastewater and human health risks. The various matter discussed in this book jointly address the important and updated material on antimicrobial resistance in wastewater and human health risks.

Chapter 1

Commonly found bacteria and drug-resistant gene in wastewater

Nidhi Dewangan
School of Studies in Life Sciences, Pt. Ravishankar University, Raipur, Chhattisgarh, India

1.1 Introduction

Wastewater serves as the hotspot for the growth of pathogenic microorganisms, with its rich nature and collection of organic and inorganic nutrients. It provides a perfect homage for the active mutation of genes and formation of "resistant gene," which can be easily passed between strains via "horizontal gene transfer (HGT)" and also to the progeny via "vertical gene transfer." There are certain species like *Escherichia coli, Enterococcus, Pseudomonas, Staphylococcus, Klebsiella*, and *Enterobacter*, which are of great interest as they are highly ubiquitous and have adapted several mechanisms to combat challenging environments very efficiently by constantly altering their genome to produce more and more pathogenic strains. These bacterial species have several highly "mutational hotspots" where they perform alterations in the genetic sequences required for their survival. Due to mutations in their genome, scientists find it difficult to keep their pathogenicity at bay. The formation of resistant genes are failing traditional antibiotics to cure bacterial infections. But with new techniques and efficient methods, it does not require to wait long periods of time to find a cure.

1.2 Bacteria : An Overview

Bacteria constitute one of the first forms of life to inhabitant Earth's atmosphere. Their abundance is quite irresistible to notice. Survival of the bacterial species is of great interest to the researchers worldwide. Billions of years of evolution, many life forms and organisms created and gone extinct, yet the bacteria with its primitive structure, still prevails and continues to conquer Earth. Their presence is so ubiquitous that you can take any minute sample of matter, from anywhere and check under a microscope, you'll discover millions of bacterial species in

Antimicrobial Resistance in Wastewater and Human Health.
DOI: https://doi.org/10.1016/B978-0-323-96124-0.00010-6

that small sample. The sources where they can extracted ranges from normal habitable environments (by humans and animals) to the extreme environments like terrestrial hot springs, hydrothermal vents, antarctica, and deep vents, which led to the placing of these species in a category called "**Extremophiles**."

A rough estimated number of Archeal and Bacterial cells present on earth is said to be approximately 10^{30} cells and an uncertainty of approximately tenfold (Bar-On, Phillips, & Milo, 2018). They can be either free-living, in colony or in association with other microbes forming "Biofilms" to produce diverse microbiota with the respect to the surroundings.

1.3 Wastewater characteristics

Wastewater can be defined as the collection of the wastes discharged from different water bodies such as commercial properties, domestic residences, agricultural facilities or land and industrial plants that have been negatively affected and exploited in the water quality by humans. It comprises of solid, liquid, and gaseous wastes and also contains a wide range of contaminants at various concentrations.

The environmental conditions of the wastewater (as experienced by the bacteria and other living organisms) are very different from other water sources and the fight for survival depends on its special characteristics. These specific characteristics will help us to understand their nature and why several pathogenic bacteria and other infectious agents manage to survive in such a harsh environment. There are basically three main characteristics of wastewater:

Physical characteristics

The physical characteristics of wastewater or sewage mainly comprises of temperature, color, odor, and turbidity. These characteristics can easily help us determine wastewater from a freshwater or other aquatic sources.

Temperature of the sewage water is seen to maintain a constant range from 15 to 21°C which is always slightly higher than the groundwater due to heat evolved during decomposition of the organic matter by the microorganisms present in the wastewater.

Color of the wastewater characteristically appears between dark-brownish and black due to mixing of loads of organic and inorganic wastes.

Odor: Emission of fetid smell via generation of "hydrogen-sulfide" from anaerobic decomposition of organic products.

Turbidity: Wastewater is highly turbid due to presence of suspended particles.

Chemical characteristics

It mainly comprises the chemical properties of a wastewater such as organic matter, biological oxygen demand (BOD), dissolved oxygen (DO), pH, and chemicals like chloride and nitrogen (Figs. 1.1–1.5).

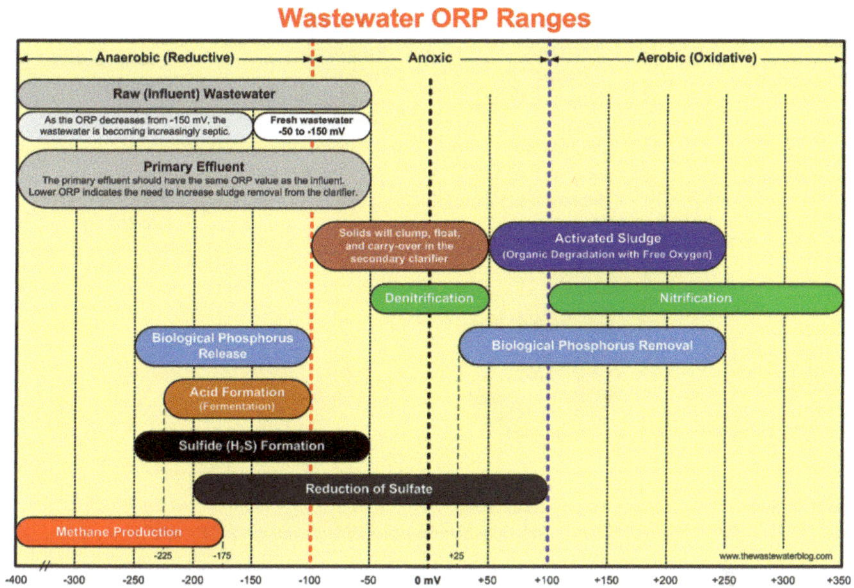

FIGURE 1.1 Wastewater ORP ranges. (*Source: The Wastewater Blog, https://www.thewastewaterblog.com/single-post/2016/12/18/orp.*)

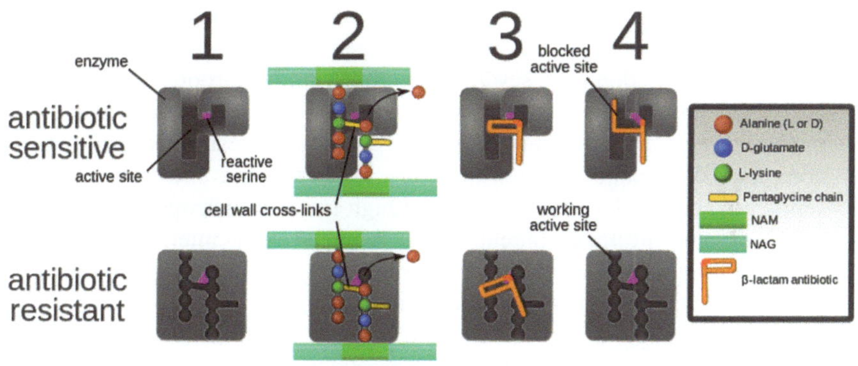

FIGURE 1.2 mecA resistance. (*Source: https://commons.wikimedia.org/wiki/File:MecA_Resistance.svg.*)

Organic matter: In general, sewage has a high concentration of organic materials. The amount of organic stuff, however, is determined by the type and condition of the sewage. Organic matter in sewage can be found as dissolved chemicals, colloidal matter, suspended matter, or sedimented matter.

Biological oxygen demand (BOD): Due to the presence of a substantial amount of organic matter, sewage often has a high BOD. BOD concentrations range from 100 mg/L for very dilute sewage to 600 mg/L or higher for concentrated sewage that contains industrial effluent mix.

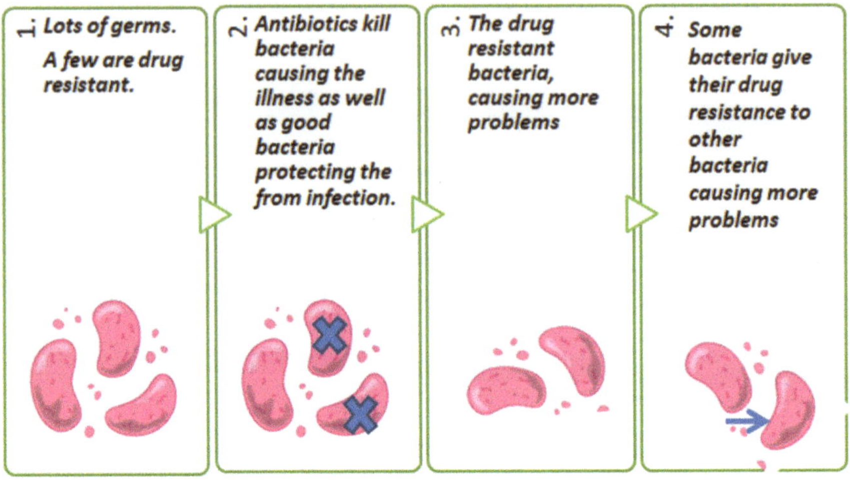

FIGURE 1.3 Evolution of drug resistant gene. (*Source: https://ib.bioninja.com.au/standard-level/topic-5-evolution-and-biodi/52-natural-selection/antibiotic-resistance.html.*)

Dissolved oxygen (DO): Sewage has extremely little DO due to the large concentration of microbial cells and biodegradable organic materials.
pH: Sewage has an alkaline pH.
Chloride: Humans excrete a substantial quantity of chloride in the form of NaCl (8–15 gm/day), primarily through urine and perspiration. As a result, chloride levels in domestic sewage from toilets and bathrooms are higher.
Nitrogen: Nitrogen is found in sewage in a variety of forms such as organic nitrogen, ammonia, nitrite, nitrate, and so on. Fresh sewage primarily comprises of organic nitrogen and very little inorganic nitrogen. Organic septic sewage, on the other hand, has a high inorganic nitrogen content and a low organic nitrogen content. Because nitrite is an intermediary result of the conversion of ammonia to nitrate (NO_3), it never accumulates in concentrations more than 1 mg/L in sewage.
Sulfite: In sewage, anaerobic bacteria produce sulfite in the form of H_2S (hydrogen sulfite) during the anaerobic breakdown of organic waste. Sewage has a horrible odor due to H_2S.

Biological characteristics

These characteristics of any water body is the presence of microbes such as bacteria, algae, fungi, viruses, aquatic plants, and aquatic animals.

Additional factors of wastewaters of utmost consideration for bacterial populations to infiltrate the sewage and survive are:

1. **Sedimentation**: The increased concentration of organic, inorganic, and chemical solutes contribute to the solid mass being settled as sediments in

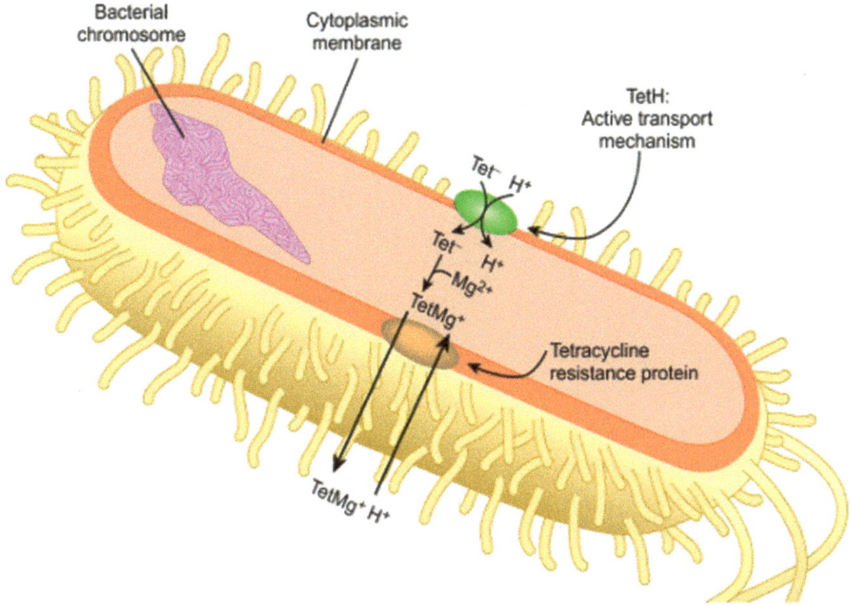

FIGURE 1.4 Tetracycline efflux mechanism by bacteria. (*Source: Tetracycline—an overview | ScienceDirect Topics https://www.sciencedirect.com/science/article/pii/B9780128132883000239.*)

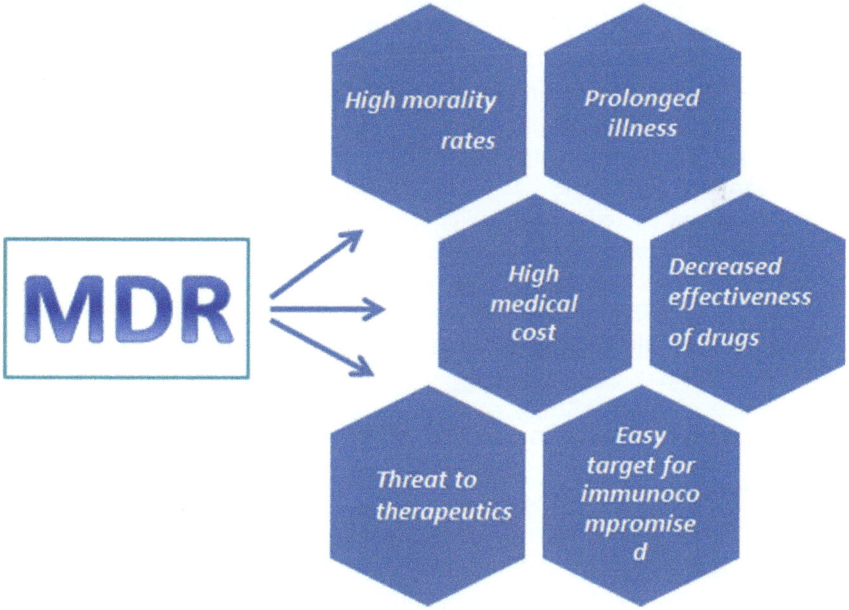

FIGURE 1.5 Problems caused due to MDR. (*Source: Tanwar, Jyoti, et al. (2014) "Multidrug resistance: an emerging crisis." Interdisciplinary perspectives on infectious diseases 2014.*)

TABLE 1.1 Typical concentration of selective enteric and coliform bacteria in sewage (Cabral, 2010).

Bacterial organism	Concentration in sewage (cells/mL)
Salmonella	10^1
Escherichia coli	10^5
Shigella	10^3
Enterococci	10^4
Enterobacter	10^4
Klebsiella	10^5
Clostridium perfringes	10^3

the wastewater. These sediments acts as a nutrient-bed and provide protective shields to the bacteria from the solar radiation (Gerba & McLeod, 1976).

2. **Salt tolerance:** The amount of solute is higher compared to the solvent in sewage which can lead to dehydration of cells and to avoid dehydration, cells absorb osmoprotectants like trehalose, glutamate, etc., which acts as a osmotic balancing agents.
3. **Light attenuation:** Since wastewater face greater turbidity due to the dissolved matter and sediments, light penetration is prevented by these suspended matter, decreasing the bactericidal effect of solar radiation.
4. **Electron–proton activity:** It was seen there is a direct correlation between electron–proton activity and bacterial metabolism (Selvarajan et al., 2018). Oxidation–reduction potential (ORP) is the measurement of breakdown of waste-products by microorganism. ORP of a healthy water source is between 300 and 500 mV (Table 1.1). The tendency of bacterial species to take up electrons and be reduced helps generate energy that is used by them for metabolic activities. Since wastewater has low electron acceptors such as oxygen, it limits the ability of bacteria to utilize organic matter for energy production, inhibiting bacterial growth and accumulation of end-products, contributing to the drop in ORP values (Gray & Gest, 1965).

1.4 Bacterial population based on wastewater source

1.4.1 Domestic sewage

It consists of the wastewater collected from the residential area which includes waste from toilets, laundry washing, and kitchen wastes. Domestic sewage contains mainly sweat, fecal and urine contaminant, and detergents from washing and cleaning. These contaminants serve as the organic and inorganic nutrients and rich environment for the bacterial pathogens survival and growth. The contents of domestic sewage mainly comprises of basic salts derived from detergents. Graywater from domestic laundry is one of the major contributors of surfactants contained in domestic wastewater. These surfactants are metabolized by the microorganisms present in the sewage providing a safer, more efficient, and less expensive physicochemical method to curb water pollution (Herbes & Schwall, 1978). Soaps and detergents commonly have high contents of "linear alkyl-benzene sulphonates" (LAS) (anionic surfactant) and "non-linear alkyl benzene sulphonate" (ABS), which require several days to biodegrade (Gledhill, 1975). These organic surfactants are observed to be a major factor in the spike of bacterial populations. They provide the carbon source and other inorganic matter for the production of energy and hence growth of bacteria (Kertesz et al., 1994).

1.4.1.1 Bacterial populations

Due to nutrient-rich rich environment provided by domestic sewage, it attracts lots of infectious bacterial species to proliferate. These bacteria are present in three different forms in wastewater: (1) free-living, (2) bacteria attached to organic matter, and (3) bacteria settled in the sediment (Giovanetti et al., 2003). Fecal coliforms are generally seen to be attached to the organic matter which proves to be a "survival strategy," helping in metabolism and protection against "grazing" by zooplanktons (Goulder, Bent, & Boak, 1981). Sewage sediment serves as a reservoir for the nutrients present in the sewage, protector against solar radiation and a large sum of enteric and coliform bacteria are present their adhered to the suspended particles but they are observed to be "metabolically active but are not culturable," with structural modifications and enzymatic growth but it turns out most of the nutrient are refractory content and are labile providing limited energy (Brown, Ellwood, & Hunter, 1977). Survival of enteric bacteria can vary from several days to several weeks in the sediment. For *Escherichia coli,* apparent mortality varies between 6 and 20 days (Le Guyader, 1989) and for *Salmonella*, it can be upto several weeks (Gudding & Krogstad, 1975). Solar irradiation can have bactericidal effect but in wastewater, light penetration is reduced due to turbidity which proves beneficial for enteric bacterial survival.

1.4.2 Industrial sewage

Since industrialization, the wastewater effluents from these industries and factories have contributed as the major source of water pollution which serve in alteration of microbial ecological landscape. Manufacture of steel wire, commercial vehicle washers, batteries, and electrical products produces wastewater effluents comprising loads of metals, polyaromatic hydrocarbons, and strong acids (Tekere, Sibanda, & Walter Maphangwa, 2016). Around 10,000 commercial dyes mostly comprising of "Azo dyes," containing double or triple nitrogen bonds which are recalcitrant and xenobiotic in nature, are being produced by the textile industries out of which 10%–15% of these pollutants are discharged during manufacturing and processing procedures as effluents. Industrial effluent has "total dissolved solids" levels of 4611 mg/L and maintains a temperature range between 18 and 25.5°C (Selvarajan et al., 2018).

1.4.2.1 Bacterial populations

The bacterial profile that is able to survive in these harsh conditions is dominated by *Proteobacteria* (44.44%–75.86%), followed by *Bacteriodetes, Fermicutes* and *Actinobacteria.* Since there is a large influx of metals like sulfur, iron, etc., it is observed these industrial effluents to harbor high populations of "sulfur-reducing" and "iron-reducing" bacteria. These pollutants are mostly toxic to all forms of life but a number of bacterial populations have been found to thrive in such environments either directly by utilizing the pollutants as a source of carbon or indirectly by "biotransformation" of the organic and inorganic matter present (Bassin et al., 2017). Synthetic dyes such as azo dyes, sulfur dyes, and others are degraded by these bacteria by the "oxidoreductive activated enzymes" (such as oxidases and azoreductases) via decolorization and mineralization which enables them to metabolize and utilize these complex xenobiotic compounds as substrates (Jamee & Siddique, 2019). The metabolites produced after degradation are mutagenic and carcinogenic. These dyes reduce the DO concentration, creating anoxic environment creating toxicity which is handled by mixed bacterial cultures of aerobic and facultative anaerobic bacteria for dye decolorization. The presence of high amounts of heavy metal and other active solid matter leads to the development of resistance in bacteria turning into pathogens and harmful to the humans.

1.4.3 Hospital sewage

Multiple medical practices occurring in a hospital such as drug treatment, radiology, laundry, surgery, chemical, and biological laboratories, serve as a potent source of hazardous pollutant discharge into the surrounding. Wastewater from hospitals have similar properties as any other swage environments but due to activities carried out in the hospital facilities have led to the rise of potential hazardous agents that includes disinfectants, lethal chemical compositions derived from medical treatment, radioactive isotopes, pharmaceuticals that are partially metabolized and antibiotic-resistant microbial populations. Hospital effluent heavy metals include mercury (Hg), cadmium (Cd), lead (Pb), chromium (Cr), cobalt (Co), nickel (Ni), and zinc (Z) with maximum concentration value of 2.5 mg/L. The toxic compounds include cyanide, phenol, and other drug components (Hocquet, Muller, & Bertrand, 2016). The pH of hospital effluent is detected between the range of 6.9–8.5 with an average value of 7.5.

1.4.3.1 Bacterial populations

Hospital wastewater provides an ecological niche and a perfect homage for emergence of **"antibiotic-resistant bacteria" (ARB)** and **"antibiotic-resistant genes" (ARG)** as patients in hospitals receive an antibiotic or antimicrobial treatment during their stay in the hospitals. These ARB in the presence of disinfectants and heavy metals have contributed to their persistence and added strength for survival in wastewater microbiome (Oberlé et al., 2012). Perseverance of ARG can be attributed to the fact of "horizontal resistance gene transfer" among intraspecies and interspecies, favored by selective pressure of antimicrobe.

1.4.3.1.1 Antibiotic-resistant Bacteria

1. Multidrug resistant ***Pseudomonas aeruginosa***—*P. aeruginosa* is strict aerobic, Gram-negative, encapsulated pathogenic bacteria found ubiquitously in the surroundings present as biofilms, but can be easily extracted from hospital water network such as taps, toilets, medical equipments, and sink because they thrive in moist conditions. In human host, they cause infections in the respiratory and urinary tracts by forming bacterial colonies. *P. aeruginosa* is a medically important species as a "multidrug resistant" pathogen with its intrinsically advanced mechanism of antibiotic resistance. They are found to be antibiotically resistant from antibiotics such as ciprofloxacin, tobramycin, beta-lactams, and many more (Slekovec et al., 2012). They produce VIM-type beta-lactamases enzyme to block the action of beta-lactam antibiotic.
2. Methicillin resistant ***Staphylococcus aureus*** **(MRSA)**—*Staphylococcus aureus* belongs to the family of Gram-positive bacteria and is responsible for causing infections in the upper respiratory tracts and soft tissues. MRSA are indistinct from the other *S. aureus* in combating resistance against antibiotics methicillin and other beta-lactams family. Terms like HA-MRSA (hospital-acquired - MRSA), LA-MRSA (Livestock acquired - MRSA), and CA-MRSA (Community acquired - MRSA) reflects this species. Alterations acquired by the extrachromosomal genetic elements like plasmids, transposon, etc., confers to the resistance against antibiotics and these "resistant-genes" are transferred among bacteria via "HGT" (Jensen & Lyon, 2009). They possess "mecA" genetic biomarker as the antibiotic resistant gene which is responsible for resistance against methicillin. mecA genetic segment codes for "penicillin binding protein 2a" (PBP2a), which possess genetic inability to bind to beta-lactam antibiotics leading to continuity of catalysis of "transpeptidation process," necessary for peptidoglycan cross-linking and enabling cell wall synthesis even in the presence of antibiotics (Faridi et al., 2018).

1.4.4 Agricultural sewage

Bacterial pollution of the soil caused by manure and sludge land application can also have an impact on indigenous soil populations and influence decomposition and nutrient recycling rates. Several studies have revealed that agricultural runoff contains significant levels of total coliforms and faecal streptococci independent of whether the soil has been grazed with subsequent contamination by animal faecal contents (Doran & Linn, 1979). Cutting the pastures lowered bacterial survival periods on the above-mentioned grass species, most likely due to the effect on drying rates and increased exposure to sun radiation. Environmental factors such as temperature, exposure to sunlight, humidity, and rainfall. According to Brown et al. primary components are involved in affecting the bacterial die-off rate on vegetation (Brown, Jones, & Donnelly, 1980).

Once applied to the field, manure becomes a possible non-point source of pollution from the agricultural sector. Fecal coliform populations tend to be more responsive to waste application than faecal streptococci or total coliform groups, which is likely due to these bacteria's greater vulnerability to die-off in terrestrial systems. Total coliforms, faecal streptococci, and enterococci population densities in runoff remained steady throughout the research at levels similar to those before treatment. This was attributable to exceptionally high levels of background contamination for these species. It is unknown if the high bacteria populations of agricultural run-off indicate a health risk or are simply caused by indigenous soil species. The longer residence time provides for more contact between soil components and microorganisms. Adsorption and fixation are enhanced through ion exchange, attractive forces, surface charge, and polymer bridging between solids and bacterial surfaces. Improperly handled livestock faeces can contaminate rivers receiving agricultural runoff. The risk to people is that these faecal microbes will invade water and food systems. One of the most serious clinical concerns is the transmission and evolution of antibiotic resistance genes, as well as their acquisition by bacterial pathogens (Arias & Murray, 2009). Large volumes of antibiotic compounds are used in animal husbandry, which increases the percentages of resistance isolates from animals.

1.4.5 Antibiotic resistant gene

While antibiotic use will primarily dictate bacterial resistance patterns in manure, the amount of antibiotics excretion and the fate of antibiotics in the soil and manure will decide the selective advantage for resistant bacteria in the environment. In manure, "beta lactams" and "macrolides" such as "tylosin" can degrade during storage with half-lives in the order of days (Kolz et al., 2005), but many antibiotic compounds in spread manure are transferred to soil. Overall, bioactive concentrations of several antibiotics are typically detected in manure, and recent studies from different regions always found maximum levels of many mg/Kg. The prevalence of bacteria containing antimicrobial resistance genes appears to be higher in pigs than in cattle or sheep, which correlates with the quantity of the antibiotics used in these animal species' husbandry. When "tetracycline" residues were evident in pig dung samples from 120 farms, the genes "tetM" and "tetO" showed greater presence. The presence of the resistance genes sul1, sul2, and blaTEM, as well as resistance (R) plasmids, in field-scale manures was shown to be associated with high antibiotic use in large farms and for weaning piglets, and low use in small and organic farms. Manure increased the total and "sulfadiazine" (SDZ) resistant cultivable bacteria quotients in soil. While selective pressure from environmentally persistent antibiotics is likely a role, manure gene load can also contribute significantly to the environmental pollution by including resistance genes, where resistance levels appear to be determined by the emission of resistance genes from farms rather than local resistance development due to selective pressure (Storteboom et al., 2010). Metals, such as As, Cu, and Zn, which accumulate in the soil as a result of manure application, are one of the reasons that may indirectly contribute to the emergence of antibiotic resistance. Using an ingenious cultivation-independent technique, Cu was demonstrated to coselect for tetracycline and vancomycin resistance in microbial soil communities under field settings (Berg et al., 2010). The incorporation of "[3H] bromodeoxyuridine" into microbial DNA in response to different antibiotic concentrations in soil was examined. Because of the high prevalence of environmental contamination with transferable antibiotic resistance genes, agricultural antibiotic use is extremely likely to have a large impact on the spread of resistance in the human community.

1.5 Drug resistant gene

One of medicine's biggest successes was the discovery and manufacture of (synthetic) antibiotics in the first part of the previous century. Upon introduction antibiotics, bacteria soon began developing resistance to these drugs and started transferring these resistant genes between bacteria and produced resistant strains.

1.5.1 Tetracycline resistance genes

The first tetracycline resistant gene was isolated from *Shigella dysenteriae* in 1953. Tetracycline is a family of medicines of oral antibiotics useful in the treatment of several bacterial infections such as acne, brucellosis, cholera, syphilis, malaria, and plague. Tetracycline is observed to inhibit the growth of bacteria by interfering with the "ribosomal machinery" that assembles amino-acids and produces vital proteins. Rather than killing bacteria, this antibiotic is used to tether their growth. This drug is sold under the name of "**Sumycin**." The noticeable side-effects in humans include diarrhea, vomiting, loss of appetite, and rash.

Tetracycline resistance genes, found in bacteria are regulated by "**tet**" genes, that help the bacteria in reducing the effect of the antibiotic via active removal of the drug, mutations within ribosomal binding sites, or chromosomal mutations leading to increased expression of intrinsic resistance mechanisms (Giovanetti et al., 2003). Till date, 40 different tet genes have been isolated and characterized in different strains, among which gram-negative bacteria have been reported to have tet(A), tet(B), tet(D), tet(E) and tet(G), while gram-positive significantly reported to have tet(K), tet(L), tet(M), tet(O). and tet(S) (Jones et al., 2006).

1.5.2 Mechanism of action

1.5.2.1 Uptake

Tetracycline can passively diffuse through porins present in the outer membrane like OmpF and OmpC of Gram-negative bacteria bound to Mg^{2+} which are now accumulated in the periplasm after dissociation from the Mg^{2+}, enabling weak lipophilic form that can easily diffuse through the inner membrane into the cytoplasm via energy-dependent processes like proton-motive force, passive diffusion (Pugsley & Schnaitman, 1978). Tetracycline antibiotics are considered to be "bacteriostatic" and once inside the cytoplasm, they prey on highly conserved bacterial 16S rRNA (Ribosomal RNA) in 30S ribosomal subunit, which then interferes with the translational process by sterically tethering the docking of "amino-acyl transferase RNA" (tRNA) of elongation step (Brodersen et al., 2000).

1.5.2.2 Resistance

Bacteria have developed resistance against antibiotics through various pathways such as:

- **Active efflux**—There are a group of transporters called major facilitator superfamily (MFS) that contains membrane traversing pumps. These pumps operate at the expense of proton energy. The pumps are classified into groups: Group 1 includes H^+ antiporter pump with 12 transmembrane segment composed of alpha and beta domains. This group governed by tet(A) and tet(B) genes in gram-negative bacteria. Group 2 includes pumps with 14 membrane traversing segments governed by tet(K) and tet(L) in Gram-positive bacteria (Guillaume et al., 2004). Groups 3–7 are not so clinically prevalent.
- **Enzymatic inactivation of the antibiotic**—Enzymatic modification of the tetracycline was first isolated in *Bacteriodetes* encoded plasmids. The class of enzymes involved in this mechanism is "Flavin-dependent monooxygenases" encoded by tet(X). The enzyme blocks the action of the drug by covalently binding "hydroxyl group" at the C-11a position detected between the C and B rings of tetracycline core (Speer & Bedzyk, 1991).
- **Alterations in the Binding Sites and Ribosomal Protection Proteins**—Tetracycline targets Ribosomal RNA (rRNA) binding sites, alteration in the genetic sequence of "tet" gene to provide resistance against Tetracycline leads to the mutations at the binding sites. Example: triple mutations of codon "AGA" (965–967 located in primary, tet1 gene) in the h31 look and codon "G" at position 942 confers to the resistance of Tetracycline in *Escherichia coli* (Trieber & Taylor, 2002).

Ribosomal protection proteins (RPPs) are a class of "GTPases," originally described in the *Streptococcus spp.*, and *Campylobacter jejuni*. These GTPases have significant sequence and are structurally similar to the elongation factors EF-Tu and EF-G (Kobayashi et al., 2007). Tet(M) and Tet(O) are the most evident and best-characterized RPP. These proteins help with the catalysis of GTP-dependent release of tetracycline from the ribosome via conformational changes induced by RPP that promote rapid formation of the EF-Tu-GTP-aa-tRNA ternary complex, which enables protein synthesis to continue even in the presence of tetracycline (Dönhöfer et al., 2012).

1.5.3 Beta-lactam resistant gene

Beta-lactams are widely used to treat bacterial illnesses around the world. These medications hinder cell wall formation and have a beta-lactam central circle (Voulgari et al., 2015). Several "extended-spectrum beta-lactamases" (ESBLs) have been identified in gram negative bacteria over the last few decades and have been categorized into four groups: penicillinases, metallo beta-lactamases, oxacillinases, and cephalosporinases. AmpC is an enzyme in Class C. *Pseudomonas aeruginosa (P. aeruginosa)* strains all possess an attenuated transcript of the ampC gene on the chromosome, which can be activated by first generation cephalosporines and produces AmpC enzymes. *P. aeruginosa* also produces a diverse set of beta-lactamases, which are passed on via plasmids and other mobile genetic elements (Easwaran, Yerat, & Ramaswamy, 2016). Another possibility is that AmpC-producing *P. aeruginosa* has natural noninduced resistance to particular drugs. Although "**uridopenicillins**" and third-generation "**cephalosporines**" are ineffective against AmpC, they do not stimulate AmpC production (Tahmasebi et al., 2018). As a result, unless AmpC is activated, these antibiotics are active; nonetheless, spontaneous mutants would be created at an intermediate frequency.

Bacterial ability to release many virulence factors is one of its clinical implications. Many virulence factors in bacteria are acquired horizontally, as evidenced by the species' high diversity. Furthermore, virulence factors are likely to serve a useful function for any organism within their natural environment, and because *P. aeruginosa* is not an obligate parasite of humans, many of the factors harmful to humans expressed by this organism will most likely serve a more innocuous function within the organism's natural habitat (Khalil et al., 2015).

1.5.4 Beta-lactamase

"**IMP-1**" is a beta-lactamase from Class B with extended spectrum metallo beta-lactamases that hydrolyze all beta-lactams except aztreonam and are resistant to clavulanic acid and sulbactam (Easwaran et al., 2016). These enzymes are known as carbapenemases because they hydrolyze imipenem and meropenem. IMP was first identified in *P. aeruginosa* as a plasmid-borne gene, and it was later shown that it could be transported via integrons. As a result, this could explain why IMP spreads in some entrobacteraceae genera (Tam et al., 2007). "**Oxacillinases**" are the other type of beta-lactamase. Except for OXA-48, they are vulnerable to clavulanic acid and hydrolyze ceftazidime and aztreonam. Except for OXA-18, oxacillinases are carried on plasmids. However, due to genetic circulation and "HGT" mechanisms identified in bacteria, oxacillinases and related variations such as OXA-51, OXA-23, and OXA-58 are more abundant in enterobacteriaceae (Li et al., 2008).

There was a link between the presence of beta-lactamase genes and of AmpC enzymes. In other circumstances, the genes responsible for resistance to lactamase-resistant enzymes, such as AmpC and ESBLs, are transferred onto the plasmid. This results in the transfer of genes across various Gram-negative bacteria, resulting in the development of resistant and harmful strains. Resistance to carbapenems is caused by AmpC activity in Gram-negative bacteria. Antibiotic resistance patterns indicated that AmpC could impact genotypic and phenotypic resistance patterns in a way that could lead to enhanced IMP gene activation. All IMP genes were found in *P. aeruginosa* that produces AmpC. Although other research claims that there is no association between genotypic and phenotypic resistance in AmpC-producing strains, the current investigation found no evidence of this (Geyer & Hanson, 2014).

1.5.5 Quinoline resistant gene

Quinolones are a common class of synthetic antimicrobials. The first member of the class, "**nalidixic acid**", was discovered as a byproduct of chloroquine manufacture in 1962 and had limited clinical usage due to its low efficacy in treating urinary tract infections and the early evolution of resistance (Slekovec et al., 2012). Chemical changes of the core quinolone and related chemical scaffolds, on the other hand, have been widely investigated and have resulted in molecules with increased potency, larger activity spectrum, improved pharmacokinetics, and a decreased incidence of resistance development. With the introduction of "ciprofloxacin" in 1987 and "norfloxacin" in 1986, a key modification of a fluorine substituent at position-8 resulted in the development of many members of what became known as the "**fluoroquinolone class**," which exhibited significantly greater potency against Gram-negative bacteria. Following that, more fluoroquinolones with greater effectiveness against Gram-positive bacteria, such as levofloxacin and moxifloxacin, were created. Fluoroquinolone s have been widely used for a variety of clinical conditions around the world due to their potency, spectrum of activity, oral absorption, and overall good safety profile.

1.5.6 Resistance due to chromosomal mutation

Quinolones inhibit "DNA gyrase" and "DNA topoisomerase IV," two important bacterial type II topoisomerase enzymes. Each enzyme is a heterotetramer, with gyrase being made up of two GyrA and two GyrB subunits and topoisomerase IV being made up of two ParC and two ParE subunits. GyrA is related to ParC, while GyrB is related to ParE. Both enzymes work by catalyzing a double-strand break in DNA, then transferring another DNA strand through the break and resealing it

(Aldred, Kerns, & Osheroff, 2014). The enzymes' DNA strand-passing domains are found in GyrA and ParC, while the enzymes' ATPase activity, which drives the catalytic cycle, is found in GyrB and ParE domains. Quinolones inhibit enzyme activity by preventing the resealing of the DNA double-strand break and by stabilizing catalytic intermediate covalent complexes of enzyme and DNA that act as a barrier to the movement of the DNA replication fork and can be converted to double-strand DNA breaks, which correlate with quinolone bactericidal activity (Hiasa, Yousef, & Marians, 1996). Quinolone resistance can be caused by single amino acid alterations in either gyrase or topoisomerase IV. These resistance mutations have typically been found in the N-terminal domains of "GyrA" (residues 67 to 106 for *E. coli* numbering) or "ParC" (residues 63 to 102) and are close to the active site tyrosines (Tyr122 for GyrA, Tyr120 for ParC), which are covalently linked to DNA in an enzyme intermediate in both enzymes (Storteboom et al., 2010). Many quinolones in clinical usage have gyrase as the more sensitive enzyme in the Gram-negative bacteria and topoisomerase IV as the more sensitive enzyme in the Gram-positive bacteria, although there are exceptions (Blanche et al., 1996). The magnitude of the resistance increase caused by such a first-step mutation can be defined by either the size of the mutation's influence on enzyme sensitivity or the intrinsic level of sensitivity of the secondary target enzyme.

1.5.7 Antibiotic efflux

Staphylococcus aureus has been the most frequently investigated Gram-positive bacteria for quinolone resistance via enhanced efflux. Overexpression of each of the three efflux pumps, NorA, NorB, and NorC, has been found to increase quinolone resistance by four to eightfold (Schindler et al., 2015). All three pumps are part of the transporter superfamily known as the "MFS," which are secondary transporters powered via proton gradient across the cytoplasmic membrane. Resistance to hydrophilic quinolones, such as ciprofloxacin and norfloxacin, is conferred by "NorA" expression, whereas resistance to hydrophobic quinolones, such as moxifloxacin and sparfloxacin, is conferred by NorB and NorC expression (Truong-Bolduc, Strahilevitz, & Hooper, 2006). The expression levels of efflux pumps, most of which are members of the "**resistance-nodulation-division**" (RND) superfamily, have been demonstrated to impart greater quinolone resistance in Gram-negative bacteria. The RND pumps are made up of three structural components: (1) a cytoplasmic membrane pump protein; (2) An outer membrane channel protein; (3) and a membrane fusion protein that connects the pump and the outer membrane protein (Li, Plésiat, & Nikaido, 2015). Fluoroquinolones are thought to cross the outer membrane via the porin diffusion channels OmpC and OmpF, and downregulation of these channels or mutations in their structural genes may also contribute as a resistance mechanism. Notably, quinolone resistance mutations in the MarR regulator boost acrB expression while decreasing OmpF expression.

1.5.8 Resistance via mutations in plasmids

The discovery of plasmid-mediated quinolone efflux pumps QepA and OqxAB added a third mechanism of "plasmid-mediated quinolone resistance" (PMQR). PMQR genes have been discovered in bacterial isolates all over the world in the previous decade. The responsible resistance gene found in plasmids was named *qnr*, later updated to *qnrA*. Qnr homologs can be found on the chromosome of many γ-Proteobacteria, Actinomycetales, and Firmicutes, including species of *Bacillus, Enterococcus, Listeria*, and *Mycobacterium*, as well as anaerobes such as *Clostridium perfringens* and *Clostridium difficile* (Sánchez et al., 2008). *Qnr* genes are commonly found in the multiresistance plasmids linked to other resistance determinants, β-lactamase genes, including genes for extended spectrum β-lactamases (ESBLs), carbapenemases, and AmpC enzymes. *Aeromonas hydrophila* chromosomally encoded AhQnr, plasmid-mediated QnrB1 and, *Enterococcus faecalis chromosomally* encoded EfsQnr. All of them are rod-like dimers. The monomers of QnrB1 and AhQnr feature 8 and 12 amino acid projecting loops that are essential for their function (Xiong et al., 2011). The deletion of the smaller A loop lowers quinolone protection, whereas the loss of the bigger B loop or both loops completely eliminates protective action. Protection is jeopardized even if a single amino acid is removed from the bigger loop (Jacoby et al., 2013).

1.5.9 Macrolide resistant gene

Macrolides are a class of antibiotic that includes erythromycin, roxithromycin, azithromycin and clarithromycin. They are useful in treating respiratory, skin, soft tissue, sexually transmitted, *H. pylori,* and a typical mycobacterial infections. Macrolides share a similar spectrum of antimicrobial activity with "benzylpenicillin" making them useful alternatives for people with a history of penicillin (and cephalosporin) allergy. Protein synthesis is inhibited by macrolide antibiotics, which target the bacterial ribosome. They bind to the peptide's nascent exit tunnel and partially block it. As a result, macrolides have been referred to as "tunnel plugs' that prevent the creation of all proteins. Recent research, on the other hand, shows that macrolides impede translation of a subset of cellular proteins selectively, and that their activity is highly dependent on the nascent protein sequence as well as the antibiotic structure (Portillo et al., 2000). As a result, macrolides appear to be translation modulators rather than worldwide inhibitors of protein synthesis. Macrolides interfere with bacterial protein synthesis and, depending on concentration and bacterial species, are either bactericidal (kill bacteria), or bacteriostatic (inhibit the growth of bacteria). Macrolides also have immunomodulatory and anti-inflammatory effects, which can be beneficial in some situations, for example, when they are used in the treatment of cystic fibrosis.

1.5.10 Antibiotic alteration

Bacteria often display cross-resistance between the macrolides. The development of unique mechanisms of antibiotic resistance, particularly in the species *Enterococcus faecium*, has made infections with these germs difficult to cure (Murray, 1990). The antibiotics "macrolide-lincosamide-streptogramin" (MLS) are an alternate therapy for the treatment of creeping enterococcal infections. Three mechanisms explain for Gram-positive bacteria's acquired resistance to MLS antibiotics: macrolides interfere with bacterial protein synthesis and, depending on concentration and bacterial species, are either bactericidal (kill bacteria), or bacteriostatic (inhibit growth of bacteria). A single 23S rRNA mutation causes broad cross-resistance to the macrolide-lincosamide-streptogramin B (MLSB) antibiotics, but the inactivation mechanism confers resistance solely to the structurally related MLS drugs. In terms of pump mechanisms, the genes mefA, mefE, msrA, and mreA have been implicated in the active efflux of macrolides in gram-positive bacteria. The genes mef and mreA have been linked to macrolide resistance, whereas the msrA gene has been linked to streptogramin B and macrolide resistance.

1.5.11 Antibiotic efflux

Macrolide-resistance genes were studied in 103 macrolide-resistant *Streptococcus pyogenes* strains obtained from pharyngotonsillitis patients. PCR was used to determine the presence of the mef(A), erm(B), and erm(TR) genes. In 48 out of 103 (46.6 percent) strains, mef(A) was discovered, while erm(B) was found in 43 isolates (41.7 percent). All mef(A) strains exhibited the M phenotype (resistance to 14- and 15-membered macrolides, and sensitivity to lincosamides and streptogramin B), whereas erm(B) strains had the MLSB phenotype (Tait-Kamradt et al., 1997). These resistant efflux genes (mef) encode a proton-motive force pump and a putative ATP-binding cassette transporter homolog and are transcribed as an operon (Masuda et al., 2000).

1.5.12 Multidrug resistant gene

Multidrug resistance (MDR) in bacteria is caused by the accumulation of genes, each of which codes for resistance to a specific agent, on resistance (R) plasmids or transposons, and/or the action of multidrug efflux pumps, each of which may pump out more than one drug type. Antibiotics used in large quantities for human therapy, farm animals, and even aquaculture fish, resulted in the selection of harmful bacteria resistant to numerous drugs. Bacterial MDR can be caused by one of two methods. First, inside a single cell, these bacteria may acquire many genes, each coding for resistance to a single antibiotic. This buildup is most common on resistance (R) plasmids. Second, increased expression of genes that code for multidrug efflux pumps, which extrude a wide spectrum of medicines, may result in MDR. The continuous use of antimicrobial medications in the treatment of infections has resulted in the emergence of resistance among distinct strains of microbes. MDR is characterized as a microorganism's insensitivity or resistance to antimicrobial treatments (which are structurally unrelated and have diverse molecular targets) notwithstanding previous the sensitivity to them (Méndez-Vilas, 2013). Although MDR is a natural phenomenon, an increase in the number of immunocompromised conditions, such as HIV infection, diabetic patients, organ transplant recipients, and severe burn patients, makes the body an easy target for hospital acquired infectious diseases, contributing to the spread of MDR.

1.5.13 Mechanism of action

Resistance is defined as a microbe's insensitivity to an antimicrobial drug when compared to other isolates of the same species. Antimicrobial medications often operate on microorganisms by inhibiting a metabolic pathway such as nucleotide synthesis, which in turn inhibits DNA/RNA synthesis, subsequent protein synthesis, and cell membrane rupture, or by competing with the substrate of any enzyme involved in cell wall formation (e.g., chitin synthase). Bacteria have developed several mechanism of action to curb and resist the effects of antibiotics.

1.5.13.1 R plasmids and Tn21

Most drug resistance genes have been shown to be effective when expressed from plasmids. Surprisingly, multiple of these genes are frequently found on a single R plasmid, allowing MDR to be given to susceptible bacteria in a single conjugation event. Many R plasmids already had resistance genes for aminoglycosides, chloramphenicol, tetracycline, and sulfonamides when they were discovered in Japan in the 1950s. The majority of the resistance genes in the early generation R plasmid sequence are components of transposons, which can transfer the genes to any piece of DNA. This is shown in "plasmid R100."

Tn21 is an outstanding example of a huge, complex, massively composite transposon (Liebert, Hall, & Summers, 1999). Surprisingly, it contains genes for mercury resistance. Tn21 contains sulfonamide resistance and aminoglycoside resistant genes. The revelation that many resistance genes in R plasmids have a unique "59-base 3′-sequence tag" led to the discovery of an integron, a piece of remarkable machinery (Hall & Stokes, 1993). An integron contains a gene that codes for an integrase, which catalyzes the insertion of resistance genes downstream from a strong promoter. Once integrated, the resistance gene is indicated by the tag, allowing it to be easily incorporated into another integron that may include a different collection of resistance genes. In addition to the benefit of great mobility, when resistance genes are placed into an integron, they are ordered into a single operon with the same transcription direction under a strong promoter given by the integron structure. Tn21 contains an integron that already contains a sulfonamide resistance gene "sul1" and a shortened version of a multidrug efflux gene "qacE," and it has integrated an aminoglycoside resistance gene "aadA1" at the specific integration site "attI." Up to eight resistance genes can be found in an integron (Rowe-Magnus & Mazel, 2002).

1.5.13.2 Horizontal gene transfer

R plasmids are not only stably maintained, but they are also typically transported between bacterial cells with high efficiency, often approaching 100%. The R plasmid DNA is entirely transported into the recipient cells, and this strand is replaced in the donor cell by a rolling-circle replication mechanism, resulting in complete R plasmids in both donor and recipient cells. The entry of foreign-origin DNA would be expected to be identified by the recipient cells' strain-specific "restriction endonuclease." However, because the first bit of DNA that enters is single-stranded (albeit the complementary strand will be created soon), it may escape this mechanism.

1.5.13.3 Efflux pumps

Drug active efflux has been shown to have a significant role in the development of resistance to specific medicines, such as tetracycline. Analysis of *S. aureus* strains resistant to multiple cationic bacteriocides and causing hospital-acquired infections revealed the presence of plasmids coding for the multidrug efflux transporter QacA (or QacB), which belongs to the "MFS" and was the first multidrug efflux pump identified in bacteria. Each protein has 14 transmembrane segments (TMSs), each of which contains multiple acidic amino acid residues. These pumps aggressively extrude monocationic biocides and dyes like benzalkonium chloride, ethidium bromide, and cetyltrimethylammonium bromide. QacA also extrudes dicationic biocides like chlorhexidine and pentamidine isethionate.

S. aureus's chromosomally coded "NorA," on the other hand, belongs to a branch with 12 TMS transporters. NorA causes resistance to cationic dyes, fluoroquinolones, and cationic inhibitors such as "puromycin" and "tetraphenylphosphonium" (Neyfakh, Borsch, & Kaatz, 1993). Because reserpine inhibits all of these efflux pumps, sensitization of bacteria to substrate medicines in the presence of reserpine can be utilized as an effective technique to assess the efflux process's contribution to resistance.

1.6 MDR effects

MDR appears to be a significant concern for researchers who rely on a single therapy, with its continual updating in the bacterial genome induced by changes in the external environment. Because of the rapid emergence of new resistance mechanisms and the decrease in the efficacy of treating common infectious diseases, there is a failure of microbial response to traditional therapy, resulting in longer sickness, increased healthcare costs, and an increased risk of mortality. Although the establishment of MDR is a natural process, the incorrect use of antimicrobial medications, insufficient hygienic conditions, improper food handling, and poor infection prevention and control procedures all contribute to the emergence and spread of MDR.

1.7 Conclusion

The cycle of "bacterial and human diseases developing antidotes to cure those infections, which develops another resistant strain with greater toxicity" will never end. The increased pollution allows pathogenic stains to thrive and transmit diseases. The good news is that researchers are using healthy bacteria to treat and clean up the environment from tainted pollutes existing in soil and groundwater via a technique known as Bioremediation. The process of bioremediation is plainly seen in wastewater treatment plants.

References

Aldred, K. J., Kerns, R. J., & Osheroff, N. (2014). Mechanism of quinolone action and resistance. *Biochemistry, 53*(10), 1565–1574.

Arias, C. A., & Murray, B. E. (2009). Antibiotic-resistant bugs in the 21st century—a clinical super-challenge. *New England Journal of Medicine, 360*(5), 439–443.

Bar-On, Y. M., Phillips, R., & Milo, R. (2018). The biomass distribution on Earth. *Proceedings of the National Academy of Sciences of the USA, 115*, 6506–6511.

Bassin, J. P., et al. (2017). Revealing the bacterial profile of an anoxic-aerobic moving-bed biofilm reactor system treating a chemical industry wastewater. *International Biodeterioration & Biodegradation, 120*, 152–160.

Berg, J., et al. (2010). Cu exposure under field conditions coselects for antibiotic resistance as determined by a novel cultivation-independent bacterial community tolerance assay. *Environmental Science & Technology, 44*(22), 8724–8728.

Blanche, F., et al. (1996). Differential behaviors of Staphylococcus aureus and Escherichia coli type II DNA topoisomerases. *Antimicrobial Agents and Chemotherapy, 40*(12), 2714–2720.

Brodersen, D. E., et al. (2000). The structural basis for the action of the antibiotics tetracycline, pactamycin, and hygromycin B on the 30S ribosomal subunit. *Cell, 103*(7), 1143–1154.

Brown, C. M., Ellwood, D. C., & Hunter, J. R. (1977). Growth of bacteria at surfaces: Influence of nutrient limitation. *FEMS Microbiology Letters, 1*(3), 163–166.

Brown, K. W., Jones, S. G., & Donnelly, K. C. (1980). The influence of simulated rainfall on residual bacteria and virus on grass treated with sewage sludge. *American Society of Agronomy, Crop Science Society of America, and Soil Science Society of America, 9*(2), 1–10.

Cabral, J. P. S. (2010). Water microbiology. Bacterial pathogens and water. *International Journal of Environmental Research and Public Health, 7*(10), 3657–3703.

Clifton, C. E. (1937). A comparison of the metabolic activities of Aerobacter aerogenes, Eberthella typhi and Escherichia coli. *Journal of Bacteriology, 33*(2), 145–162.

Dönhöfer, A., et al. (2012). Structural basis for TetM-mediated tetracycline resistance. *Proceedings of the National Academy of Sciences, 109*(42), 16900–16905.

Doran, J. W., & Linn, D. M. (1979). Bacteriological quality of runoff water from pastureland. *Applied and Environmental Microbiology, 37*(5), 985–991.

Easwaran, S., Yerat, R., & Ramaswamy, R. (2016). A study on detection of extended-spectrum beta-lactamases (ESBLs) and comparison of various phenotypic methods of AmpC detection in Pseudomonas aeruginosa from various clinical isolates in a tertiary care teaching hospital. *Muller Journal of Medical Sciences and Research, 7*(1), 35.

Eriksson, E., et al. (2002). Characteristics of grey wastewater. *Urban Water, 4*(1), 85–104.

Faridi, A., et al. (2018). Detection of methicillin-resistant Staphylococcus aureus (MRSA) in clinical samples of patients with external ocular infection. *Iranian Journal of Microbiology, 10*(4), 215.

Flemming, H.-C., & Wuertz, S. (2019). Bacteria and archaea on Earth and their abundance in biofilms. *Nature Reviews Microbiology, 17*(4), 247–260.

Gerba, C. P., & McLeod, J. S. (1976). Effect of sediments on the survival of Escherichia coli in marine waters. *Applied and Environmental Microbiology, 32*(1), 114–120.

Geyer, C. N., & Hanson, N. D. (2014). Multiplex high-resolution melting analysis as a diagnostic tool for detection of plasmid-mediated AmpC β-lactamase genes. *Journal of Clinical Microbiology, 52*(4), 1262–1265.

Giovanetti, E., et al. (2003). Presence of the tet (O) gene in erythromycin-and tetracycline-resistant strains of Streptococcus pyogenes and linkage with either the mef (A) or the erm (A) gene. *Antimicrobial Agents and Chemotherapy, 47*(9), 2844–2849.

Gledhill, W. E. (1975). Screening test for assessment of ultimate biodegradability: Linear alkylbenzene sulfonates. *Applied Microbiology, 30*(6), 922–929.

Goulder, R., Bent, E. J., & Boak, A. C. (1981). Attachment to suspended solids as a strategy of estuarine bacteria. *Feeding and Survival Srategies of Estuarine Organisms* (pp. 1–15). Boston, MA: Springer.

Gray, Clarke T., & Gest, Howard (1965). Biological Formation of Molecular Hydrogen: A" hydrogen valve" facilitates regulation of anaerobic energy metabolism in many microorganisms. *Science, 148*(3667), 186–192.

Gudding, R., & Krogstad, O. (1975). The persistence of Escherichia coli and Salmonella typhimurium in fine-grained soil. *Acta Agriculturae Scandinavica, 25*(4), 285–288.

Guillaume, Gilliane, et al. (2004). Phylogeny of efflux-mediated tetracycline resistance genes and related proteins revisited. *Microbial Drug Resistance, 10*(1), 11–26.

Hadidian, Z., & Hoagland, H. (1941). Chemical pacemakers: III. Activation energies of some rate-limiting components of respiratory systems. *The Journal of General Physiology, 24*(3), 339–352.

Hall, R. M., & Stokes, H. W. (1993). Integrons: Novel DNA elements which capture genes by site-specific recombination. *Genetica, 90*(2-3), 115–132.

Herbes, S. E., & Schwall, L. R. (1978). Microbial transformation of polycyclic aromatic hydrocarbons in pristine and petroleum-contaminated sediments. *Applied and Environmental Microbiology, 35*(2), 306–316.

Hiasa, Hiroshi, Yousef, Diana O., & Marians, Kenneth J. (1996). DNA strand cleavage is required for replication fork arrest by a frozen topoisomerase-quinolone-DNA ternary complex. *Journal of Biological Chemistry, 271*(42), 26424–26429.

Hocquet, Didier, Muller, Allison, & Bertrand, Xavier (2016). What happens in hospitals does not stay in hospitals: Antibiotic-resistant bacteria in hospital wastewater systems. *Journal of Hospital Infection, 93*(4), 395–402.

Hunter, J. V., & Heukelekian, H. (1965). The composition of domestic sewage fractions. *Journal (Water Pollution Control Federation), 37*, 1142–1163.

Jacoby, G. A., et al. (2013). Mutational analysis of quinolone resistance protein QnrB1. *Antimicrobial Agents and Chemotherapy, 57*(11), 5733–5736.

Jamee, R., & Siddique, R. (2019). Biodegradation of synthetic dyes of textile effluent by microorganisms: An environmentally and economically sustainable approach. *European Journal of Microbiology and Immunology, 9*(4), 114–118.

Jensen, S. O., & Lyon, B. R. (2009). Genetics of antimicrobial resistance in Staphylococcus aureus. *Future Microbiology, 4*(5), 565–582.

Jones, C. H., et al. (2006). Identification and sequence of a tet (M) tetracycline resistance determinant homologue in clinical isolates of Escherichia coli. *Journal of Bacteriology, 188*(20), 7151–7164.

Kertesz, M. A., et al. (1994). Desulfonation of linear alkylbenzenesulfonate surfactants and related compounds by bacteria. *Applied and Environmental Microbiology, 60*(7), 2296–2303.

Khalil, M. Abd El F., et al. (2015). Comparative study of virulence factors among ESβL-producing and nonproducing Pseudomonas aeruginosa clinical isolates. *Turkish Journal of Medical Sciences, 45*(1), 60–69.

Klock, J. W. (1971). Survival of coliform bacteria in wastewater treatment lagoons. *Journal (Water Pollution Control Federation), 43*, 2071–2083.

Kobayashi, T., et al. (2007). Molecular evidence for the ancient origin of the ribosomal protection protein that mediates tetracycline resistance in bacteria. *Journal of Molecular Evolution, 65*(3), 228–235.

Kolz, A. C., et al. (2005). Degradation and metabolite production of tylosin in anaerobic and aerobic swine-manure lagoons. *Water Environment Research, 77*(1), 49–56.

Laponogov, I., et al. (2013). Structure of an 'open'clamp type II topoisomerase-DNA complex provides a mechanism for DNA capture and transport. *Nucleic Acids Research, 41*(21), 9911–9923.

Le Guyader, F. (1989). Colonisation bactérienne et implantation de E. Coli dans le sédiment d'origine littorale. *Diss. Université de Rennes, 1*.

Lesher, G. Y., et al. (1962). 1, 8-Naphthyridine derivatives. A new class of chemotherapeutic agents. *Journal of Medicinal Chemistry, 5*(5), 1063–1065.

Li, X.-Z., Plésiat, P., & Nikaido, H. (2015). The challenge of efflux-mediated antibiotic resistance in Gram-negative bacteria. *Clinical Microbiology Reviews, 28*(2), 337–418.

Li, Yi, et al. (2008). Prevalence of plasmid-mediated AmpC β-lactamases in a Chinese university hospital from 2003 to 2005: First report of CMY-2-type AmpC β-lactamase resistance in China. *Journal of Clinical Microbiology, 46*(4), 1317–1321.

Liebert, C. A., Hall, R. M., & Summers, A. O. (1999). Transposon Tn 21, flagship of the floating genome. *Microbiology and Molecular Biology Reviews, 63*(3), 507–522.

Masuda, N., et al. (2000). Substrate specificities of MexAB-OprM, MexCD-OprJ, and MexXY-oprM efflux pumps in Pseudomonas aeruginosa. *Antimicrobial Agents and Chemotherapy, 44*(12), 3322–3327.

Méndez-Vilas, A. (2013). Microbial pathogens and strategies for combating them: Science, technology and education Ed. Formatex Research Center.

Munro, P. M., Gauthier, M. J., & Laumond, F. M. (1987). Changes in Escherichia coli cells starved in seawater or grown in seawater-wastewater mixtures. *Applied and Environmental Microbiology, 53*(7), 1476–1481.

Murray, Barbara E. (1990). The life and times of the Enterococcus. *Clinical Microbiology Reviews, 3*(1), 46–65.

Neyfakh, A. A., Borsch, C. M., & Kaatz, G. W. (1993). Fluoroquinolone resistance protein NorA of Staphylococcus aureus is a multidrug efflux transporter. *Antimicrobial Agents and Chemotherapy, 37*(1), 128–129.

Nomura, Y., et al. (1998). Application of a linear alkylbenzene sulfonate biosensor to river water monitoring. *Biosensors and Bioelectronics, 13*(9), 1047–1053.

Oberlé, K., et al. (2012). Evidence for a complex relationship between antibiotics and antibiotic-resistant Escherichia coli: From medical center patients to a receiving environment. *Environmental Science & Technology, 46*(3), 1859–1868.

Plummer, D. H., Owens, N. J. P., & Herbert, R. A. (1987). Bacteria—particle interactions in turbid estuarine environments. *Continental Shelf Research, 7*(11-12), 1429–1433.

Pommepuy, M., et al. (1992). Enteric bacteria survival factors. *Water Science and Technology, 25*(12), 93–103.

Portillo, A., et al. (2000). Macrolide resistance genes in Enterococcus spp. *Antimicrobial Agents and Chemotherapy, 44*(4), 967–971.

Pugsley, A. P., & Schnaitman, C. A. (1978). Outer membrane proteins of Escherichia coli VII. Evidence that bacteriophage-directed protein 2 functions as a pore. *Journal of Bacteriology, 133*(3), 1181–1189.

Roberts, M. C. (1996). Tetracycline resistance determinants: Mechanisms of action, regulation of expression, genetic mobility, and distribution. *FEMS Microbiology Reviews, 19*(1), 1–24.

Rowe-Magnus, D. A., & Mazel, D. (2002). The role of integrons in antibiotic resistance gene capture. *International Journal of Medical Microbiology, 292*(2), 115–125.

Sánchez, M. B., et al. (2008). Predictive analysis of transmissible quinolone resistance indicates Stenotrophomonas maltophilia as a potential source of a novel family of Qnr determinants. *BMC Microbiology, 8*(1), 1–14.

Schindler, Bryan D., et al. (2015). Analyses of multidrug efflux pump-like proteins encoded on the Staphylococcus aureus chromosome. *Antimicrobial Agents and Chemotherapy, 59*(1), 747–748.

Selvarajan, R., et al. (2018). Industrial wastewaters harbor a unique diversity of bacterial communities revealed by high-throughput amplicon analysis. *Annals of Microbiology, 68*(7), 445–458.

Sizer, I. W. (1939). Temperature activation of the urease-urea system using crude and crystalline urease. *The Journal of General Physiology, 22*(6), 719–741.

Slekovec, C., et al. (2012). Tracking down antibiotic-resistant Pseudomonas aeruginosa isolates in a wastewater network. *Plos One, 7*(12), E49300.

Smith, M. C., & Chopra, I. (1984). Energetics of tetracycline transport into Escherichia coli. *Antimicrobial Agents and Chemotherapy, 25*(4), 446–449.

Speer, B. S., Bedzyk, L., & Salyers, A. A. (1991). Evidence that a novel tetracycline resistance gene found on two Bacteroides transposons encodes an NADP-requiring oxidoreductase. *Journal of Bacteriology, 173*(1), 176–183.

Storteboom, H., et al. (2010). Tracking antibiotic resistance genes in the South Platte River basin using molecular signatures of urban, agricultural, and pristine sources. *Environmental Science & Technology, 44*(19), 7397–7404.

Tahmasebi, H., et al. (2018). Investigation of the relationship between the presence of chromosomal and plasmid-encoded ampc genes and type of clinical specimen in pseudomonas aeruginosa. *Journal of Babol University of Medical Sciences, 20*(3), 36–43.

Tait-Kamradt, A., et al. (1997). mefE is necessary for the erythromycin-resistant M phenotype in Streptococcus pneumoniae. *Antimicrobial Agents and Chemotherapy, 41*(10), 2251–2255.

Tam, V. H., et al. (2007). Prevalence of AmpC over-expression in bloodstream isolates of Pseudomonas aeruginosa. *Clinical Microbiology and Infection, 13*(4), 413–418.

Tekere, M., Sibanda, T., & Walter Maphangwa, K. (2016). An assessment of the physicochemical properties and toxicity potential of carwash effluents from professional carwash outlets in Gauteng Province, South Africa. *Environmental Science and Pollution Research, 23*(12), 11876–11884.

Trieber, C. A., & Taylor, D. E. (2002). Mutations in the 16S rRNA genes of Helicobacter pylori mediate resistance to tetracycline. *Journal of Bacteriology, 184*(8), 2131–2140.

Truong-Bolduc, Q. C., Strahilevitz, J., & Hooper, D. C. (2006). NorC, a new efflux pump regulated by MgrA of Staphylococcus aureus. *Antimicrobial Agents and Chemotherapy, 50*(3), 1104–1107.

Voulgari, E., et al. (2015). Emergence of OXA-162 carbapenemase-and DHA-1 AmpC cephalosporinase-producing sequence type 11 Klebsiella pneumoniae causing community-onset infection in Greece. *Antimicrobial Agents and Chemotherapy, 60*(3), 1862–1864.

Woodcock, S., & Sloan, W. T. (2017). Biofilm community succession: A neutral perspective. *Microbiology (Reading, England), 163*(5), 664–668.

Xiong, X., et al. (2011). Structural insights into quinolone antibiotic resistance mediated by pentapeptide repeat proteins: Conserved surface loops direct the activity of a Qnr protein from a gram-negative bacterium. *Nucleic Acids Research, 39*(9), 3917–3927.

Zavala, L., Ángel, M., & Estrada, Eunice Espinoza (2016). The contribution of the type of detergent to domestic laundry graywater composition and its effect on treatment performance. *Water, 8*(5), 214.

Zürrer, D., Cook, A. M., & Leisinger, T. (1987). Microbial desulfonation of substituted naphthalenesulfonic acids and benzenesulfonic acids. *Applied and Environmental Microbiology, 53*(7), 1459–1463.

Chapter 2

Development and spread of drug resistance through wastewater

Sonia Chadha and Prerna Tandon
Amity University Uttar Pradesh, Lucknow Campus, Amity Institute of Biotechnology, Lucknow, India

2.1 Introduction

Antibiotics are broadly used for controlling bacterial infection as well as livestock breeding. The rigorous use and abuse of antibiotics as well as their inappropriate disposal have lead to the loss of antibiotic effectiveness. This has lead to the development of antibiotic-resistant bacteria and antibiotic resistance genes. The increase in resistance to antibiotics is being depicted as a major global health threat by WHO and European Centre for Disease Prevention and Control (Baker, Hobman, Dodd, Ramsden, & Stekel, 2016). Due to antibiotic-resistance hundreds of thousands of deaths are being reported annually. Transmission of drug-resistant genes occurs through horizontal gene transfer and vertical gene transfer methods such as transformation, transduction, and conjugation. Plasmids, transposons, and integrons mediate gene transfer between bacteria thereby promoting antibiotic-resistant genes. Antibiotic-resistant bacteria also spread through wind, water, soil, human, and animal activities. Hospitals, farms, private households, and industrial plants have largely contributed to spread of heavy metals, antibiotic-resistant genes, and antibiotic-resistant bacteria (Behera, Kim, Oh, & Park, 2011).

Wastewater treatment plants are major hotspots for the progress and propagation of antibiotic-resistant genes to the environment. As microbes like bacteria from diverse sources are in close proximity during the purification process, they promoting horizontal gene transfer (Biswal, Mazza, Masson, Gehr, & Frigon, 2014). The fecal bacteria like *E. coli* are largely found in wastewater treatment plants that act as a source of resistance and contributes to serious diseases like MDR *Mycobacterium tuberculosis*, vancomycin-resistant enterococcus, and methicillin-resistant *Staphylococcus aureus*. Wastewater moving into the wastewater treatment plants contains mixture of pollutants, drug-tolerant

Antimicrobial Resistance in Wastewater and Human Health.
DOI: https://doi.org/10.1016/B978-0-323-96124-0.00011-8

bacteria, antibiotics, heavy metals, and antibiotic-resistant genes that penetrate the environment from hospitals, industrial plants, aquaculture, animal farms, and private households (Andersson & Hughes, 2014). There are various strategies for wastewater treatment which are aimed at expulsion of drug-resistant bacteria and genes (Table 2.1).

A majority of the wastewater treatment plants are not equipped with effective technologies for eliminating antibiotics, antibiotic-resistant genes, or antibiotic-resistant bacteria. As a result treated effluents containing resistant genes or bacteria are spread when they are discharged to lakes, rivers, oceans, and are reused for irrigation and agricultural practices (Auerbach, Seyfried, & McMahon, 2007; Gao, Munir, & Xagoraraki, 2012). Constructed wetlands are another major contributor of antibiotic- resistant genes as there is a constant release of microbes to water (Bairán, Rebollar-Pérez, Chávez-Bravo, & Torres, 2020). In addition to antibiotics, toxic elements like heavy metals and pesticides are also involved in the genetic exchange of antibiotic-resistant genes (Ardern & Lockett, 1914). Heavy metals are nondegradable and tend to accumulate in the environment. Contamination of the natural environment with a high concentration of heavy metals induces a coselection mechanism which occurs due to a single gene or multiple genes conferring resistance to antibiotics and heavy metals. The process of cross-resistance also leads to antibiotic and metal resistance as both antibiotics and metals undergo identical biochemical pathway. Prolonged usage of antibiotics in the protection of human and animals have created resistance in bacterial community (Bairán et al., 2020). Therefore, identifying sources of resistance genes, their environmental distribution and how anthropogenic inputs affect their spread will aid in establishing strategies to overcome antibiotic resistance.

2.1.1 Factors responsible for development of drug resistance

Antibiotics are employed for treating bacterial infections, in addition, antibacterial metals are widely used to prevent bacterial attachment and to combat biofilms (Pal et al., 2017). Antibiotics have been detected in sewage effluents, ground and surface water, sewage sludge, soil, and manure. The presence of antibiotics in the habitat may contribute to the spread of drug resistance and antibiotics may enter the human food chain. Studies carried out by Kim and Aga (Kim & Aga, 2007) have shown wastewater treatment plants to be the source of drug resistance development among pathogenic bacteria due to the persistent presence of pharmaceuticals in them. Their studies showed the accumulation of these metabolites in activated sludge. Nonsteroidal anti-inflammatory drugs were predominantly present in municipal wastewater treatment plants whereas antibiotics were dominant in livestock wastewater treatment plants (Benedetti et al., 2013). Widespread use of antibiotics in the livestock industry resulted in resistance of antibiotics to degradation that can lead to antibiotic resistance development in the environment (Ekpeghere, Lee, Kim, Shin, & Oh, 2017).

TABLE 2.1 Table showing the various strategies used for watewater treatment.

	Treatment strategy	Characteristic	Remark	References
1	Aerobic and anaerobic treatment processes	Low energy and environmentally friendly strategies which are mostly used to treat chemical oxygen demand	Successful in removing antibiotic resistant bacteria and antibiotic resistant genes	Du et al. (2015), Christgen et al. (2015)
2	Constructed wetlands	Small semiaquatic ecosystems, supporting various physical–chemical reactions	Can efficiently remove aqueous antibiotic resistant genes; however, they can also act as reservoirs for specific antibiotic resistant genes	Chen et al. (2016), Fang et al. (2017)
3	Disinfection	Disinfection kills a significant percentage of pathogenic organisms, the most popular being chlorination and UV radiation	Chlorination if the most effective disinfection technique to remove antibiotic resistant genes	Sharma et al. (2016), Zhuang et al. (2016)
4	Use of nanomaterials	Diverse combinations of nanomaterial have proved that antimicrobial nanotechnology can be effective defenses against antibiotic resistance	Defense against multidrug resistance	Shahverdi et al. (2007), Aruguete et al. (2013)
5	Coagulation	Removal of colloidal particles	Active method for removal of antibiotic resistant genes from wastewater treatment plant effluent	Li et al. (2017), Xiao et al. (2013)
6	Biochar	Used for sorption of contaminants	A practical strategy that can treat antibiotics contaminated soils	Li et al. (2017), Ye et al. (2016), Cui et al. (2016)

The table shows the comparative account of the various watewater treatment strategies.

Improper disposal of antibiotics and incomplete metabolism in humans leads to the release of antibiotics in municipal wastewater treatment plants (Barancheshme & Munir, 2018). Hospitals and healthcare centers are the most common fertile grounds for breeding resistant bacteria and microbes as they deal with many sick patients requiring intensive antimicrobial therapy. These antibiotic-resistant bacteria enters contaminate the groundwater, surface water, and agricultural soil through wastewater reuse. Sewage treatment plants are known to contain a large quantity of resistance genes and their improper and untreated discharge into water bodies like lakes, rivers, oceans act as an important vector enabling bacterial transmission between hosts through the environment (Chen et al., 2016).

Industrial activities promote spread of antibiotic-resistant genes as they are the main sources of pollutants such as heavy metals antibiotics, toxic elements like pesticides. The use of antimicrobial agents in animals for promoting growth or for therapeutic purposes can cause spread of drug-resistant bacteria (Behera et al., 2011).

Lack of proper healthcare facilities, hygiene, and poor sanitation creates a breeding ground for resistant bacteria to proliferate (Bergeron, Boopathy, Nathaniel, Corbin, & LaFleur, 2015). Misuse and excessive consumption of antibiotics cause selection pressure due to which microorganisms that were earlier sensitive to antibiotics now become resistant to the antibiotics. Anthropogenic activities have largely contributed to the incident and outspread of antibiotic-resistant genes. Genotoxic substances are often mutagenic and carcinogenic and are, therefore, the potential suspect in the origin of antibiotic-resistant organisms (Devarajan et al., 2015). Increased concentration of chemicals near ultrafiltration and reverse osmosis treatment systems membrane surface can act as mutagens and can result in the occurrence of resistance in the bacteria in gel/cake layer and surroundings via mutation. It has also been observed that physical forces such as wind and water environment also aids in spread of resistant genes as bacteria can mover over large distances. Bacterial species like micrococcus, bacillus, staphylococcus, and aeromonas have been isolated from air. Wild birds and animals living close to humans have been known to cater to bacteria carrying resistance genes and may contribute to spreading those genes across large areas.

Poor hygiene and lack of proper preventive measures have contributed to the spread of resistant strains of bacteria. Human beings are exposed to resistant pathogens through body contact or indirect contact transmission, aerosols, and food prepared by persons carrying the pathogen. It may also be transferred through a community setting (Fig. 2.1).

2.1.1.1 Wastewater and development of drug resistance

Wastewater treatment plants are important sources for spread of drug-resistant bacteria particularly those present in activated sludge. In the absence of selection pressure, antibiotic resistance can occur naturally (intrinsic) due to the presence of spontaneous gene mutation (Bengtsson-Palme et al., 2016). Resistance is

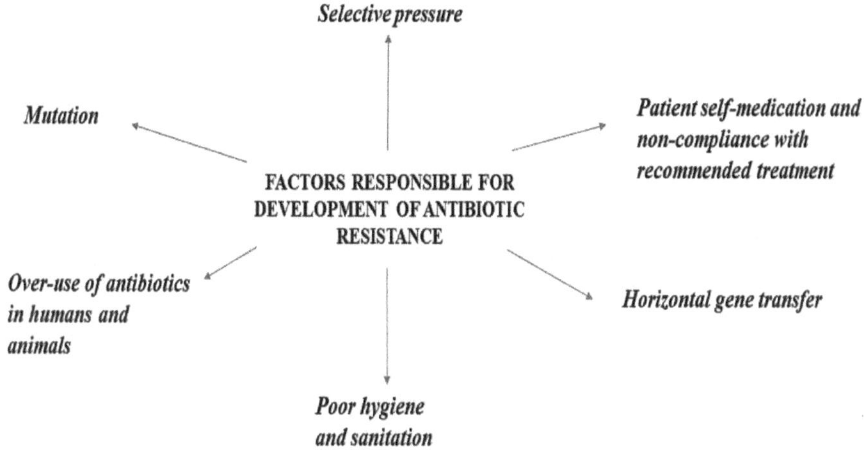

FIGURE 2.1 Factors responsible for the development of antibiotic resistance. No Permission Required. The above figure gives the major factors involved in the development of drug resistance.

developed when at least one bacterium in a heterogeneous colony carries the genetic determinants capable of expressing resistance to the antibiotic. As these treatment plants are constantly being polluted by discharging of detergents, disinfectants, and antimicrobial agents, their long-term prevalence in wastewater and pressure selection at subinhibitory concentrations of antibiotics may lead to the development of antibiotic resistance gene in bacteria. Hospitals and animal husbandry have largely contributed to the outspread of antibiotic resistance genes and antibiotic-resistant bacteria. The occurrence of these genes has also been detected in aquatic ecosystems as they are constantly being polluted by antimicrobial compounds. Most of the antibiotic-resistant genes present in wastewater are linked to clinical pathogens (Smith & Coast, 2013). A large number of antibiotics are eliminated through feces and urine which are then released into municipal sewers, sewage sludge, and the soil. The municipal wastewater contains high proportion of organic and inorganic matter including microorganisms (including pathogenic, commensal, and environmental bacteria) thereby contributing to spread of drug-resistant bacteria which then contaminate groundwater and agricultural soil.

Antibiotics and their associated antibiotic resistance genes and antibiotic-resistant bacteria enter the environment through discharges into rivers, wastewater reuse, and irrigation. Improper wastewater treatment and lack of hygiene allow for the perpetuation of antibiotic-resistant genes in the environment. Wastewater treatment plants are not equipped with effective technologies for removing mobile genetic elements. Therefore, wastewater treatment plants are potential hotspots and meeting points of most of antibiotic-resistant bacteria particularly those present in activated sludge or biological filters (Dcosta et al., 2011). As bacterial density at these locations is very high and due to increasing availability of nutrients and selective pressure for resistance, antibiotic resistance genes are transmitted to the new bacterial population by horizontal gene transfer method. Studies carried out by Du et al. (2015) have shown that anaerobic and anoxic sludge treatment methods are better at removing antibiotic-resistant genes than aerobic treatment methods as microorganisms have lower reactivity under anaerobic conditions and thus propagation of antibiotic resistance is inhibited.

2.1.1.2 Development of drug resistance

There are mainly four types of resistance mechanism that bacteria have adopted against antibiotics (Fig. 2.2). These are (1) efflux pumps where antibiotics are excreted by the cell. They are five different types of efflux pump families like ATP-binding cassette, multidrug and toxic compound extrusion, major facilitators, resistance nodulation cell division, and small multidrug resistance. (2) Hydrolysis of the antibiotic leading to antibiotics inactivation. (3) By-passing targets by the creation of new pathways, prevention of antibiotic binding and changes in structure of the cell wall, (4) target modification by modifying the antibiotic targets.

Resistance to antibiotics can be obtained by mutations, vertical inheritance, or horizontal gene transfer (Jury, Vancov, Stuetz, & Khan, 2010). The spread of drug-resistant genes via the horizontal gene transfer method is more effective than mutation. Horizontal gene transfer mediates the transfer of genetic information by motile genetic elements like plasmids and transposons (Chiang, Penadés, & Chen, 2019). The mobility of genes between these genetic elements is facilitated by integrons (Davies & Davies, 2010; Gillings, 2014; Mazel, 2006). Multiple resistance occurs due to the presence of different genes, encoding for specific antibiotics are positioned at the same motile genetic element or chromosome. The transfer of antibiotic resistance genes takes place from phages, free DNA, donor bacteria to the recipient cell. They are four different mechanisms of horizontal gene transfer.

Conjugation—It is a process wherein DNA is transferred from donor cell to recipient cell via a conjugation tube or sex pilus. As this process requires cell to cell contact the recipient cell that was earlier sensitive to antibiotics now shows resistance against the antibiotic.

Transformation—Uptake of free extracellular DNA by bacteria.

Transduction—It involves the use of bacteriophages to pass on DNA from one bacterium to another. Transduction includes generalized transduction where only a small fragment of bacterial DNA is packaged into the bacteriophage head, and specialized transduction where both phage and specific bacterial DNA are packaged into the head (Chiang et al., 2019).

Gene transfer agents (GTAs) are bacteriophage-like elements made by several bacteria. These GTAs act as a carrier of resistance in the environment and carry random segments of DNA present in the host bacterium, which are transduced to a recipient cell.

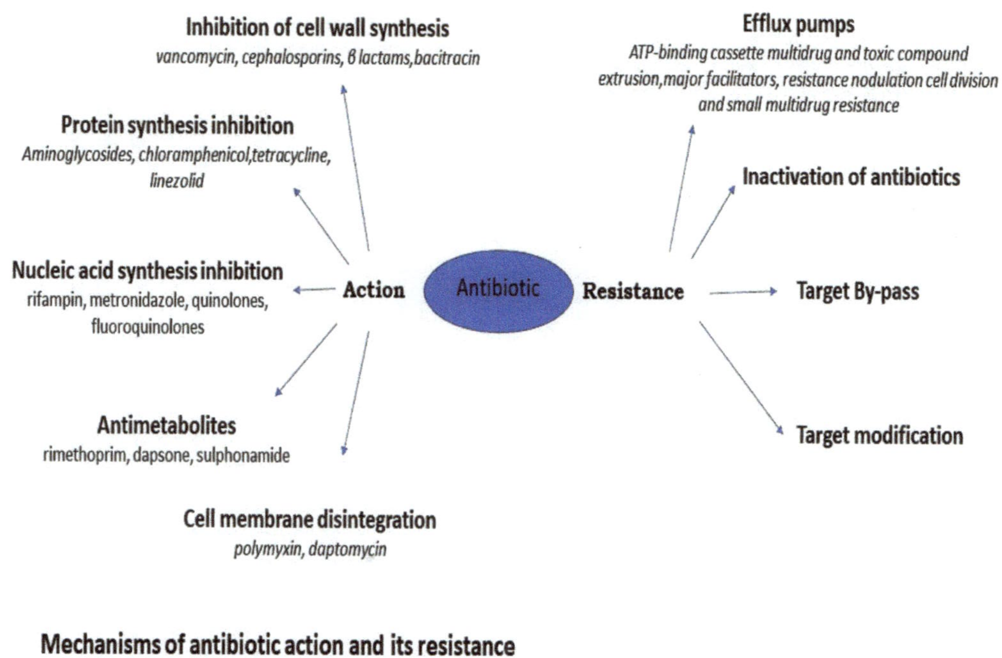

FIGURE 2.2 Figure showing mechanisms of action of antibiotics and mechanisms of antibiotic resistance. No Permission Required. The figure depicts the various mechanisms of action of antibiotics and the various ways adopted by the microorganisms to develop drug resistance.

Integrons include resistance cassettes encoding antibiotic-resistant genes as well as efflux pump genes. These multigene cassettes are controlled by a common mutual promoter and helps in the coselection of antibiotic-resistant genes (Alexander, Bollmann, Seitz, & Schwartz, 2015, 2016). The presence of heavy metals and organic compounds may cause coselection of antibiotic and heavy metal resistance. Exposure to heavy metals acts as a selective pressure on antibiotic resistance. Antibiotic-resistant selection also depends on coresistance and cross-resistance. Coresistance gives resistance against several classes of antibiotics. It is caused by physically linking different resistance genes. Cross-resistance occurs when one single resistance mechanism provides resistance against an entire class of compounds, antibiotics, and other toxicants (Amador, Fernandes, Prudêncio, Barreto, & Duarte, 2015).

The antibiotic-resistant genes present in wastewaters reside most often in the pathogenic bacteria. Hospital wastewater contains pathogenic drug-resistant bacteria, and high concentrations of antibiotics. Since many of the administered antibiotics are not completely metabolized in human beings, they are excreted into the sewerage (Done & Halden, 2015; Sabri et al., 2020). Studies have shown that incompletely degraded compounds present in the wastewaters exert biological selection pressure for the development of drug-resistant genes and provide a breeding ground for horizontal gene transfer between bacteria (Bouki, Venieri, & Diamadopoulos, 2013; McKinney & Pruden, 2012; Sharma, Siskova, Zboril, & Gardea-Torresdey, 2014; Zhang, Marrs, Simon, & Xi, 2009) and propagation of resistance genes (Davies & Davies, 2010).

2.1.1.3 Dissemination of antibiotic resistance

The rise in the global human population has lead to an increase in the water requirement for agricultural, food production, and household purposes. A practical solution for this is the reuse of wastewater. This leads to the contamination of the environment and spread of antibiotic-resistant genes and bacteria. Studies carried out by Zhang et al. (2009) and Szczepanowski et al. (2009) confirm that final effluents from wastewater treatment plants are responsible for disseminating antibiotic resistance. Antibiotics, pharmaceutical residues, and heavy metals are in continuous contact with bacteria during wastewater treatment strategies, and this leads to build-up of selection pressure for resistance genes (Bouki et al., 2013; Ding & He, 2010; Zhang et al., 2009). A guarded concentration of antibiotics in wastewater is difficult to predict as there is general disagreement on the fact that lower than minimum inhibitory antibiotic concentrations may cause the selection of antibiotic resistance genes (Gullberg et al., 2011; Klümper et al., 2019). Heavy metals and organic compounds can enhance the selective pressure for antibiotic-resistant genes by coselection (Cesare et al., 2016; Schlüter, Szczepanowski, Pühler, & Top, 2007; Tuckfield & McArthur, 2008). Coselection occurs through coresistance or cross-resistance. Coresistance indicates resistance to more than one class of antibiotics in the same bacterial strain.

Resistance genes against zinc and copper have been found to increase antibiotic-resistant gene dissemination by coresistance (Yazdankhah, Rudi, & Bernhoft, 2014). In cross-resistance, a single resistance mechanism is responsible for conferring resistance to an entire class of drugs or other toxicants (Baker-Austin, Wright, Stepanauskas, & McArthur, 2006). An example of cross-resistance is multidrug resistance pumps that export both metals and antibiotics (Baker-Austin et al., 2006). Therefore due to coselection antibiotic-resistant genes may persist even in the absence of antibiotics (Zhang, Song, Yang, Li, & Wang, 2018) and both co- and cross-resistance effect the antibiotic resistance selection in different environments (Knapp et al., 2017; Stepanauskas et al., 2005).

2.1.1.3.1 Drug-resistant bacterial population associated with wastewater

The wide use of antibiotics in hospitals and the livestock industry has lead to the spread of antibiotic resistance in the environment. Molecular and phenotypic approaches have been used to characterize the antibiotic-resistant bacteria in wastewaters. Enterobacteriaceae members, including Escherichia coli, Klebsiella spp., Shigella spp., Salmonella spp., Vibrio spp., Acinetobacter spp., and Enterococcus spp., were among the most common resistant bacteria identified in the wastewater samples. Additionally, high levels of multidrug-resistant bacteria and antibiotic-resistant genes conferring resistance to varied classes of antimicrobial drugs, including beta-lactams, carbapenems, tetracyclines, aminoglycosides, fluoroquinolones, sulfonamides, macrolides, vancomycin and erythromycin, were found (Bootsma, van der Horst, Guryeva, ter Kuile, & Diekmann, 2012). Bacteria like E. coli and Salmonella are resistant to multiple antibiotics. Antibiotics such as tetracycline, ciprofloxacin, norfloxacin, ofloxacin, trimethoprim, and sulfamethoxazole are present in high concentrations in the sludge of wastewater treatment plants. Tet(x), tet(w), tet(g), sul(1) intI(1)were detected in influent and effluent of municipal wastewater treatment plants (Du et al., 2015).

Resistance to aztreonam, pefloxacin, trimethoprim-sulfamethoxazole, gentamicin, oxacillin, penicillin, piperacillin, and ampicillin are very common bacterial resistant population. Betaproteobacteria and flavobacteria found in wastewater show multidrug resistance against carbapenems and cephalosporins. Aeromonas and Pseudomonas aeruginosa isolates obtained from some water reservoirs were found to express 50% and 100% multidrug resistance, respectively.

A study was conducted wherein presence of specific resistant genes to tetracyclines (tetQ, tetA, tetO),sulphonamide (sul1, sul2), erythromycin (emb), quinolone (qnrD, qnrS), beta-lactams (cepA, cfxA, bla TEM), methicillin (mecA), vancomycin (vanA) and aminoglycoside (aac(3)-II, aacA4, aadA, aadB, aadE, aphA1, aphA2, strA and strB) were analyzed and was confirmed that wastewater treatment plants are the main sources of antibiotic-resistant transmission (Amador et al., 2015).

Wastewater treatment plants constitute certain enterococcal resistant strains (Enterococcus faecalis and Enterococcus faecium) against erythromycin, tetracycline, amipicillin, fluoroquinolones and aminoglycosides. Fluoroquinolones and bla_{TEM}, qnr(S), erm(B), sul(1), and tet(W) are detected at the highest concentration in wastewater treatment plants. The reason for the presence of these antibiotic resistance genes is due to the use of penicillin and aminoglycosides.

2.1.1.3.2 Human exposure to drug resistance via water and its health impact

Aquatic environments are one of the reservoirs and transmission routes for the spread of antibiotic resistance as drinking water and wastewater treatment plants are incapable of completely removing antibiotic resistance genes. Antibiotics, as well as antibiotic resistance genes, enter the wastewater treatment plants through various pathways such as discharge of municipal sewage, pharmaceutical plants, animal husbandry, agricultural field containing livestock manure, and aquaculture ponds. Antibiotic resistance is defined as the ability of the bacteria to resist effects of antibiotic resistance to which the bacteria once was sensitive to (Behera et al., 2011). Prolonged exposure of bacterial population to antibiotics or pharmaceutical substances can mediate horizontal gene transfer of drug resistance genes phylogenetically distant bacteria and between nonpathogenic bacteria. Resistance against antibiotics can lead to enhanced virulence, prolonged morbidity, hospitalization, disease outbreaks, and even mortality. The use of treated wastewater can be beneficial for agricultural irrigation and recreational activities which can further lead to the occurrence and spread of antibiotic-resistant bacteria and genes to the environment. Humans can be exposed these through activities like drinking, bathing, aquatic sports, occupational exposure during agricultural irrigation, and consumption of food produce irrigated with reclaimed water.

Human and animal microbiota is a reservoir for antibiotic-resistant genes. Antibiotic residues on entering the human body interact with human flora which contains diverse microorganisms inhabiting the human body. Bacteria entering the human gastrointestinal tract acquire antibiotic resistance via conjugation and thereby gets accumulated in human and animal feces. Daily intake of antibiotic residues can create a selective pressure (a biotic or abiotic factor that alters the intestinal microbiome composition of an organism in a given environment) inside the gut, leading to the development of antibiotic resistance in enteric bacteria. Bacterial diversity belonging to actinobacteria, proteobacteria, or bacteroidetes phyla promotes the exchange of antibiotic-resistant genes and antibiotic-resistant bacteria to humans. Imbalance in intestinal microbiota may lead to growth, proliferation, and penetration of harmful and opportunistic pathogens or bacteria into tissues. This can further lead to acute immune reactions, allergic reactions, and various infectious diseases like colorectal cancer, pseudomembranous colitis, and intestinal disorders. In some cases, diseases caused by these bacteria can also lead to the death of an individual (Smith & Coast, 2013).

Mechanisms by which the antibiotic resistance in drinking water can affect human health include (1) direct infection by an antibiotic-resistant pathogen following consumption of contaminated water, without any human-to-human transmission; (2) sustained human-to-human transmission due to direct infection with an antibiotic-resistant bacteria that is pathogenic or opportunistic; and (3) horizontal gene transfer of environmental antibiotic-resistant genes in drinking water with the antibiotic-resistant pathogens selected by human antibiotics and/or coselecting agents. This may result in certain diseases, for example, Klebsiella pneumoniae is one of the major global causes of urinary tract, lower respiratory tract, and bloodstream infections that are due to carbapenem resistance. In African and Asian regions with poor sanitation and unsafe drinking water facilities, ciprofloxacin-resistant Shigella spp. continues to cause diarrhea and dysentery, which is sometimes fatal in infants.

2.2 Conclusion

The present chapter highlights the contribution of wastewater treatment plants toward the spread of antibiotic resistance genes and antibiotic-resistant bacteria. The current spread of antibiotic resistance genes is due to anthropogenic activities. Poor hygiene, lack of sanitization facilities, high density and diversity of microorganisms sustained by a nutrient-rich environment in activated sludge, constant selective pressure due to antibiotic residues at subinhibitory concentrations, absence of restrictions, and strict monitoring has contributed to the spread of antibiotic-resistant genes and antibiotic-resistant bacteria.

Both conventional and advanced treatment plants have been unsuccessful in removing antibiotic-resistant bacteria from water environments. Advanced treatment strategies, like UV, ozonation, and oxidation of water effluents, and heat drying, lime stabilization, and pyrolysis of biosolids, are effective in removing antibiotic resistance genes and antibiotic-resistant bacteria.

They are some ways through which we can limit the spread of resistant genes:

- There should be a threshold limit for the release of antibiotics from hospitals, pharmaceuticals, and animal husbandries.
- The economy and ecology of the waterway should be examined before setting up advanced treatment processes. Smaller-scale treatment plants with disinfection units at prove to be more cost-effective than redesigning and rebuilding larger municipal wastewater treatment plants.

Mathematical modeling should assist experimental studies to assess the mechanisms for the spread of resistant and nonresistant bacteria in wastewater treatment environments.

References

Alexander, J., Bollmann, A., Seitz, W., & Schwartz, T. (2015). Microbiological characterization of aquatic microbiomes targeting taxonomical marker genes and antibiotic resistance genes of opportunistic bacteria. *Science of the Total Environment, 512–513*, 316–325. https://doi.org/10.1016/j.scitotenv.2015.01.046.

Alexander, J., Knopp, G., Dötsch, A., Wieland, A., & Schwartz, T. (2016). Ozone treatment of conditioned wastewater selects antibiotic resistance genes, opportunistic bacteria, and induce strong population shifts. *Science of the Total Environment, 559*, 103–112. https://doi.org/10.1016/j.scitotenv.2016.03.154.

Amador, P. P., Fernandes, R. M., Prudêncio, M. C., Barreto, M. P., & Duarte, I. M. (2015). Antibiotic resistance in wastewater: Occurrence and fate of Enterobacteriaceae producers of Class A and Class C β-lactamases. *Journal of Environmental Science and Health - Part A Toxic/Hazardous Substances and Environmental Engineering, 50*(1), 26–39. https://doi.org/10.1080/10934529.2015.964602.

Andersson, D. I., & Hughes, D. (2014). Microbiological effects of sublethal levels of antibiotics. *Nature Reviews Microbiology, 12*(7), 465–478. https://doi.org/10.1038/nrmicro3270.

Ardern, E., & Lockett, W. T. (1914). Experiments on the oxidation of sewage without the aid of filters. *Journal of the Society of Chemical Industry, 33*(10), 523–539. https://doi.org/10.1002/jctb.5000331005.

Aruguete, D. M., Kim, B., Hochella, M. F. Jr, Ma, Y., Cheng, Y., Hoegh, A., et al. (2013). Antimicrobial nanotechnology: its potential for the effective management of microbial drug resistance and implications for research needs in microbial nanotoxicology. *Environmental Science: Processes & Impacts, 15*, 93–102. doi:10.1039/c2em30692a.

Auerbach, E. A., Seyfried, E. E., & McMahon, K. D. (2007). Tetracycline resistance genes in activated sludge wastewater treatment plants. *Water Research, 41*(5), 1143–1151. https://doi.org/10.1016/j.watres.2006.11.045.

Bairán, G., Rebollar-Pérez, G., Chávez-Bravo, E., & Torres, E. (2020). Treatment processes for microbial resistance mitigation: The technological contribution to tackle the problem of antibiotic resistance. *International Journal of Environmental Research and Public Health, 17*(23), 1–20. https://doi.org/10.3390/ijerph17238866.

Baker, M., Hobman, J. L., Dodd, C. E. R., Ramsden, S. J., & Stekel, D. J. (2016). Mathematical modelling of antimicrobial resistance in agricultural waste highlights importance of gene transfer rate. *FEMS Microbiology Ecology, 92*(4). doi:10.1093/femsec/fiw040.

Baker-Austin, C., Wright, M. S., Stepanauskas, R., & McArthur, J. V. (2006). Co-selection of antibiotic and metal resistance. *Trends in Microbiology, 14*(4), 176–182. https://doi.org/10.1016/j.tim.2006.02.006.

Barancheshme, F., & Munir, M. (2018). Strategies to combat antibiotic resistance in the wastewater treatment plants. *Frontiers in Microbiology, 8*. doi:10.3389/fmicb.2017.02603.

Behera, S. K., Kim, H. W., Oh, J. E., & Park, H. S. (2011). Occurrence and removal of antibiotics, hormones and several other pharmaceuticals in wastewater treatment plants of the largest industrial city of Korea. *Science of the Total Environment, 409*(20), 4351–4360. https://doi.org/10.1016/j.scitotenv.2011.07.015.

Benedetti, L., Langeveld, J., Comeau, A., Corominas, L., Daigger, G., Martin, C., et al. (2013). Modelling and monitoring of integrated urban wastewater systems: Review on status and perspectives. *Water Science and Technology, 68*(6), 1203–1215. https://doi.org/10.2166/wst.2013.397.

Bengtsson-Palme, J., Hammarén, R., Pal, C., Östman, M., Björlenius, B., Flach, C. F., et al. (2016). Elucidating selection processes for antibiotic resistance in sewage treatment plants using metagenomics. *Science of the Total Environment, 572*, 697–712. https://doi.org/10.1016/j.scitotenv.2016.06.228.

Bergeron, S., Boopathy, R., Nathaniel, R., Corbin, A., & LaFleur, G. (2015). Presence of antibiotic resistant bacteria and antibiotic resistance genes in raw source water and treated drinking water. *International Biodeterioration and Biodegradation, 102*, 370–374. https://doi.org/10.1016/j.ibiod.2015.04.017.

Biswal, B. K., Mazza, A., Masson, L., Gehr, R., & Frigon, D. (2014). Impact of wastewater treatment processes on antimicrobial resistance genes and their co-occurrence with virulence genes in Escherichia coli. *Water Research, 50*, 245–253. https://doi.org/10.1016/j.watres.2013.11.047.

Bootsma, M. C. J., van der Horst, M. A., Guryeva, T., ter Kuile, B. H., & Diekmann, O. (2012). Modeling non-inherited antibiotic resistance. *Bulletin of Mathematical Biology, 74*(8), 1691–1705. https://doi.org/10.1007/s11538-012-9731-3.

Bouki, C., Venieri, D., & Diamadopoulos, E. (2013). Detection and fate of antibiotic resistant bacteria in wastewater treatment plants: A review. *Ecotoxicology and Environmental Safety, 91*, 1–9. https://doi.org/10.1016/j.ecoenv.2013.01.016.

Cesare, Di, A. , Eckert, M. , E., D'Urso, S., Bertoni, R., Gillan, D. C., et al. (2016). Co-occurrence of integrase 1, antibiotic and heavy metal resistance genes in municipal wastewater treatment plants. *Water Research, 94*, 208–214. https://doi.org/10.1016/j.watres.2016.02.049.

Chen, J., Ying, G. G., Wei, X. D., Liu, Y. S., Liu, S. S., Hu, L. X., et al. (2016). Removal of antibiotics and antibiotic resistance genes from domestic sewage by constructed wetlands: Effect of flow configuration and plant species. *Science of the Total Environment, 571*, 974–982. https://doi.org/10.1016/j.scitotenv.2016.07.085.

Chiang, Y. N., Penadés, J. R., & Chen, J. (2019). Genetic transduction by phages and chromosomal islands: The new and noncanonical. *PLoS Pathogens, 15*(8). https://doi.org/10.1371/journal.ppat.1007878.

Christgen, B., Yang,, Y., Ahammad, S. Z., Li, B., Rodriquez, D. C., Zhang, T., et al. (2015). Metagenomics shows that low-energy anaerobic-aerobic treatment reactors reduce antibiotic resistance gene levels from domestic wastewater. *Environmental Science & Technology, 49*, 2577–2584. doi:10.1021/es505521w.

Cui, E., Wu, Y., Zuo, Y., & Chen, H. (2016). Effect of different biochars on antibiotic resistance genes and bacterial community during chicken manure composting. *Bioresource Technology, 203*, 11–17. doi:10.1016/j.biortech.2015.12.030.

Davies, J., & Davies, D. (2010). Origins and evolution of antibiotic resistance. *Microbiology and Molecular Biology Reviews, 74*(3), 417–433. https://doi.org/10.1128/MMBR.00016-10.

Dcosta, V. M., King, C. E., Kalan, L., Morar, M., Sung, W. W. L., Schwarz, C., et al. (2011). Antibiotic resistance is ancient. *Nature, 477*(7365), 457–461. https://doi.org/10.1038/nature10388.

Devarajan, N., Laffite, A., Graham, N. D., Meijer, M., Prabakar, K., Mubedi, J. I., et al. (2015). Accumulation of clinically relevant antibiotic-resistance genes, bacterial load, and metals in freshwater lake sediments in central europe. *Environmental Science and Technology, 49*(11), 6528–6537. https://doi.org/10.1021/acs.est.5b01031.

Ding, C., & He, J. (2010). Effect of antibiotics in the environment on microbial populations. *Applied Microbiology and Biotechnology, 87*(3), 925–941. https://doi.org/10.1007/s00253-010-2649-5.

Done, H. Y., & Halden, R. U. (2015). Reconnaissance of 47 antibiotics and associated microbial risks in seafood sold in the United States. *Journal of Hazardous Materials, 282*, 10–17. https://doi.org/10.1016/j.jhazmat.2014.08.075.

Du, J., Geng, J., Ren, H., Ding, L., Xu, K., & Zhang, Y. (2015). Variation of antibiotic resistance genes in municipal wastewater treatment plant with A2O-MBR system. *Environmental Science and Pollution Research, 22*(5), 3715–3726. https://doi.org/10.1007/s11356-014-3552-x.

Ekpeghere, K. I., Lee, J. W., Kim, H. Y., Shin, S. K., & Oh, J. E. (2017). Determination and characterization of pharmaceuticals in sludge from municipal and livestock wastewater treatment plants. *Chemosphere, 168*, 1211–1221. https://doi.org/10.1016/j.chemosphere.2016.10.077.

Fang, H., Zhang, Q., Nie, X., Chen, B., Xiao, Y., Zhou, Q., et al. (2017). Occurrence and elimination of antibiotic resistance genes in a long-term operation integrated surface flow constructed wetland. *Chemosphere, 173*, 99–106. doi:10.1016/j.chemosphere.2017.01.027.

Gao, P., Munir, M., & Xagoraraki, I. (2012). Correlation of tetracycline and sulfonamide antibiotics with corresponding resistance genes and resistant bacteria in a conventional municipal wastewater treatment plant. *Science of the Total Environment, 421–422*, 173–183. https://doi.org/10.1016/j.scitotenv.2012.01.061.

Gillings, M. R. (2014). Integrons: Past, present, and future. Microbiology and. *Molecular Biology Reviews, 78*(2), 257–277. https://doi.org/10.1128/MMBR.00056-13.

Gullberg, E., Cao, S., Berg, O. G., Ilbäck, C., Sandegren, L., Hughes, D., et al. (2011). Selection of resistant bacteria at very low antibiotic concentrations. *PLoS Pathogens, 7*(7), e1002158. https://doi.org/10.1371/journal.ppat.1002158.

Jury, K. L., Vancov, T., Stuetz, R. M., & Khan, S. J. (2010). Antibiotic resistance dissemination and sewage treatment plants. *Current Research, Technology and Education. Topics in Applied Microbiology and Microbial Biotechnology, 2*, 509–519.

Kim, S., & Aga, D. S. (2007). Potential ecological and human health impacts of antibiotics and antibiotic-resistant bacteria from wastewater treatment plants. *Journal of Toxicology and Environmental Health, 10*(8), 559–573. Part B https://doi.org/10.1080/15287390600975137.

Klümper, U., Recker, M., Zhang, L., Yin, X., Zhang, T., Buckling, A., et al. (2019). Selection for antimicrobial resistance is reduced when embedded in a natural microbial community. *ISME Journal, 13*(12), 2927–2937. https://doi.org/10.1038/s41396-019-0483-z.

Knapp, C. W., Callan, A. C., Aitken, B., Shearn, R., Koenders, A., & Hinwood, A. (2017). Relationship between antibiotic resistance genes and metals in residential soil samples from Western Australia. *Environmental Science and Pollution Research, 24*(3), 2484–2494. https://doi.org/10.1007/s11356-016-7997-y.

Li, N., Sheng, G. P., Lu, Y. Z., Zeng, R. J., & Yu, H. Q. (2017). Removal of antibiotic resistance genes from wastewater treatment plant effluent by coagulation. *Water Research, 111*, 204–212. doi:10.1016/j.watres.2017.01.010.

Mazel, D. (2006). Integrons: Agents of bacterial evolution. *Nature Reviews Microbiology, 4*(8), 608–620. https://doi.org/10.1038/nrmicro1462.

McKinney, C. W., & Pruden, A. (2012). Ultraviolet disinfection of antibiotic resistant bacteria and their antibiotic resistance genes in water and wastewater. *Environmental Science and Technology, 46*(24), 13393–13400. https://doi.org/10.1021/es303652q.

Pal, C., Asiani, K., Arya, S., Rensing, C., Stekel, D.J., Larsson, D.G. J., et al. (2017). Metal resistance and its association with antibiotic resistance. In: *Advances in Microbial Physiology*, 70, 261–313. Academic Press (US). https://doi.org/10.1016/bs.ampbs.2017.02.001.

Sabri, N. A., Schmitt, H., Van der Zaan, B., Gerritsen, H. W., Zuidema, T., Rijnaarts, H. H. M., et al. (2020). Prevalence of antibiotics and antibiotic resistance genes in a wastewater effluent-receiving river in the Netherlands. *Journal of Environmental Chemical Engineering, 8*(1), 102245. https://doi.org/10.1016/j.jece.2018.03.004.

Schlüter, A., Szczepanowski, R., Pühler, A., & Top, E. M. (2007). Genomics of IncP-1 antibiotic resistance plasmids isolated from wastewater treatment plants provides evidence for a widely accessible drug resistance gene pool. *FEMS Microbiology Reviews, 31*(4), 449–477. https://doi.org/10.1111/j.1574-6976.2007.00074.x.

Shahverdi, A. R., Fakhimi, A., Shahverdi, H. R., & Minaian, S. (2007). Synthesis and effect of silver nanoparticles on the antibacterial activity of different antibiotics against Staphylococcus aureus and Escherichia coli. *Nanomedicine: Nanotechnology, Biology, and Medicine, 3*(2), 168–171. https://doi.org/10.1016/j.nano.2007.02.001.

Sharma, V. K., Johnson, N., Cizmas, L., McDonald, T. J., & Kim, H. (2016). A review of the influence of treatment strategies on antibiotic resistant bacteria and antibiotic resistance genes. *Chemosphere, 150*, 702–714. doi:10.1016/j.chemosphere.2015.12.084.

Sharma, V. K., Siskova, K. M., Zboril, R., & Gardea-Torresdey, J. L. (2014). Organic-coated silver nanoparticles in biological and environmental conditions: Fate, stability and toxicity. *Advances in Colloid and Interface Science, 204*, 15–34. https://doi.org/10.1016/j.cis.2013.12.002.

Smith, R., & Coast, J. (2013). The true cost of antimicrobial resistance. *BMJ, 346*(mar11 3), f1493. https://doi.org/10.1136/bmj.f1493.

Stepanauskas, R., Glenn, T. C., Jagoe, C. H., Tuckfield, R. C., Lindell, A. H., & McArthur, J. V. (2005). Elevated microbial tolerance to metals and antibiotics in metal-contaminated industrial environments. *Environmental Science and Technology, 39*(10), 3671–3678. https://doi.org/10.1021/es048468f.

Szczepanowski, R., Linke, B., Krahn, I., Gartemann, K. H., Gützkow, T., Eichler, W., et al. (2009). Detection of 140 clinically relevant antibiotic-resistance genes in the plasmid metagenome of wastewater treatment plant bacteria showing reduced susceptibility to selected antibiotics. *Microbiology (Reading, England), 155*(7), 2306–2319. https://doi.org/10.1099/mic.0.028233-0.

Tuckfield, R. C., & McArthur, J. V. (2008). Spatial analysis of antibiotic resistance along metal contaminated streams. *Microbial Ecology, 55*(4), 595–607. https://doi.org/10.1007/s00248-007-9303-5.

Xiao, F., Simcik, M. F., & Gulliver, J. S. (2013). Mechanisms for Removal of Perfluorooctane Sulfonate (PFOS) and Perfluorooctanoate (PFOA) from drinking water by conventional and enhanced coagulation. *Water Research, 47*, 49–56. doi:10.1016/j.watres.2012.09.024.

Yazdankhah, S., Rudi, K., & Bernhoft, A. (2014). Zinc and copper in animal feed – development of resistance and co-resistance to antimicrobial agents in bacteria of animal origin. *Microbial Ecology in Health & Disease, 25*. doi:10.3402/mehd.v25.25862.

Ye, M., Sun, M., Feng, Y., Wan, J., Xie, S., Tian, D., Zhao, Y., Wu, J., Hu, F., Li, H., & Jiang, X. (2016). Effect of biochar amendment on the control of soil sulfonamides, antibiotic-resistant bacteria, and gene enrichment in lettuce tissues. *Journal of Hazardous Materials, 309*, 219–227. https://doi.org/10.1016/j.jhazmat.2015.10.074.

Zhang, S., Song, H. L., Yang, X. L., Li, H., & Wang, Y. W. (2018). A system composed of a biofilm electrode reactor and a microbial fuel cell-constructed wetland exhibited efficient sulfamethoxazole removal but induced sul genes. *Bioresource Technology, 256*, 224–231. https://doi.org/10.1016/j.biortech.2018.02.023.

Zhang, Y., Marrs, C. F., Simon, C., & Xi, C. (2009). Wastewater treatment contributes to selective increase of antibiotic resistance among Acinetobacter spp. *Science of the Total Environment, 407*(12), 3702–3706. https://doi.org/10.1016/j.scitotenv.2009.02.013.

Zhuang, Y., Ren, H., Geng, J., Zhang, Y., Zhang, Y., Ding, L., et al. (2016). Inactivation of antibiotic resistance genes in municipal wastewater by chlorination, ultraviolet, and ozonation disinfection. *Environmental Science and Pollution Research, 22*, 7037–7044. doi:10.1007/s11356-014-3919-z.

Chapter 3

Enrichment of drug resistance genes in human pathogenic bacteria showing antimicrobial resistance

Karuna Singh[a] and Radha Chaube[b]
[a] *Department of Zoology, Mahila Maha Vidyalaya, Banaras Hindu University, Varanasi, India,*
[b] *Department of Zoology, Institute of Science, Banaras Hindu University, Varanasi, India*

3.1 Introduction

Microbial infections are responsible of morbidity and mortality worldwide. There is ample of drugs being used to combat such infectious diseases. The discovery of antibiotics has been regarded as the most important event in the history of chemotherapy. But, later on it was observed that most pathogenic bacteria have the capability of developing resistance to at least some antimicrobial agents. The genetic plasticity of bacteria and the abusive use of antibiotics both in the agriculture as well as in clinical set up contribute significantly in the emergence of the multidrug resistant (MDR) bacteria. According to WHO, the respiratory infection and diarrhea caused by the MDR bacteria are amongst top ten diseases responsible for societal morbidity and mortality (World Health Organization, 2014). In recent years, advent of antimicrobial resistance has added significantly in the number of infections as well as healthcare costs. These concerns prompted the WHO to launch a Global Action Plan on AMR in 2015 (World Health Organization, 2015).

It is only natural that organisms that produce antibiotics should also contain self-resistance mechanisms against their own antibiotics. Resistance determinants found in bacteria have generated significant attention in recent years because of their possible link with the emergence of resistance in pathogenic clinical isolates (Martinez, 2018; Surette and Wright, 2017). Indeed, with the global epidemic of antibiotic resistance (ABR) unfolding before us, it is important to understand the origin of these determinants in pathogens.

Various researchers have evidence to support the idea that feeding antibiotics to animals may result in the development of antimicrobial resistant organisms,

Antimicrobial Resistance in Wastewater and Human Health.
DOI: https://doi.org/10.1016/B978-0-323-96124-0.00008-8

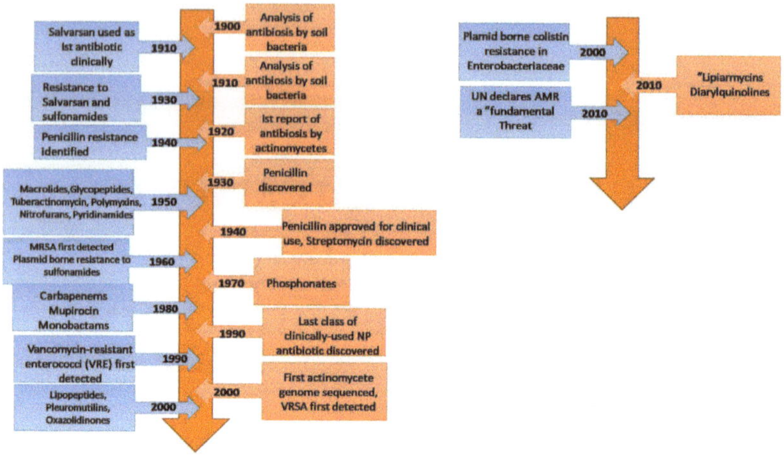

FIGURE 3.1 A journey of past, present, and future antibiotic drug development and discovery is shown in this figure with timeline.

and that those resistant organisms may be transferred to the humans who consume those animals (Wegner, 2012). For example, in salmon aquaculture antimicrobial agents used in farming are mainly administered to the fish through medicated feed, thus there is significant potential for a large proportion of the drug to enter the environment via uneaten medicated feed in addition to through urinary and faecal excretion.

3.2 History

When we look back in the history of drugs, it was Penicillin, the first natural antibiotic to be discovered accidentally by Alexander Fleming in 1928, when the *Penicillium notatum* contaminated a culture plate in his laboratory. However, penicillin was not used as an antibiotic until the late 1930s (Hopwood, 2007). Penicillin as an inhibitor of cell wall synthesis was found effective against Gram-positive but not against Gram-negative bacteria. Following the discovery of penicillin by Fleming, other microbiologists, including Rene Dubos and Selman Waksman, started a deliberate search for antibacterial agents among soil microorganisms, including bacteria and fungi.

The isolation of streptomycin (produced by *Streptomyces griseus*) in 1943 was the landmark event that pioneered the golden age of antibiotic discovery (1940–1990). Streptomycin inhibits protein synthesis by binding to the 30S subunit of the prokaryotic ribosome and was found to be effective not only against Gram-negative bacteria but also against the tubercle bacillus (Hopwood, 2007). A journey of past, present, and future antibiotic drug development and discovery is shown in Fig. 3.1 which shows the timeline, and Fig. 3.2 depicts the history of AMR.

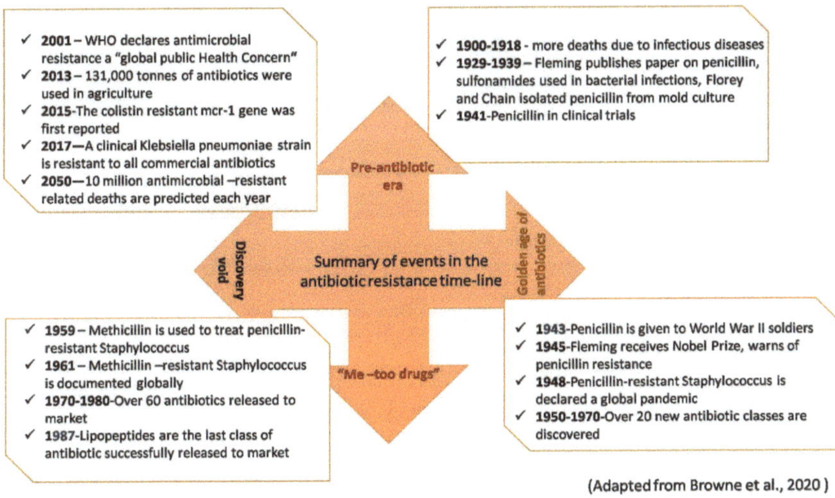

FIGURE 3.2 Depicts the history of antimicrobial resistance.

3.3 Pathogenic bacteria

Bacteria are microorganisms made of a single cell. They are diverse and have variety of shapes and features. They can survive in any (extrinsic or intrinsic) environmental condition. Most of the bacterial strains are harmless and beneficial, but some are infectious or cause various diseases termed as pathogenic bacteria. They are capable of causing disease when enters into the body and can be spread through water, air, soil, and also through physical contact. The common diseases caused by bacteria, their symptoms and treatment strategies are shown in Table 3.1A. Twelve families of bacteria are declared as the highest risk priority pathogens by WHO on Feb. 27, 2021. The antibiotic-resistant bacteria or priority pathogens are classified into critical, high, and medium priority groups (Fig. 3.3).

3.4 Drugs against pathogenic bacteria

The antibiotics have been used for treating or preventing disease since long. It has been noticed that continued increases in AMR have led to fewer treatment options for patients, and an associated increase in morbidity and mortality.

The antibacterial drugs are classified as follows:

1. Bacteriocidal, (if they kill the bacteria), or
2. Bacteriostatic (if they just prevent the bacterial growth).

Thus, based on these properties there are many types and classes of antibiotics and each class inhibits a process that is different in the pathogen from that found in the host (Table 3.1A and Table 3.1B). For example—chloramphenicol and tetracycline exhibit selective toxicity thus inhibit the bacterial ribosome not the structurally different eukaryotic ribosome.

TABLE 3.1A List of some selected pathogenic bacteria causing human diseases.

Bacteria	Genus	Disease	Symptom	Treatment
Bordatella pertussis is a gram negative, aerobic, pathogenic, encapsulated coccobacillus	Bordetella	Whooping cough, secondary bacterial pneumonia	Highly contagious bacterial disease, initial sympoms are usually similar to those of the common cold with a runny nose, fever and mild cough, but these are followed by weeks of severe coughing fits, following a fit of coughing, a high-pitched whoop sound or gasp may occur as the person breathes in	Macrolides such as erythromycin, before paroxysmal stage **Prevention:** Pertussis vaccone, such as in DPT vaccine.
Yersinia pestis is a gram negative, non-motile, odd-shaped, coccobacillus bacterium without spores that is related to both Yersinia pseudotuberculosis and Yersinia entero-olitica	Yersinia	Bubonic plague, plague takes three main forms: pneumonic, septicemic, bubonic	One to seven days after exposure to the bacteria, flu-like symptoms develop. These symptoms include fever, headaches and vomiting, as well as swollen and painful lymph nodes occurring in the area closest to where the bacteria entered the skin. Occassionally the swollen lymph nodes known as buboes may break open.	Streptomycin primarily, teracyclin, supportive therapy for shock **Prevention:** Plague vaccine, minimize exposure to rodents and fleas.
Mycobacterium tuberculosis: It requires oxygen to grow, it is debated whether it produces spores and is non-motile. M. tuberculosis has an unusual, waxy coating on its cell surface primarily due to the presence of mycolic acid. This coating makes the cells impervious to Gram staining, and as a result, *M. tuberculosis* can appear either Gram-negative or Gram-positive. Acid-fast stains such as Ziehl-Neelsen or fluorescent stains such as auramine are used instead to identify *M. tuberculosis* with a microscope.	Mycobacterium	TB(Tuberculosis)	Tuberculosis: chronic cough with blood-containing sputum, fever, night sweats and weight loss	Standard "short" course: First 2 months combination: Isoniazid, Rifampicin, Pyrazinamide, Ethambutol Further 4 months combination: Isoniazid, Rifampicin **Prevention:** BCG vaccine, Isoniazid

TABLE 3.1B List of some selected pathogenic bacteria causing human diseases.

Bacteria	Genus	Disease	Symptom	Treatment
Vibrio cholerae is a species of Gram negative, facultative anaerobe and comma-shaped bacteria.	Vibrio	Cholera: severe "rice water"diarrhea	The classic symptom is large amounts of watery diarrhea that lasts a few days. Vomiting and muscle cramps may also occur. Diarrhea can be so severe that it leads within hours to severe dehydration and electrolyte imbalance. This may result in sunken eyes, cold skin decreased skin elasticity and wrinkling of the hands and feet. Dehydration can cause the skin to turn bluish.	Fluid and electrolyte replacement, Doxycycline **Prevention:**Proper sanitation, adequate food preparation.
S. typhi is a subspecies of *Salmonella enterica*, the rod shaped, flagellated, aerobic, gram negative bacterium.	Salmonella	Typhoid fever type salmonellasis (fever, abdominal pain, hepatos plenomegaly, rose spots, chronic carrier state	This is commonly accompanied by weakness, abdominal pain, constipation, headaches and mild vomiting. Some people develop a skin rash with rose colored spots. In severe cases, people may experience confusion. Without treatment, symptoms may last weeks or months.	Ceftriaxone, Fluoroquinolones e.g ciprofloxacin Prevention: Ty21a and ViCPS vaccines, Hygiene and food preparation.
Clostridium botulinum is a gram –positive, rod shaped, anaerobic, spore-forming, motile bacterium with the ability to produce the neurotoxin botulinum.	Clostridium	Botulism: Mainly muscle weakness and paralysis	The disease begins with weakness, blurred vision, feeling tired and trouble speaking. This may then be followed by weakness of the arms, chest muscles and legs. Vomiting, swelling of the abdomen and diarrheamay also occur. The disease does not usually affect consciousness or causes a fever.	Antitoxin, Penicillin, Hyperbaric oxygen, Ventilation Preventon: Proper foodpreservation technique.
Clostridium tetani is a commn soil bacterium and the causative agent of tetanus. It is a rod shaped, gram –positive bacterium, up to 0.5 μm wide and 2.5 μm long. It is motile by way of Various flagella that surrounds its body.	Clostridium	Tetanus: muscle spasms	Spasms occur frequently for three to four weeks. Some spasms may be severe enough to fracture bones. Other symptoms of tetanus may include fever, sweating, headache, trouble swallowing, high blood pressure, and afast heart rate. Onset of symptoms is typically three to twenty one days following infection	Tetanus immune globulin, sedatives, muscle relaxants, mechanical ventilation, Penicilin or metronidazole Prevention: Tetanus vaccine (such as in the DPT vaccine)

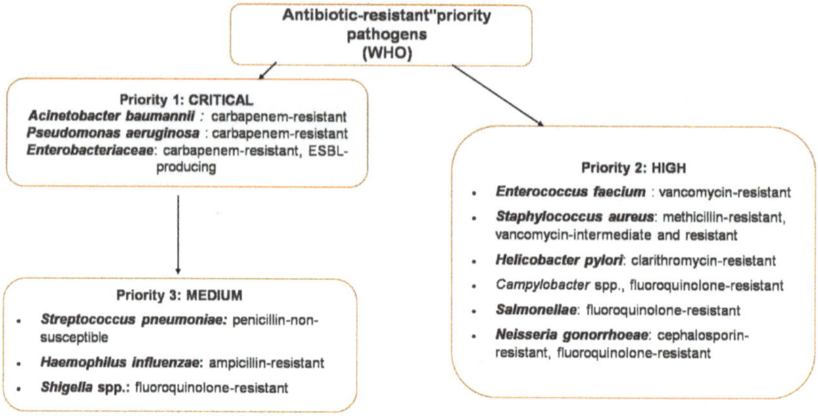

FIGURE 3.3 The antibiotic-resistant bacteria or priority pathogens are classified into critical, high, and medium priority groups.

Antibacterial agents can be divided into following groups based on their bacterial targets:

- agents that inhibit cell wall synthesis,
- cell membrane depolarizing agents,
- agents that inhibit protein synthesis,
- inhibitors of nuclei acid synthesis, and
- agents that inhibit metabolic pathways in bacteria.

Currently, antibiotics affecting almost every process in the bacterial cell are known. Based on their structure and mode of action, at least seven major groups of antibiotics have been described. These include

1. β-lactams (inhibit cell wall synthesis),
2. aminoglycosides (protein synthesis),
3. macrolides (protein synthesis),
4. tetracyclines (protein synthesis),
5. daptomycin (cell membrane function),
6. platensimycin (fatty acid biosynthesis), and
7. glycopeptides (cell wall synthesis).

Antibiotics are used both in treating human disease and in intensive farming to promote animal growth. Both uses may be contributing to the rapid development of ABR in bacterial populations.

3.5 Drug resistance

Bacteria may also acquire resistance genes from other related organisms, and the level of resistance will vary depending on the species and the genes acquired

(Coculescu, 2009; Martinez, 2014). Pathogenic bacteria exhibit two mechanisms of ABR- intrinsic (always expressed in the species), or acquired (the genes are naturally occurring in the bacteria, but are only expressed to resistance levels after exposure to an antibiotic). The intrinsic mechanism is a species-specific trait that neither requires prior exposure of antibiotics nor is associated with horizontal gene transfer (HGT) (Cox & Wright, 2013; Martinez, 2014). The nonpathogenic commensals can convert into opportunistic pathogens owing to the intrinsic mechanism. Because of mobile genetic elements (MGEs), acquired resistance has more severe consequences. Bacteria acquire ABR through the following methods.

3.6 Enzymatic modification and inactivation

β-lactamases are a family of enzymes found in many bacterial pathogens. AmpC β-lactamase of *E. coli* was the first enzyme found to have resistance to a β lactam antibiotic, penicillin (Abraham & Chain, 1940). Since then, a number of β-lactamase have been reported to produce against β-lactam antibiotics (Table 3.2). The β-lactamase degrades the β-lactam ring thus inhibiting the binding of a drug to its target. Carbapenemases are the most problematic β-lactamases because they break down all members of the β-lactam family of antibiotics (including carbapenems), severely limiting treatment options.

3.7 Antibiotic target site alterations

Alterations in antibiotic target sites can lead to low binding affinity of ABDs to their binding sites thus resulting in drug resistance. Both Gram-positive and Gram-negative bacteria gain ABR through this mechanism. The steric hindrance caused by acetylases, adenylases, and phosphorylases decreases the binding affinity of chloramphenicol, aminoglycosides, and lincosamides to their respective targets (Munita & Arias, 2016). The resistance for zorbamycin, tallysomycin, bleomycin, and phleomycin is achieved through acetylation (Peterson & Kaur, 2018). The binding affinity of the drug for its target is also reduced by mutational changes in the target site (ex- rifampin resistance), enzymatic alteration of the target site (Cfr-mediated linezolid resistance), target protection (tetracycline resistance determinants Tet(M) and Tet(O)) and target modification (Munita & Arias, 2016).

3.8 Antibiotic efflux and change in the permeability of bacterial cell wall

Efflux of antibiotic drugs from the bacterial cells leads to ABR. In the 1980s, the mechanism of efflux was first time reported in *E. coli*. The bacterium was

TABLE 3.2 Types of β-lactamases.

S.no.	Charcteristics/basis of classification	Groups				References
		1	2	3	4	
1.	Biochemical	Serine-β-lactamase	Serine-β-lactamase	Metello- β-lactamase	Serine-β-lactamase	Wright (2005)
2.	Amino acid homology	Ambler A β-lactamase	Ambler B metallic enzyme with Zn cofactor	Ambler C	Ambler D	Ambler (1980); Poirel, Naas, and Nordmann (2010)
3.	Hydrolyzing activity	Narrow	Moderate	Broad; penicillin- and cephalosporin-resistant; not inhibited by β-lactamase inhibitors (Tazobactam and clavulanic acid)	Extended spectrum β-lactamase (ESBL); formed by the point mutation in the parent enzyme (*BlaTEM*); resistant to the first, second, and third generations of cephalosporins, penicillin, and aztreonam; inhibited by β-lactamase inhibitors; CTX-M, a major cause of concern	Canton et al. (2012)

seen to efflux tetracycline, an antibiotic drug (Li, Sharma, & Kaur, 2014). Since then, it has been reported in many Gram-positive and Gram-negative bacteria. In general, the bacterial efflux pumps are substrate-specific, but sometimes they may exhibit broad-spectrum substrate activity (Poole, 2005, 2015). Moreover, any change (even small) in porins-mediated permeability can reduce the importation of ABD in the bacterial cell. Tetracyclines, β-lactams, and other hydrophilic antibiotics are affected by this strategy (Munita & Arias, 2016). Based on the substrate/s, conformational changes, energy source, and bacterium type, the efflux pumps of bacteria can be classified into the following five types:

a) Major facilitator superfamily transporters.
b) ATO-binding cassette transporters.
c) The multidrug and toxic compound extrusion.
d) The small drug resistance.
e) Resistance-nodulation-cell division (Peterson & Kaur, 2018; Piddock, 2006).

3.9 Degradation of antibiotic drugs (ABDs)

This is the most common mechanism of ABR. The bacteria are known to have several enzymes which can modify or degrade the antibiotic molecules. The modification enzymes were first time reported in *Streptomyces* species in the 1970s (Walker & Walker, 1970). The well-known examples of antibiotic modification enzymes are aminoglycoside modification enzymes (AMEs).

1. Overproduction:
   The bacteria *Haemophilus influenza* and *E. coli* utilize this strategy to confer the resistance to trimethoprim (TMP). The bacteria outnumber the antibiotic molecules by their overproduction. TMP along with sulfamethoxazole (TMP-SMX) hampers DNA synthesis by inhibiting two sequential enzymes dihydrofolate reductase and dihydropteroate synthase resulting in a lower number of antibiotics compared to their targets (Kakoullis et al., 2021; Walsh, 2016).
2. Bifunctional enzymes:
   Tp47, the first enzyme having two active sites (one for PBP activity and the second for β-lactamase activity) was isolated from syphilis causing bacterium *Treponema palladium*. These enzymes are encoded by two linked genes, thus causing a notable increase in substrate specificity and resistance against many ABDs (Egorov, Ulyashova, & Rubtsova, 2018). The other enzymes are bifunctional β-lactamase, blaLRA-13, bifunctional AAC(6')-Ie/APH(2")-Ia, ANT(3")-Ii/AAC(6')-IId, AAC(6')-30/AAC(6') - Ib, AAC(3)-Ib/ AAC(6')-Ib' and AAC(6')-Ib-cr acetyltransferase (Boehr, Daigle, & Wright, 2004; Cha, Ishiwata, & Mobashery, 2004; Robicsek et al., 2006; Allen et al., 2009).

3.10 Antibiotic resistance genes (ABRs)

Gene plasticity allows bacteria to acquire resistance against antibiotics which can be achieved by mutations or by HGT (Sharma, Johnson, Cizmas, McDonald, & Kim, 2016). ABR in bacteria is originated either as the antibiotic selection pressure posed by the antibiotic therapy or as natural selection during the course of phylogeny. Some nonantibiotic toxic compounds like detergents are also contributed to the emergence of multidrug resistance genes (MDRGs) in bacteria (Alonso, SaÂnchez, & MartõÂnez, 2001). It has been evident that firstly apart from antibiotic producers, the nonproducers also have ARGs and secondly, the genes involved in the other process can act as ARGs during an exaptation (natural selection) mechanism. Ex- heavy metal (silver)/ABR gene linkage in clinical practices. The ABR gene families and their characteristics are summarized in Table 3.3.

(1) AMR databases and ARGs prediction tools—Several ARG databases have been developed in the recent past which have comprehensive information about ARGs. They facilitate the accurate identification and characterization of antimicrobial genes (AMGs), thus expedite the process of rational designing of antibacterial drugs.
   a. Databases
      i. National database of antibiotic resistant organisms (NDARO)—NDARO was developed in 2015, as a part of National Action Plan of White house for combating ABR. This curated database maintains by National Center for Biotechnology Information (NCBI) together with other outside agencies like FDA, USDA, WHO, CDC, PHE and is widely available for public. An AMRFinderPlus has also developed for the identification of AMR genes in bacterial genome. The "Isolate browser" of this user friendly database permits rapid identification of bacterial genome having ARGs. NCBI's browsers such as Pathogen Detection Reference Gene Catalog, Pathogen Detection Reference HMM Catalog, Microbial Browser for Identification of Genetic and Genomic Elements (MicroBIGG-E), Genomes with AMR genotypes or phenotypes, and AMR-related sequence and phenotype data submission tool of NDARO ease real-time surveillance of microorganisms having AMGs (https://www.ncbi.nlm.nih.gov/pathogens/antimicrobial-resistance).
      ii. The Comprehensive Antibiotic Resistance Database (CARD; http://arpcard.mcmaster.ca)—This database is a collection of ABGs and their products and ABG-associated phenotypes. CARD curates three branches of resistance determinants, namely, ABR, its mechanism and molecules having antibiotic property. The addition of mobile genetic and virulence elements is in progress. The quick

TABLE 3.3 Antibiotic resistance gene families and their characteristics.

S. no.	ARG family	Description	Resistant genera	Antibiotic drug	Source	Mode of action	References
1	aadA	*aadA* family of *genes* encode aminoglycoside-3″-adenylyltransferases (AAD)	*Pseudomonas, Riemerella*		Natural or Semi synthetic	Inhibition of cell membrane and protein syntheses	van Hoek et al. (2011), Kwiecień, Ilona, Chrobak-Chmie, Agnieszka, and Magdalena (2020)
2	cat	encodes chloramphenicol acetyltransferases (CATs)	*Acinetobacter, Escherichia, Klebsiella, Salmonella, Serratia, Shigella*	Chloramphenicol	*Streptomyces venezuelae*	Inhibition of protein synthesis	van Hoek et al. (2011), Sultan et al. (2018)
3	qnrA	QnrA (now termed QnrA1), is a protein of 218 amino-acids that belongs to the repeat protein family	*Citrobacter, Enterobacter, Escherichia, Klebsiella, Shigella*	Quinolone	Several plant, animal and microbial species	Inhibit replication of DNA in bacteria by blocking ligase domain of bacterial DNA gyrase	van Hoek et al. (2011), Sultan et al. (2018)
4	Erm (B)	ErmB confers the MLSb phenotype. Its expression is inducible by erythromycin	*Aggregatibacter, Acinetobacter, Aerococcus, Arcanobacterium, Bacillus, Bacteroides, Citrobacter, Corynebacterium, Clostridium, Enterobacter, Escherichia, Eubacterium, Enterococcus, Fusobacterium, Gemella, Haemophilus, Klebsiella, Lactobacillus, Micrococcus, Neisseria, Pantoea, Pediococcus, Peptostreptococcus, Porphyromonas, Proteus, Pseudomonas, Ruminococcus, Rothia, Serratia, Staphylococcus, Streptococcus, Treponema, Wolinella*	Macrolide–lincosamide–streptogramin B (MLS)	*Streptomyces* spp.	rRNA methylase	van Hoek et al. (2011), Pernodet, Fish, Blondelet-Rouault, and Cundliffe (1996)

(*continued on next page*)

TABLE 3.3 Antibiotic resistance gene families and their characteristics—cont'd

S. no.	ARG family	Description	Resistant genera	Antibiotic drug	Source	Mode of action	References
5	tet(A)	The TET enzymes catalyze the hydroxylation of DNA 5-methylcytosine (5mC) to 5-hydroxymethylcytosine (5hmC), and can further catalyse oxidation of 5hmC to 5-formylcytosine (5fC) and then to 5-carboxycytosine (5caC).	*Acinetobacter, Aeromonas, Bordetella, Chryseobacterium, Citrobacter, Edwardsiella, Enterobacter, Escherichia, Flavobacterium, Klebsiella, Laribacter, Plesiomonas, Proteus, Pseudomonas, Salmonella, Serratia, Shigella, Variovorax, Veillonella, Vibrio*	Tetracycline	fermentation of *Streptomyces aureofaciens e Streptomyces rimosus*	Efflux	Chopra and Roberts (2001), van Hoek et al. (2011), Sultan et al. (2018)
6	dfrA1	*dfrA1* is an integron-encoded dihydrofolate reductase	*Actinobacter, Enterobacter, Escherichia, Klebsiella, Morganella, Proteus, Pseudomonas, Salmonella, Serratia, Shigella, Vibrio*	trimethoprim	synthetic	Inhibits bacterial dihydrofolate reductase (DHFR)	van Hoek et al. (2011), Wüthrich et al. (2019)
7	Sul genes	*sul genes* have been identified on both chromosomes and plasmids and are often associated with MGEs such as transposons, integrons and insertion etc.	*Escherichia coli and Salmonella*	Sulfonamide	synthetic	Inhibit dihydropteroate synthetase which leads to suppressed DNA replication and bacteriostatic activity against both (aerobic) Gram-positive and -negative bacteria	Fair and Tor (2014), Sultan et al. (2018)

(*continued on next page*)

8	ESBL genes	Located on plasmids, Some are mutant derivatives of plasmid-mediated β-lactamases (e.g., $bla_{TEM/SHV}$), and mobilized from environmental bacteria($bla_{CTX\text{-}M}$).	Members of Enterobacteriaceae	β-lactams	natural or semisynthetic	Interference with cell wall synthesis	van Hoek et al. (2011), Navon-Venezia, Kondratyeva, and Carattoli (2017), Sultan et al. (2018)
9	CRDG genes	carbapenem resistance determining genes (CRDG): bla_{VIM}, bla_{IMP}, bla_{KPC}, $bla_{OXA\text{-}48}$, and bla_{NDM}	*Klebsiella pneumoniae, E. coli*	Carbepenems	developed at from the carbapenem thienamycin, a natural derivative of *Streptomyces cattleya*	Penicillin binding proteins	Meletis (2016), Navon-Venezia et al. (2017), Sultan et al. (2018)
10	mcr	The mobilized colistin resistance (mcr) gene confers plasmid-mediated resistance to colistin mcr-1, the original variant has capability of HGT.	*Pseudomonas aeruginosa, Acinetobacter baumannii,* Enterobacteriaceae members, such as *E. coli, Salmonella* spp., and *Klebsiella* spp.	Colistin	produced by the soil bacterium *Bacillus polymyxa*	LPS (lipopolysaccahride layer of bacteria)	Liu et al. (2016), Sultan et al. (2018)
11	*van*	Van genes are the most common phenotypes observed in nosocomial environment.	VanA-enterococci and staphylococci, and VanB –enterococci	Glycopeptide	produced by *Amycolatopsis orientalis*, a soil bacterium and semisynthetic	Inhibits peptidoglycan synthesis thus inhibiting bacterial cell wall formation	Sultan et al. (2018), van Hoek et al. (2011), Yushchuk, Binda, and Marinelli (2020)

identification of putative ARGs, ontogenic curation of sequences and molecular data in antibiotic resistance ontogeny (ARO) are the unique features of CARD (McArthur et al., 2013). CARD2017 permits advance data sharing and resistome prediction using resistance gene identifier (RGI) (Jia et al., 2017). CARD is updating its genome analysis-related literature and computation text mining monthly. The tools and resources of CARD are as follows:

(1) Browser facilitates browsing of ARO and detection of AMR genes.
(2) The analysis tools BLAST and RGI (based on homology and SNP models).
(3) Ontogenies and data download.
(4) Predictors of computer-generated resistance and prevalence status of AMRGs.
(5) CARD: Live project which provides list of AMR genes along with submission details.

The Antibiotic Resistance Gene Online (ARGO)—ARGO is a curation of specifically vancomycin and βlactam resistance genes (Scaria, Chandramouli, & Verma, 2005). These two classes of antibiotics target cell wall synthesis in Gram-positive bacteria.

1. Deep learning models—Almost all the existing databases use "best hit" approach for the prediction or identification of ARGs which often gives false negative results. This limitation of ARG databases can be overcome by creating deep learning approaches.
   a. DeepARG—This deep learning approach has been used to predict ABRGs from metagenomic data. Arango-Argoty et al. (2018) constructed two deep learning models—DeepARG-SS for short-read sequences and DeepARG-LS for full gene length sequences. These models can predict ARGs with >0.97 (high) precision and >0.90 recall, thus giving lower rates of false-negative in comparison to best hit approach. DeepARG can identify a wide range of ARGs because the strict cutoffs are not required. Both the models of DeepARG are available at http://bench.cs.vt.edu/deeparg.
   b. Hierarchial multi-task deep learning of annotating antibiotic resistance genes (HMD-ARG)—It is the most recently developed deep learning framework for ARGs. Li et al. (2021) proposed an end-to-end HMD framework for ARG in Feb. 2021. Detailed annotations from resistant antibiotic class, gene mobility, and resistance mechanism of ARGs are provided in HMD framework, thus it can simultaneously identify the multiple properties of ARG. Database is available at http://www.cbrc.kaust.edu.sa/HMDARG/.
   c. Structured antibiotic resistance genes (SARGs)—An online analysis pipeline ARGs-OAPv2.0 including SARG database with ARG hierarchy

(AR gene-type-subtype-sequence) and Hidden Markov Models have been developed (Yin et al., 2018). The database allows quick quantification and characterization of ARGs environmental metagenic datasets. This online tool uses SARGfam (HMM) for model-based identification of sequences.

d. NanoARG-HGT and antibiotic-mediated selection pressure play important role in ABR. NanoARG is an online computational tool that can identify ARGs mobile genetic elements (MRGs) and coselection pressure indicators like metal resistance genes (MRGs), thus provides useful information on ABR mobility, virulence and co-selective forces (Arango-Argoty et al., 2019). NanoARG also allows the analysis of long sequences along with data processing and visualization.

e. ARGAnalyzer (ARGA)—In 2019, Wei et al. updated the sequence database of antibiotic resistance (SDARG), an online pipeline ARG analyzer. ARGA can analyze existing ARGs primer. As it annotates ARGs from metagenomic datasets, thus can be better alternative of environmental survey of ARGs (Wei et al., 2019).

f. Prediction of Antimicrobial resistant via Game theory (PARGT)—This bioinformatic tool is an open-source package that can predict AMRGs in both Gram-positive and Gram-negative bacteria. PARGT implements a machine-learning model and a game-theory algorithm for the identification of ARGs in the bacteria (Chowdhury, Call, & Broschat, 2020).

g. Antibiotic Resistance Gene-ANNOTation (ARG-ANNOT)—This was developed to identify putative new and existing ARGs in the bacterial genome. The Bio-Edit interface of this database can identify new genetic determinants of ARG in bacterial genomes (Gupta et al., 2014).

h. ResFinder 4.0- ResFinder 4.0 uses whole genome sequencing (WGS)-based antimicrobial susceptibility testing for the prediction of phenotypes. This database has species-specific *in-silico* antibiograms and genotype-phenotype key connecting determinants of AMP with phenotype (Bortolaia et al., 2020).

i. Multiple Antibiotic Resistance Annotator (MARA) and database—MARA database curates ABRs and some important mobile elements from Gram-negative bacteria only. MARA has the facility of sequence annotation and is a comprehensive database for ABR and ARGs of Gram-negative bacteria (Partridge & Tsafnat, 2018).

3.11 Future perspectives

The rise in bacterial infections that are resistant to almost all known antibiotics is alarming and scientifically, the identification of new chemical matter with the unique physicochemical characteristics required for antibiotic discovery and development is a key challenge. Various alternative approaches have been explored by scientists and researchers in past and recent times. Researchers look for ways to inhibit the formation of biofilm (Khan et al., 2019) by different agents isolated from plants, animals, and microbes. Recently chitosan and its derivatives have got more attention due to its properties like biodegradability, biocompatibility, nonallergic and nontoxicity. Researchers are trying to employ chitosan and its derivatives as effective agents to inhibit biofilm formation and attenuate virulence properties by various pathogenic bacteria. The antibiotic delivery system has been improving by the use of nanotechnology (Eleraky, Allam, Hassan, & Omar, 2020). With passive targeting efficiency and tailored physicochemical properties, the antibiotic nanosystems offer better alternative for combating the threats of AMR.

3.12 Conclusions

Bacteria are very versatile and adaptive organisms. In order to survive, they need to be capable of dealing with the toxic substances. Generally, bacteria that infect humans are able to defend themselves against antimicrobial agents. Recently, alarming increase in AMR among pathogenic bacteria has provoked scientists and clinicians to find out ways to combat these pathogens. Unfortunately, till date there is no answer to this problem. Now it is time demand to rethink and look for alternative how we design new antimicrobial agents or may start looking toward natural, recombinant, biological, DNA, RNA based substances for clues what could be used to overcome current challenges.

Acknowledgment

Authors are thankful to Ms Gunjan Uttam for helping in the compilation of Table 3.3.

References

Abraham, E., & Chain, E. (1940). An enzyme from bacteria able to destroy penicillin. *Nature, 146*, 837. https://doi.org/10.1038/146837a0.

Allen, H. K., Moe, L. A., Rodbumrer, J., Gaarder, A., & Handelsman, J. (2009). Functional metagenomics reveals diverse beta-lactamases in a remote Alaskan soil. *The ISME Journal, 3*(2), 243–251. Epub 2008 Oct 9. PMID: 18843302. doi:10.1038/ismej.2008.86.

Alonso, A., SaÂnchez, P., & MartõÂnez, J. L. (2001). Environmental selection of antibiotic resistance genes. *Environmental Microbiology, 3*(1), 1–9.

Ambler, R. P. (1980). The structure of beta-lactamases. *Philosophical Transactions of the Royal Society B: Biological Sciences, 289*, 321–331. doi:10.1098/rstb.1980.0049.

Aminov, R. I. (2010). A brief history of the antibiotic era: Lessons learned and challenges for the future. *Frontiers in Microbiology, 1*, 134. doi:10.3389/fmicb.2010.00134.

Arango-Argoty, G., Garner, E., Pruden, A., Heath, L. S., Vikesland, P., & Zhang, L. (2018). DeepARG: A deep learning approach for predicting antibiotic resistance genes from metagenomic data. *Microbiome, 6*, 23. https://doi.org/10.1186/s40168-018-0401-z.

Arango-Argoty, G. A., Dai, D., Pruden, A., Vikesland, P., Heath, L. S., & Zhang, L. (2019). NanoARG: A web service for detecting and contextualizing antimicrobial resistance genes from nanopore-derived metagenomes. *Microbiome, 7*(1), 88. PMID: 31174603; PMCID: PMC6555988. doi:10.1186/s40168-019-0703-9.

Boehr, D. D., Daigle, D. M., & Wright, G. D. (2004). Domain-domain interactions in the aminoglycoside antibiotic resistance enzyme AAC(6′)-APH(2″). *Biochemistry, 43*(30), 9846–9855. PMID: 15274639. doi:10.1021/bi049135y.

Bortolaia, V., Kaas, R. S., Ruppe, E., Roberts, M. C., Schwarz, S., Cattoir, V., et al. (2020). ResFinder 4.0 for predictions of phenotypes from genotypes. *Journal of Antimicrobial Chemotherapy, 75*(12), 3491–3500. PMID: 32780112; PMCID: PMC7662176. doi:10.1093/jac/dkaa345.

Canton, R, Maria, GAJ, & Carlos, GJ (2012). CTX-M enzymes: Origin and diffusion. *Frontiers in Microbiology, 3*, 110. doi:10.3389/fmicb.2012.00110.

Cha, J. Y., Ishiwata, A., & Mobashery, S. (2004, 9). A novel beta-lactamase activity from a penicillin-binding protein of Treponema pallidum and why syphilis is still treatable with penicillin. *Journal of Biological Chemistry, 279*(15), 14917–14921. Epub 2004 Jan 27, PMID: 14747460. doi:10.1074/jbc.M400666200.

Chopra, I., & Roberts, M. (2001). Tetracycline antibiotics: Mode of action, applications, molecular biology, and epidemiology of bacterial resistance. *Microbiology and Molecular Biology Reviews, 65*(2), 232–260. doi:10.1128/MMBR.65.2.232-260.2001. PMID: 11381101; PMCID: PMC99026.

Chowdhury, A. S., Call, D. R., & Broschat, S. L. (2020). PARGT: A software tool for predicting antimicrobial resistance in bacteria. *Science Reports, 10*, 11033. https://doi.org/10.1038/s41598-020-67949-9.

Coculescu, B. I. (2009). Antimicrobial resistance induced by genetic changes. *Journal of Medicine and Life, 2*, 114–123.

Cox, G., & Wright, G. D. (2013). Intrinsic antibiotic resistance: Mechanisms, origins, challenges and solutions. *International Journal of Medical Microbiology, 303*, 287–292.

Egorov, A. M., Ulyashova, M. M., & Rubtsova, M. Y. (2018). Bacterial enzymes and antibiotic resistance. *Acta Naturae, 10*(4), 33–48.

Eleraky, N. E., Allam, A., Hassan, S. B., & Omar, M. M. (2020, February 8). Nanomedicine Fight against Antibacterial Resistance: An Overview of the Recent Pharmaceutical Innovations. *Pharmaceutics, 12*(2), 142. PMID: 32046289; PMCID: PMC7076477. doi:10.3390/pharmaceutics12020142.

Fair, R., & Tor, Y. (2014). Antibiotics and Bacterial Resistance in the 21st Century. *Perspectives in Medicinal Chemistry, 6*, 25–64. doi:10.4137/PMC.S14459.

Gupta, S. K., Padmanabhan, B. R., Diene, S. M., Lopez-Rojas, R., Kempf, M., Landraud, L., et al. (2014). ARG-ANNOT, a new bioinformatic tool to discover antibiotic resistance genes in bacterial genomes. *Antimicrobial Agents and Chemotherapy, 58*(1), 212–220. Epub 2013 Oct 21. PMID: 24145532; PMCID: PMC3910750. doi:10.1128/AAC.01310-13.

Hopwood, D. A. (2007). *Streptomyces in Nature and Medicine: The Antibiotic Makers* p. 2007. New York, NY: Oxford University Press.

Jia, B., Raphenya, A. R., Alcock, B., Waglechner, N., Guo, P., Tsang, K. K., et al. (2017). CARD 2017: Expansion and model-centric curation of the comprehensive antibiotic resistance database. *Nucleic Acids Research, 45*(D1), D566–D573. Jan 4; E pub 2016 Oct 26. PMID: 27789705; PMCID: PMC5210516. doi:10.1093/nar/gkw1004.

Kakoullis, L, Papachristodoulou, E, Chra, P, & Panos, G. (2021). Mechanisms of antibiotic resistance in important Gram-positive and Gram-negative pathogens and novel antibiotic solutions. *Antibiotics, 10*(4), 415. https://doi.org/10.3390/antibiotics10040415.

Khan, F., Pham, D. T. N., Oloketuyi, S. F., Manivasagan, P., Oh, J., & Kim, Y-Mog (2019). Chitosan and their derivatives: Antibiofilm drugs against pathogenic bacteria. *Colloids and Surfaces B: Biointerfaces*. https://doi.org/10.1016/j.colsurfb.2019.110627.

Kwiecień, E., Ilona, S., Chrobak-Chmie, D., Agnieszka, G., & Magdalena, R. (2020). New determinants of aminoglycoside resistance and their association with the class 1 integron gene cassettes in Trueperella pyogenes. *International Journal of Molecular Sciences, 21*, 4230. doi:10.3390/ijms21124230.

Li, W., Sharma, M., & Kaur, P. (2014). The DrrAB efflux system of Streptomyces peucetius is a multidrug transporter of broad substrate specificity. *Journal of Biological Chemistry, 289*, 12633–12646. doi:10.1074/jbc.M113.536136.

Li, Y., Xu, Z., Han, W., Cao, H., Umarov, R., Yan, A., et al. (2021, February 8). HMD-ARG: Hierarchical multi-task deep learning for annotating antibiotic resistance genes. *Microbiome, 9*(1), 40. PMID: 33557954; PMCID: PMC7871585. doi:10.1186/s40168-021-01002-3.

Liu, Y. Y., Wang, Y., Walsh, T. R., Yi, L. X., Zhang, R., Spencer, J., et al. (2016). Emergence of plasmid-mediated colistin resistance mechanism MCR-1 in animals and human beings in China: A microbiological and molecular biological study. *The Lancet Infectious Diseases, 16*(2), 161–168. Epub 2015 Nov 19. PMID: 26603172. doi:10.1016/S1473-3099(15)00424-7.

Martinez, J. L. (2018). Ecology and evolution of chromosomal gene transfer between environmental microorganisms and pathogens. *Microbiology Spectrum, 6*, 1–16. doi:10.1128/microbiolspec.MTBP-0006-2016.

Martinez, JL (2014). General principles of antibiotic resistance in bacteria. *Drug Discov Today, 11*, 33–39.

McArthur, A. G., Waglechner, N., Nizam, F., Yan, A., Azad, M. A., Baylay, A. J., et al. (2013). The comprehensive antibiotic resistance database. *Antimicrobial Agents and Chemotherapy, 57*(7), 3348–3357. 13 Epub 2013 May 6. Jul; E pub 2013 May 6. PMID: 23650175; PMCID: PMC3697360. doi:10.1128/AAC.00419-13.

Meletis, G. (2016). Carbapenem resistance: overview of the problem and future perspectives. *Therapeutic Advances in Infectious Disease, 3*(1), 15–21 PMID: 26862399; PMCID: PMC4735501. doi:10.1177/2049936115621709.

Munita, J. M., & Arias, C. A. (2016). Mechanisms of antibiotic resistance. *Microbiology Spectrum, 4*(2) VMBF00162015; VMBF-0016-2015. doi:10.1128/microbiolspec.

Navon-Venezia, S., Kondratyeva, K., & Carattoli, A. (2017). Klebsiella pneumoniae: a major worldwide source and shuttle for antibiotic resistance. *Fems Microbiology Review, 41*(3), 252–275. PMID: 28521338. doi:10.1093/femsre/fux013.

Partridge, S. R., & Tsafnat, G. (2018). Automated annotation of mobile antibiotic resistance in Gram-negative bacteria: the multiple antibiotic resistance annotator (MARA) and database. *Journal of Antimicrobial Chemotherapy, 73*, 883–890. doi:10.1093/jac/dkx513.

Pernodet, J. L., Fish, S., Blondelet-Rouault, M. H., & Cundliffe, E. (1996). The macrolide-lincosamide-streptogramin B resistance phenotypes characterized by using a specifically deleted, antibiotic-sensitive strain of Streptomyces lividans. *Antimicrobial Agents and Chemotherapy, 40*(3), 581–585. PMID: 8851574; PMCID: PMC163161. doi:10.1128/AAC.40.3.581.

Peterson, E., & Kaur, P. (2018). Antibiotic resistance mechanisms in bacteria: Relationships between resistance determinants of antibiotic producers, environmental bacteria, and clinical pathogens. *Frontiers in Microbiology, 9*, 2928. doi:10.3389/fmicb.2018.02928.

Piddock, L. J. (2006, 19). Clinically relevant chromosomally encoded multidrug resistance efflux pumps in bacteria. *Clinical Microbiology Reviews*, (2), 382–402.

Poirel, L., Naas, T., & Nordmann, P. (2010). Diversity, epidemiology, and genetics of class D beta-lactamases. *Antimicrobial Agents and Chemotherapy, 54*(1), 24–38, Epub 2009. PMID: 19721065; PMCID: PMC2798486. doi:10.1128/AAC.01512-08.

Poole, K. (2005). Efflux-mediated antimicrobial resistance. *Journal of Antimicrobial Chemotherapy, 56*(1), 20–51.

Poole, K., Lau, C. H., Gilmour, C., Hao, Y., & Lam, J. S. (2015). Polymyxin susceptibility in Pseudomonas aeruginosa linked to the MexXY-OprM multidrug efflux system. *Antimicrobial Agents and Chemotherapy, 59*(12), 7276–7289. doi:10.1128/AAC.01785-15.

Robicsek, A., Strahilevitz, J., Jacoby, G. A., Macielag, M., Abbanat, D., Park, C. H., et al. (2006). Fluoroquinolone-modifying enzyme: a new adaptation of a common aminoglycoside acetyltransferase. *Nature Medicine, 12*(1), 83–88, Epub 2005 Dec 20. PMID: 16369542. doi:10.1038/nm1347.

Scaria, J., Chandramouli, U., & Verma, S. K. (2005). Antibiotic resistance genes online (ARGO): A database on vancomycin and beta-lactam resistance genes. *Bioinformation, 1*(1), 5–7, Published 2005 Mar 17. doi:10.6026/97320630001005.

Sharma, V. K., Johnson, N., Cizmas, L., McDonald, T. J., & Kim, H. (2016). A review of the influence of treatment strategies on antibiotic resistant bacteria and antibiotic resistance genes. *Chemosphere, 150*, 702–714. https://doi.org/10.1016/j.chemosphere.2015.12.084.

Sultan, I., Rahman, S., Jan, A. T., Siddiqui, M. T., Mondal, A. H., & Haq, Q. M. R. (2018). Antibiotics, resistome and resistance mechanisms: A bacterial perspective. *Frontiers in Microbiology, 9*, 2066, PMID: 30298054; PMCID: PMC6160567. doi:10.3389/fmicb.2018.02066.

Surette, M. D., & Wright, G. D. (2017). Lessons from the environmental antibiotic resistome. *Annual Review of Microbiology, 71*, 309–329. doi:10.1146/annurevmicro-090816-093420.

van Hoek, A. H., Mevius, D., Guerra, B., Mullany, P., Roberts, A. P., & Aarts, H. J. (2011). Acquired antibiotic resistance genes: An overview. *Frontiers in Microbiology, 2*, 203 Published 2011 Sep 28. doi:10.3389/fmicb.2011.00203.

Walker, M. S., & Walker, J. B. (1970). Streptomycin biosynthesis and metabolism. Enzymatic phosphorylation of dihydrostreptobiosamine moieties of dihydro-streptomycin-(streptidino) phosphate and dihydrostreptomycin by Streptomyces extracts. *Journal of Biological Chemistry, 245*, 6683–6689.

Walsh, C. (2016). Major classes of antibiotics and their modes of action. *Antibiotics: Challenges, Mechanisms, Opportunities* (pp. 16–32). Washington, DC: ASM Press.

Wegener, H. C. (2012). Antibiotic resistance—Linking human and animal health. *Improving Food Safety Through a One Health Approach* (pp. 331–349). Washington, DC: National Academy of Sciences.

Wei, Z., Wu, Y., Feng, K., Yang, M., Zhang, Y., Tu, Q., et al. (2019). ARGA, a pipeline for primer evaluation on antibiotic resistance genes. *Environment International, 128*, 137–145, Epub 2019 May 3. PMID: 31054477. doi:10.1016/j.envint.2019.04.030.

World Health Organization (2014). Antimicrobial resistance: Global report on Survillance 2014. World Health Statistics 2014.Geneva, 30th April, 2014.

World Health Organization (2015). Global action plan on antimicrobial resistance.

Wright, G. D. (2005). Bacterial resistance to antibiotics: Enzymatic degradation and modification. *Advanced Drug Delivery Reviews, 57*, 1451–1470. doi:10.1016/j.addr.2005.04.002.

Wüthrich, D., Brilhante, M., Hausherr, A., Becker, J., Meylan, M., & Perreten, V. (2019). A novel trimethoprim resistance gene, dfrA36, characterized from Escherichia coli from calves. *mSphere., 4*(3) American Society for Microbiology. doi:10.1128/mSphere.00390-19.

Yin, X., Jiang, X. T., Chai, B., Li, L., Yang, Y., Cole, J. R., et al. (2018, July 1). ARGs-OAP v2.0 with an expanded SARG database and Hidden Markov Models for enhancement characterization and quantification of antibiotic resistance genes in environmental metagenomes. *Bioinformatics, 34*(13), 2263–2270. PMID: 29408954. doi:10.1093/bioinformatics/bty053.

Yushchuk, O., Binda, E., & Marinelli, F. (2020). Glycopeptide antibiotic resistance genes: Distribution and function in the producer Actinomycetes. *Frontiers in Microbiology, 11*, 1173. doi:10.3389/fmicb.2020.01173.

Chapter 4

Direct reuse of wastewater: problem, challenge, and future direction

Sandeep Dharmadhikari[a], Amit Jain[a], Nitin Pawar[b] and Parmesh Kumar Chaudhari[c]

[a]*Department of Chemical Engineering, School of Studies of Engineering and Technology, Guru Ghasidas Vishwavidyalaya (A Central University), Bilaspur, Chhattisgarh, India,* [b]*Department of Chemical Engineering, Visvesvaraya National Institute of Technology, Nagpur, Maharashtra, India,* [c]*Department of Chemical Engineering, National Institute of Technology Raipur, Raipur, Chhattisgarh, India*

4.1 Introduction

Wastewater treatment and reuse are not newly developed ideas, and knowledge regarding the subject has evolved. Human waste has been diverted away from urban areas for centuries by using untreated municipal wastewater. Using domestic wastewater as fertilizer has been a common practice for decades (Paranychianakis, Salgot, Snyder, & Angelakis, 2015). The result has been a better understanding of treatment processes and water quality standards. Since 3200–1100 BC, that is, the bronze era of the civilization, water from domestic sources has been utilized for agriculture purposes by primeval civilizations like Mesopotamia, the Indus Valley civilization, the Minoans, and many more. Later on, the Hellenic era of civilization and the cities like Athens, Rome, and their adjacent area used recycled wastewater for irrigation, fertilization, and disposal (Tzanakakis, Paranychianaki, & Angelakis, 2007). Sewage was applied to the land for agricultural use and disposal in Bunzlau of Poland during the 16th century, and Edinburgh of Scotland in the mid of 17th century (Tzanakakis et al., 2014). Most of Europe and America's rapidly growing cities followed this trend. As cities grew in size and population, sewage farms rose to popularity as a means of disposing of large amounts of wastewater and these techniques are still in use today in several large cities in Europe and the United States. It was in Gennevilliers in 1872 that the first sewage farms were established, eventually treating the entire town's sewage. Similarly, Australia's Melbourne established a large "sewage farm" at the end of the nineteenth century

Antimicrobial Resistance in Wastewater and Human Health.
DOI: https://doi.org/10.1016/B978-0-323-96124-0.00002-7

(Angelakis & Snyder, 2015; Crites, Ronald, Joe Middlebrooks, & Sherwood, 2010; Tzanakakis et al., 2014). Although the land treatment technique had some environmental and community health issues, it was widespread in Central Europe, the United States, and other parts of the world throughout the 20th century and utilized for many decades. Apart from environmental and health issues, these systems had some other limitations also, like the requirement of a huge area, operating challenges, and unable to attain good hygiene conditions; therefore, these systems had gone out of favor in the first half of the 21st century (Tzanakakis et al., 2014). The development of industries and cities was very fast in the fifties and sixties of the 19th century which resulted in the worsening of hygienic conditions. Therefore, the need for a new sewage system originated and modern sewage systems were developed. At the same time, in the first half of the 19th century, due to contamination of water sources, cholera took a serious form in London and thousands of civilians lost their lives. After this incident, Edwin Chadwick, then the Royal Commissioner, prepared a report on sanitation improvements, and Sir Joseph Bazalgete was appointed by the Metropolitan Commission of Sewers, and was entrusted with the responsibility of developing the extensive underground sewage system for the orderly and safe disposal of waste (Ashton & Janet, 1991).

In several countries, there is a rapid increase in the planning of wastewater treatment and effluent reuse projects. Some most prevalent (re)uses of treated wastewater are agricultural irrigation, aquifer boosting, brackish water barriers, industrial and manufacturing sectors, lavatory flushing, and many more municipal uses. The World Bank, the United Nations' Food and Agriculture Organization, and the World Health Organization (WHO) estimate that the average yearly growth in the reused volume of such water in the United States, China, Japan, Spain, Israel, and Australia is between 25% and 50%. In 2010, approximately 4300 million cubic meters of wastewater were treated in California, but only 20% of treated water (860 million cubic meters) was reused and the remaining 80% (3440 million cubic meters/year) had no use and was released into the ocean. But taking into account, the demand and supply of water, it is estimated that by 2030, 2470 million cubic meters of water will be reused per year (Levine, Tchobanoglous, & Asano, 1985). Currently, Spain is producing more than 500 million cubic meters of treated wastewater yearly and it is estimated to treat 1000 million cubic meters yearly in the coming years (Mudgal et al., 2015). In Israel, almost 80% of treated wastewater is recycled and recycled sewage is used to irrigate crops. According to NEWater Singapore alone meets around 30% of current US water requirements, which are predicted to increase to 55% by 2060 (Angelakis & Snyder, 2015; Angelakis, 2014; Hedberg, Pardo, Frontini, & Toutia Daryoush, 2015). California and Texas in the United States have inspired more research and deployment of direct potable water systems. California's governor declared in 2013 that potable reuse guidelines, including direct reuse, would be enacted, and by 2016, a strategy has been implemented. With large-scale operations in Big Spring and Wichita Falls, Texas has already made significant progress in indirect potable reuse. The only scientific difference between indirect and direct potable reuse in an identical healthcare facility is time. Groundwater injection and other environmental buffers have little, if any, effect on water quality. When high-purity water is injected into aquifers, metals like arsenic can leak out and mix with lower-quality water during subsurface storage. On the other hand, environmental buffers provide a theoretical response time based on the amount of water held.

Drinkable water recycling systems would need to respond significantly faster, necessitating the usage of the vital controller and real-time sensor devices. Although gauges are becoming more common for monitoring the treatment effectiveness, real-time systems that give operators instant feedback and/or autonomous treatment system operations are necessary. Another emerging trend is that stakeholders in water reuse are growing increasingly concerned about chemical mixes' potential health impacts (Snyder, 2014). Speedy, huge-output water toxicity screening is increased by means of in-vitro bioassays (Escher & Leusch, 2012; Escher et al., 2014).

4.1.1 Demand for water in the Asian nation

Asia has a population of 4.5 billion individuals consuming 65% of the world's water. Around 30% of Asia's population already lacks access to clean water. India and China have experienced double-digit GDP growth as well as population growth in recent years. Many river basins are already struggling to meet the demands placed on them. What will happen to water supply and demand as the effects of climate change become more visible and continued improvements in socioeconomic conditions because population numbers rise even more with wealthier societies requiring more water on average anyway? Will some areas have more problems than others? What will aggravate the situation? Is it possible to solve the issue?

As Asia's population and economy have grown, so has the water demand. Securing water for agriculture, the region's largest water-use sector has proven difficult for many countries in the region, especially because increasing water volume is required for both industries and households. The problem is also linked to the region's food security. There has been some success in lowering agricultural water usage thanks to more water-efficient irrigation technologies like drip irrigation and traditional practices like "subak" in Bali—a collaborative and participatory local water allocation method of irrigation (Cosgrove, William, & Frank, 2000). Rice paddy fields are important for flood control and groundwater recharge in addition to providing food. This fact deserves full recognition in the Asian context. In most countries, water withdrawal from industry and households is expected to increase. In 1980, industrial water consumption accounted for 10% of total water consumption in China. By the year 2000, it had risen to 25%. It grew from 10% in 1990 to 21% in 2000 in Malaysia (Kog, 2015). By 2025, industrial water withdrawals in Asia are expected to rise to 9.5%

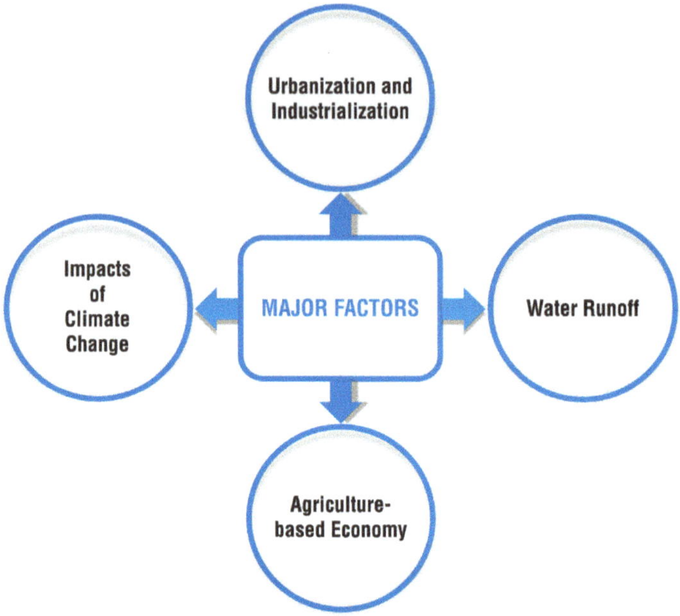

FIGURE 4.1 Factors influencing reclaimed water reuse.

of total withdrawals, up from 6.9% in 1995. Domestic water withdrawals are expected to rise from 9.9% to 15.2% during the same period (Shiklomanov, 2000). Asia has outpaced and continues to outpace other regions in terms of water consumption growth rate and absolute volume. In 2025, 2.4 billion people in the region are expected to be affected by water scarcity, nearly double the number in 1995 (Cosgrove, William, & Frank, 2000). Tensions between different water users over limited water resources are likely to worsen as demand for water rises. Conflicts between upstream and downstream users/countries may be more common in the region. In Asia, there are 53 international rivers. As urbanization progresses, the demand for water resources in cities will increase, resulting in increased competition between urban and nonurban areas. Overuse and improper development of water resources have the potential to significantly harm the environment. Dam construction and overuse of water, for example, could have a significant impact on river ecosystems. Groundwater overuse can cause a drop in the water table, which can cause land subsidence. Land subsidence is essentially irreversible, putting people at risk of flooding during high tides and making them more vulnerable to natural disasters such as tsunamis.

4.1.2 Factors influencing water reutilization

The main influencing factors for water reutilization are represented in Fig. 4.1. Water availability has become a significant problem for the government as a result of expanding population, changing rainfall patterns, decreased water availability, industrial expansion, and fast urbanization. Water shortage is a result of increased per capita demand, technical improvements, environmental and economic constraints, and other issues, all of which need the reuse of treated wastewater. It is vital to produce big amounts of food as an agro-economic country, which demands a large amount of water. A population of 1600 million people will require 450 Mega tons of food by 2050 (Central water, 2017). As a result of anthropogenic activity and extreme weather, surface water is drying up. Highly polluted water is found in 275 of 445 rivers and 76 of 85 lakes, making them unsafe for human consumption (Dutta, 2018). People thought that water was free and they used it abundantly and withdrew the adequate amount of water without recharging the underground water sources resulting in groundwater levels have plummeted by 61% in the last decade. According to a report many metro cities of India, mainly New Delhi, Chennai, Hyderabad, and many more, will face an acute shortage of groundwater by the end of 2030. Furthermore, agriculture relies heavily on groundwater, prompting the hunt for new water sources to supplement those already available. In total, 67% of the Himalayan mountains have receded as a result of global warming, resulting in less water in the streams (2018). Furthermore, the annual number of rainy days is decreasing. Climate change has affected the planning and management of water supplies, making society vulnerable to its impacts. Higher industrial output is predicted as a result of the country's development, which is directly connected with rising water consumption. Furthermore, the population is also growing at a very fast rate, especially in urban areas of India. If the population will grow in this manner, then by 2030 top five most populated cities of the world will be from India which will put pressure on water supplies (Moefc, 2018; McNabb, 2019). The key socio-economic factors affecting water resources are urbanization, agriculture, and industry.

4.1.3 Challenges associated with the recycling of wastewater

In India, wastewater reuse and reclamation are still in their infancy. Many obstacles exist in the execution of such projects which reuse the treated wastewater and restrict the nation's progress in wastewater reclamation and reuse. A large amount of wastewater or sewage is generated in India but the amount of wastewater reutilization is extremely low due to many reasons. The first reason is the inadequate quantity of plants treating the sewage, as cities of approximately two-thirds of the nation do not have such type of unit, at the same time many of the installed units are antiquated and malfunctioning. Due to these reasons about two-thirds of the wastewater is left untreated and also the amount of wastewater produced and handled varies greatly. Scarcity of dependable technology to treat sewage water for removing the requisite impurities for acceptable reuse, as well as qualified employees who understand how wastewater reuse treatment units work, are also the reason behind the low reutilization of sewage water (Mekala, Brian Davidson, & Madar Samad, 2007).

The lack of institutions and regulatory bodies to create clear and mandatory policy for reutilization of treated effluent, and inefficiency of management to coordinate among the department of freshwater supply and effluent treatment are the challenges to creating treated wastewater reuse systems. This fragmentation causes unnecessary planning and implementation delays, as well as conflicts, resource division, and interdepartmental agreements, which are responsible for making treated effluent reutilization projects very complicated. Owing to the disintegration of the department, data on the gap between water demand and supply, amount of effluent generated, and consumption of water, among other things, is restricted, which has an impact on recovered water reuse planning. Local administrations have become very reliant on the resources available due to the abundance of water in smaller cities and villages. Newly treated wastewater reuse programs have excessively high upfront costs, and finance and technology to establish such initiatives are uncommon. For effective reutilization of treated water, the enduring policy and the economic feasibility of such programs are very crucial (Recycling and reuse of treated wastewater in urban India: a proposed advisory and guidance document. CGIAR Research Program on Water, Land and Ecosystems, 2016). The rate of water supplied for household and industrial purposes is usually different in most cities, with the industrial rate being much more than the domestic rate. In certain circumstances, industrial levies are utilized to offset the expenses of municipal/domestic water supply. Local officials are wary to reuse treated wastewater since it could eliminate some of the increased money. Because of a lack of community trust in government officials, treated effluent reutilization projects have failed because of the lack of public understanding and perception, particularly when it comes to water quality and health issues. The public's resistance to recycling treated wastewater is fueled by some major issues like inadequate technologies of treatment, safety, and health issues of the common man. Perceptions of consumers' reactions to the reutilization of treated water function more as a hurdle rather than a reaction itself. All such circumstances are impeding the growth of the reutilization of wastewater. However, as a result of technological developments and economic progress, the condition is rapidly moving. Plenty of possibilities are available in the nation to reutilize the treated water, and several opportunities have been exploited (Kumar & Kirti, 2020).

4.2 Various sources of water supply and sanitation in India

Since the 1980s, India's water supply and sanitation have vastly improved. While India's population has access to toilets, on that comparison many of the population are unable to access safe drinking water or sewage infrastructure. Although the government has started various programs at different levels which resulted in significant improvements in sanitation and water supply. Some of these initiatives are still ongoing. In 1980, rural sanitation coverage was estimated to be 1%, and by 2018, it had risen to 95% (Dutta, 2018). From

72% in 1990 to 88% in 2008, the percentage of Indians who have access to improved water sources has increased significantly (Ravikumar, Rajiv Raman, & Shubhra, 2014). As India is the second-highest populated country in the world, we are moving up in the list of countries that have a scarcity of fresh drinking water. But at the same time, there is a strong possibility in India to reutilize the treated wastewater. India is a developing country and its economy is mainly agriculture-based. Therefore, its water resources, especially groundwater, are unable to fulfill the demands of the agricultural and industrial sectors. Therefore, to fulfill the need of both sectors for a long time, reclaimed water can play an important role. Recycling and reusing treated wastewater keep it away from the freshwater source and thus benefits the environment and also ensures to meet the water demand by continuous supply and at the same time minimum use of freshwater sources. Multiple nutrients such as nitrogen, phosphorus, and potassium, among others, can be recovered by reusing treated wastewater. Thus, the produced sludge is mostly used as compost to fulfill the demand of crops as it has a very good nutritional value. As India imports the majority of potassium and phosphorus. Thus, the reuse of wastewater not only saves the Indian money by reducing the import but also create new jobs. One approach to repay the entire project's cost including its processing and maintenance is to sell the water reuse plant. The treated water is sent to the industries that use it, subject to the required treatment levels. As the treated wastewater has many minerals, therefore, its use for cropping increased the yield from 30% to 40% and at the same time fertilizer consumption was reduced by 50% and thus, the farmers' income increased by a minimum of 30% (Dutta, 2018). Similarly, the use of such water for agricultural purposes minimizes the pumping cost resulting in low energy consumption and consequently lower greenhouse gas emissions (Kumar & Kirti, 2020). According to studies, it reduces the CO_2 emission by 1.5 million tons per year and saves around 1.75 million MWh of energy per year.

4.3 Wastewater management and water audits

Wastewater is such water that must be cleaned after it has been utilized. Laundry, bathing, dishwashing, toilets, waste disposals, and industrial-generated liquid waste falls under this category. Wastewater that accumulates pollutants flows into seas, lakes, and rivers with rainwater. The main aim of effluent management is to keep the water clean and safe so that the water must be safe to drink and wash with, as well as suitable for use by industry (Walmsley & Pearce, 2010).

Such water must be so safe that there should not be any harm if such water is released into water bodies like ponds, rivers, lakes, oceans, etc. Wastewater can be categorized into two categories, first point source and second nonpoint source. The effluent that enters into the natural source of water like rivers, lakes, oceans, etc., by specified areas falls under the first category of wastewater. A common example of such a category is the drain from storm and sanitary. Effluent which is not tied with a single source belongs to the second category of wastewater. Some common examples are acidic fluids from mines, as well as runoff from agricultural lands and municipal regions. Because the source and pollutants in point source wastewater are known, it is easier to handle in many ways. On the other hand, it is very tedious to identify the second category of wastewater and its treatment is also very difficult (Grafton & Hussey, 2011; Tchobanoglous, Burton, & Stensel, 2003).

A water audit is a study of a group's water usage. It examines all aspects of use from the point where water enters the premises to the point where wastewater is discharged, starting at the point where water enters the premises and ending at the point where wastewater is discharged. The audit determines the amount of water used, any wastage, existing leaks, excess use, and other factors, as well as areas where consumption can be reduced. It examines current treatment systems and practices critically and makes recommendations for changes to increase efficiency and reduce usage. An audit, based on this in-depth investigation and observations, makes recommendations on how to reduce water waste and consumption, as well as improvements in treatment practices and methods, as well as cost-benefit analyses. It also suggests that a system be set up to keep track of the amount of water that enters a system as well as how that water is distributed and used. Water auditing is a method of obtaining an objective water balance by measuring the flow of water from the point of withdrawal or treatment, through the distribution system, and into the areas where it is used before being discharged. Calculating the water balance, water use, and identifying ways to save water are all part of a water audit. A preliminary water survey and a detailed water audit are both parts of a water audit. A preliminary water survey is carried out to gather background information on plant operations, water consumption, and discharge patterns, as well as water billing, rates, and water cess (Rai, Vijay, & Alka, 2017).

Following the analysis of secondary data gathered from the industry, a thorough water audit is carried out, which includes the following steps (McMillan, Hampton, & Lennert, 2019):

- Analyze the water system.
- Baseline water map quantification.
- Pressure and flow meters, as well as other devices, are used to monitor and measure the situation.
- Inefficiencies and leaks are measured and quantified.
- Water quality loads and discharges are quantified.
- Variability inflows and quality parameters are quantified.
- Water treatment and reuse strategies, as well as direct use.

Finally, a detailed water balance is created. Water quality requirements are mapped for various user areas, which aids in the development of "recycle" and "reuse" opportunities.

The following items must be included in the detailed water audit report:

- Patterns of water consumption and wastewater generation Specific water conservation and use.
- The facility's water balance is complete.
- Opportunities to conserve water.
- Methodology for putting the proposals into action.
- Figures and detailed description.
- Investing is required.

For water conservation, industries can take the following steps:

- Industrial process modernization to reduce water consumption.
- A recirculating cooling system recycles water.
- Norms for water budgeting are being established when compared to traditional chemical treatment.
- Ozonation cooling water approach can result in a fivefold reduction in blowdown.
- Eliminating some plenum flushes, switching from a continuous to an intermittent flow system, and improving control over the use of deionized water will reduce the amount of deionized water reused.
- Using wastewater in the garden.
- Processing of effluents following disposal standards.

4.4 Various applications for wastewater treatment in India

A huge amount of wastewater is produced in India and its reutilization is possible which depends on the various parameters, as shown in Fig. 4.2. Some of them are the quantity and quality of wastewater, consciousness about the reutilization of effluent, the possible market for treated water, the type of process used for the treatment of wastewater and its processing cost, distribution of treated water, etc. Reutilization of treated water depends on its quality, as shown in Table 4.1.

4.4.1 Agricultural reutilization

The majority of the reutilized wastewater, processed, and unprocessed is used for agricultural purposes. It has gained a lot of traction, especially in developing economies. India, for example, has a large population. Agricultural reuse of sewage is a reality for a long time. Both food and nonfood crops can be irrigated with it. Untreated wastewater contains a high amount of salt, dissolved nutrients (such as phosphates) and heavy metals which are detrimental repercussions of using such water. It was observed in a study that the production of rice crops decreased by 40%–50% in the agricultural land situated near the Musi River, Hyderabad, as the farmers were using the untreated wastewater of this river which had high salt content (Devi & Madar Samad, 2008).

In addition, the Ministry of Agriculture in Varanasi recently issued a notice forbidding the utilization of treated wastewater within the city irrigation of crops

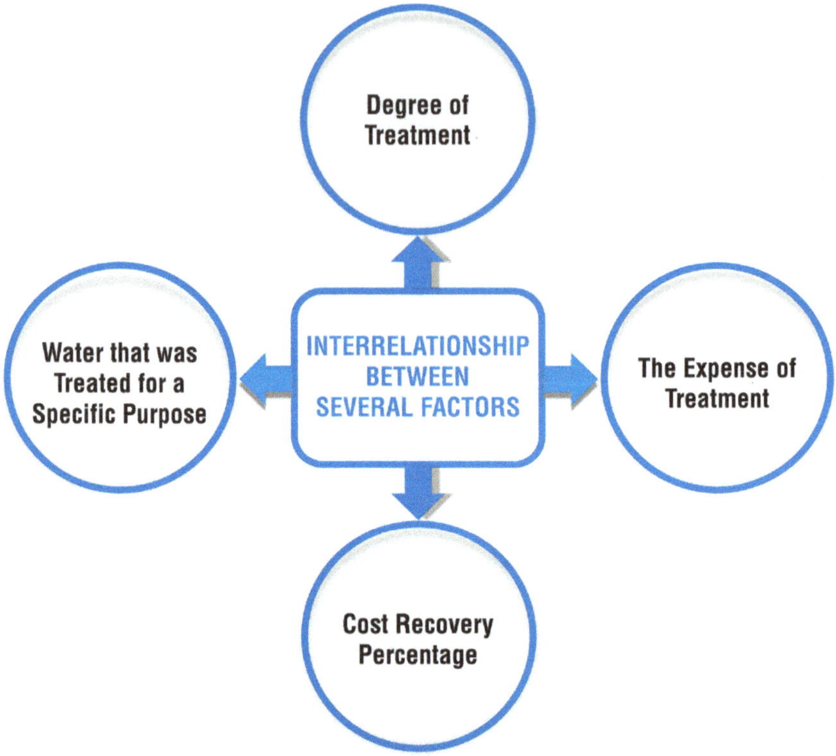

FIGURE 4.2 Interdependence of numerous elements in treatment procedure selection.

TABLE 4.1 Reuse of treated water: water quality guidelines (Rode, Shirbhate, & Shewale, 2013).

Parameter	Unit	Urban use	Agriculture use	Environmental use
pH	6.5–8.3	–	–	–
BOD_5	mg/L	≤10	≤10	≤10
Turbidity	NTU	≤2	–	≤2
TSS	mg/L	–	–	–

as the treated effluent of sewage and small textile industries running at home, as this water contains metals (2015). Various studies indicated that the quality of water sources played an important role when the reutilization of treated water is considered. If such water is to be used for agricultural purposes, then such water should be given a secondary treatment process combined with a disinfection process. But, the requirements of these processes depend on the quality of water to be treated and the available technology for treatment. The faecal coliform count is the most important parameter for water quality. To know about the minimum water quality for reutilization of wastewater for agricultural purposes in India, the manual on Sewerage and Sewage Treatment (2013) published by the Central Public Health Engineering and Environmental Organization (CPHEEO) should be preferred.

4.4.2 Urban reutilization

Urban reutilization has gained attention in the urban areas in India. It is possible to reuse urban materials in both unrestricted and regulated ways. The most worrisome water quality criteria for urban reuse are fecal coliform, residual chlorine and turbidity, necessitating considerable tertiary treatment, and sophisticated disinfection to avoid being labeled a public health issue (Barringer, 2014; Tare, 2011; Alley, Kelly, Nutan Maurya, & Sukanya, 2018; Rode, Shirbhate, & Shewale, 2013).

4.4.3 Environmental/recreational reutilization

The treated effluent or wastewater can be utilized to promote water bodies, restoration of existing water bodies, development of manmade water bodies like lakes for the entertaining purpose which boosted tourism, and livestock upkeep. Hyderabad's famous Hussain Sagar Lake one of the examples of such a case where treated municipal effluent is used to compensate the losses of water in the lack due to evaporation (Arceivala, Sol, & Shyam, 2006). Such a type of environmental reuse will be more attractive for tourism purposes and will get more consideration in the coming years. As a result, a minimum quality standard must be set for the water to be acceptable for reuse in the environment, guaranteeing that it is free of health and environmental hazards.

4.4.4 Industrial reutilization

This is one of the reasons that the industries are utilizing the treated municipal effluent or wastewater for general uses. Many regions have effectively utilized reclaimed water for secondary activities both inside and outside of the sector. Almost all the industrial sectors, like power generation units, paper industries, fabric industries, leather industries, pharmaceutical industries, oil and gas sector, metal processing units, etc., are getting benefitted from reclaimed water. The construction business has recently gained a lot of impetus in many Indian cities. In recent years, treated wastewater has been utilized by these industries to fulfill the requirement of water for processing, boiler, cooling tower, bathrooms, toilets, etc. For industrial reuse, tertiary treatment is required, which includes disinfection as well as control of minerals like phosphorus, nitrogen, etc. (Kumar & Indira Khurana, 2014; Bazza, 2003; Lautze, Stander, Drechsel, da Silva, & Keraita, 2014; Chauhan, Jensen, & Sengaiah, 2016).

4.4.5 Indirect and direct potable reutilization

In India, the use of treated wastewater for nonpotable purposes is in the early stage. To fulfill impending water demands, it is important to reuse treated wastewater to supplement potable water sources. But it needs an advanced level of treatment like microfiltration, ultrafiltration, reverse osmosis, removal of minerals, disinfection, etc., and requires careful quality monitoring also (Rode, Shirbhate, & Shewale, 2013). There are no strict standards or guidelines in India for the reutilization of treated wastewater for drinking purposes. Although, due to wastewater disposal in water bodies, like lakes, rivers and streams, and injection of groundwater, the existing reutilization of treated wastewater has been done in the entire nation for a long time. Planned potable water reuse, on the other hand, is a future growth sector that requires critical planning and research before it can be pushed and implemented (Chauhan, Jensen, & Sengaiah, 2016; Lautze, Stander, Drechsel, da Silva, & Keraita, 2014).

4.4.6 Process industry reutilization

The first step in reducing industrial water consumption is to figure out how much water is currently being consumed. Examining old water bills or conducting a water audit can help with this. An audit produces baseline data on water usage and identifies areas of the operation that use a lot of water (Table 4.2). There is no way to track success if baseline data does not exist before consumption-reduction attempts. Once a corporation understands its present water usage, it may set clear, measurable targets for the next generation. Some water-saving suggestions for businesses have been given by many researchers. Nondrinking water can be reused in a variety of ways (Rai, Vijay, & Alka, 2017). Guidelines for the reuse of treated water in industrial applications are mentioned in Table 4.2. Water can be reused after it has been utilized in one portion of an operation. For example, after given proper treatment used water may be reutilized for farming purposes and also in cooling tower in industries. Further, the condensate of cooling water may be used as "makeup" water to minimize the freshwater requirement in the cooling tower. Although high quantities of heavy metals make condensate water dangerous to drink but it is safe to use in cooling system. The amount of water in blowdown can be minimized by proper monitoring the operation of the cooling tower and optimizing the concentration cycle of the tower, as these are the one of the highest energy-consuming equipment and operation of any industries and there are a lot of chances of improvement. High-dissolved-solids wastewater departs the cooling tower and is replaced by freshwater throughout each cooling tower cycle. If industries are closely regulated, they can drastically cut their

TABLE 4.2 Guidelines for the reuse of treated water in industrial applications.

Constituent	Boiler feed	Pulp and paper	Textiles	Petroleum and coal
Calcium (mg/L)	0.01–0.40	20	–	75
Iron (mg/L)	0.05–1.00	0.30–1.00	0.10–0.30	1
Manganese (mg/L)	0.01–0.30	0.05–0.50	0.01–0.05	1
Alkalinity (mg/L)	40–350	100	–	125
Chloride (mg/L)	–	200–1000	–	300
TDS (mg/L)	200–700	–	100	1000
Hardness (mg/L)	0.07–350	100	25	350
Color (Hazen)	–	10–30	5	–

consumption of freshwater. It is advisable to check regularly the functioning of the equipment especially fittings and joints. To minimize the water requirement, use automatic valves in the pipeline assembly, use proper pressure and level switches at appropriate places and replace the old sanitary with new water-efficient accessories in washrooms.

Likewise water-cooled equipment's should be replaced with air-cooled equipment at suitable places (McMillan, Hampton, & Lennert, 2019). Water is removed from the cleaning procedure when a dry option is available. Examining each phase of a company's cleaning procedures may identify simple solutions to save lots of water by employing low- or no-water practices. Sweeping or vacuuming before employing water-based technologies, for example, could cut down on water consumption without requiring more infrastructures. Using minimum water for nonessential tasks is the way to save water and landscaping is a perfect example. When drought-resistant, drought-tolerant, and climate-adapted crops are chosen, minimum amount of water is used for landscaping tasks that aren't necessarily essential to the operation. Perhaps the organization has made some progress in terms of water consumption monitoring and reduction. Similarly, in industries, replacement of energy-consuming equipment with energy-efficient equipment is a good practice (Kumar & Indira Khurana, 2014; Chauhan, Jensen, & Sengaiah, 2016).

4.5 Various treatment systems used to treat wastewater

4.5.1 Sewage treatment plant (STP)

The direct discharge of industrial and human effluents into natural resources without treatment has the greatest impact on human health and the environment.

Treatment of sewage is required to reduce sewage toxicity, maintain a safe and healthy environment, and promote human welfare.

4.5.1.1 Treatment of sewage

Sewage contains a significant amount of toxic organic matter. In sewage treatment plants, microorganisms are commonly used to remove toxic organic matter. There are two stages in a sewage or wastewater treatment plant (Khopkar, 2007).

4.5.1.2 Primary treatment

It entails using physical processes to remove large and small components from wastewater (Khopkar, 2007).

4.5.1.3 Biological treatment

As a biological treatment, sewage treatment plants are infected with aerobic bacteria. By using the organic components of the sewage, these microorganisms minimize toxicity. BOD can be used to determine this (biological oxygen demand). The sludge is pumped from the treatment plant into a large tank after it has undergone biological treatment. This large tank contains anaerobic bacteria that aid in sludge digestion. Biogas is created during digestion and used as a source of energy. As a result, sewage treatment plant design and sewage management are essential for human wellbeing (Henze, van Loosdrecht, Ekama, & Brdjanovic, 2008).

4.5.1.4 Generation of energy

Microorganisms that produce energy are referred to as microbial fuel cells (MFCs). Among other things, MFCs are used to create biogas and energy. Agricultural waste, manure, and domestic garbage are used as raw materials in the production of biogas. The biogas is created in a biogas plant, which is a massive concrete tank. The slurry is fed by the biogas plant, which collects biomasses (biowastes). Biomasses are rich in organic stuff. Some bacteria can thrive anaerobically inside the biogas plant. These bacteria can digest the biomass contained in sludge and sewage. During digestion, a substantial amount of gas mixture is discharged inside the tank. A mixture of these gases is referred to as biogas. The biogas plant's biogas is removed through a separate outlet. MFCs are also utilized to generate heat in addition to generating power from wastewater. MFCs utilize organic debris from the wastewater treatment plant. During digestion, organic matter is broken down into a simple molecule, releasing carbon dioxide and electrons. The electrons are absorbed by the electrode and used as a source of electricity (Logan & Regan, 2006).

4.5.2 Faecal sludge treatment plant (FSTP)

Faecal Sludge Treatment Plant (FSTP) is a vermifiltration-based, comprehensive, and long-term technological solution for managing faecal sludge. FSTP can separate solid and liquid wastes and convert them into vermicompost and water for nonpotable uses like gardening, flushing, and farming with proper treatment (Cairncross, 2018).

Faecal sludge management (FSM) involves the collection, transport, and treatment of faecal sludge from pit latrines, septic tanks, or other onsite sanitation systems. In highly populated places where a section of the population is not connected to a sewerage system and pit latrines cannot be covered and rebuilt, FSM is required (Strande, 2014).

The tiger biofilter technology is used in the FSTP. It is made up of five major processes:

- Screening.
- Anaerobic digestion is a type of digestion that occurs in the absence of oxygen.
- Vermifiltration-I—separation of residual solids and liquids.
- Wastewater treatment with vermifiltration-II.
- Polishing is a type of tertiary treatment.

4.5.3 Cotreatment or combined treatment

As a pretreatment phase, a biological treatment method using fluidizing media was used on the dye wastewater. This is intended to be a replacement for the current activated sludge processes. The treated water was then pumped back into the system. Cotreatment systems are used to further process the waste. Sludge bulking is a result of traditional biological treatments. As a result, more efficient biological treatment techniques are necessary, such as the use of fluidizing media. This has been proven to minimize sludge generation and pollutant concentrations in discharged water while producing outcomes comparable to typical activated sludge procedures (Byrd, 1957).

4.5.4 Thermal hydrolysis

Wastewater treatment, waste by-product reduction, and biogas production are all applications of thermal hydrolysis technology. A huge amount of sludge is generated in a conventional water treatment plant in any industry. Sludge is viewed as a valuable source of energy rather than a waste by thermal hydrolysis facilities. After the wastewater has been treated and the sludge has been collected, biogas generation begins. The sludge is heated and crushed in enormous vats. High pressure ranging from 7 to 12 bar and temperature ranging from 160 to 165°C are necessary (Barber & Lancaster, 2012; Neyens & Baeyens, 2003).

4.5.5 Microbial fuel cells

MFCs are a type of technology that uses microbes to purify wastewater while also performing three other purposes. Even more astounding is the fact that bacteria digesting wastewater sludge may transform charged electrons into electricity as a by-product. Researchers have successfully generated considerable amounts of energy in lab conditions by transferring the electrons generated during bacterial oxidation to an electrode. The consumption of fossil fuels to create energy can be minimized by the mass production of MFC (Logan & Regan, 2006; Logan et al., 2006).

4.5.6 Solar photocatalytic wastewater treatment

Choosing what to do with the sludge produced is one of the most difficult components of wastewater treatment. Sludge is an organic content that produces in a huge quantity in conventional wastewater treatment units. If this conventional system is replaced by a photocatalytic system, the sludge formation can be reduced by 80% due to the "solar irradiation," a microbial breakdown-oxidation process. This is because that solar irradiation has a synergetic effect that lowers the quantity of carbon in the sludge when paired with hydrogen peroxide and we know that the sludge is an organic content and carbon is the principal ingredient in it (Gutierrez-Mata et al., 2017).

4.5.7 Natural techniques to treat wastewater

Roofs and streets or roadways are two of the most common wastewater sources. Strom drain is the main root to reach back the pollutants into the nature. All local and government bodies around the world have adopted natural techniques to treat the wastewater instead of throwing it into the water bodies. Natural techniques comprise sediment ponds, large-scale soil filters, and excavated wetlands with filtration systems. As a result of slowing the departure of stormwater and allowing sediments and microorganisms to settle down, it is caught in the pores of the earth crust that act as a natural filter, the water discharged back into nature is substantially cleaner (Mahmood et al., 2013).

4.6 Prospects of wastewater reutilization

Recovery of wastewater and its reutilization is not a frequent practice in India; therefore, the focus has switched in recent years to finding answers to problems and encouraging the reuse of treated wastewater. Enormous prospective is available in the wastewater sector that leads to successful treatment of wastewater and its reutilization.

4.6.1 Technological and financial viability

Continuous improvements in treatment protocols and the development of new technology-enhanced the quality of wastewater accessible for reutilization. Various factors like geographical, demographic, and financial considerations play an important role to identify the suitable treatment process so that the recovered treated water can be reutilized. Location and time are the two categories in which treatment procedures are classified. At the same time, to decide on a treatment process, financial parameters and the possible use of reutilized water are also important considerations. Cost and frequency of use are proportionate to the amount of treatment required. The greater the needed water quality, the more expensive it is to treat it. We know that there are three types of processes for the treatment of water; physical processes, chemical processes, and biological processes. Traditional wastewater treatment plant utilizes all three processes for the treatment of wastewater by removing the organic and inorganic matters, minerals, and other particles. This wastewater treatment for reuse can be accomplished using a variety of established technologies (Goyal & Arun, 2021). The activated sludge process, as well as its variations such as sequential batch reactors, extended aeration, and others, as well as up-flow anaerobic sludge blankets, fluidized aerobic bed reactors, and others, are among the most often used processes. The advanced treatment technology, like membrane bioreactor, can convert the wastewater similar to that of river water after treatment. Natural solutions of treatment like stabilization ponds and the development of wetlands are also drawing the attention of industries to use in addition to traditional treatment techniques. Natural technologies, among other things, surpass traditional procedures in terms of cost, energy, and labor needs. The requirement of less framework, energy, and capital in decentralized wastewater treatment technologies made it popular in developing and densely populated countries like India. This technology provides a holistic solution for ecological units, with proper functioning at personal, social and institutional levels (Kumar & Kirti, 2020). They have a high level of technological, social, and financial viability and are site-specific. Wastewater treatment and reuse systems are based on removal efficiency, treatment area availability, finances, electricity, and skilled labor.

The financial viability of the treatment plant is another important parameter. Installation of a new treatment plant needs a huge amount of investment. Various governments, private and public funding agencies are available for this purpose.

Government sources like central government, state governments, urban development corporation (HUDCO), and many more are the government sources.

Banks, insurance companies, and share markets are the sources of capital for commercial sectors.

World bank, Asian Development Bank, and many more are multinational organization that provides financial support for big projects. Similarly, some financial organizations cover at least two countries, for example, Japan International Cooperation Agency and Department for International Development.

Public–private partnership (PPP) is a new concept in which a long-term agreement is established between a private company and a government organization to provide services to the public. In this model, the private company bears the responsibility and risk of project execution, along with financial liability.

In India, now private agencies are taking interest in the design, operation, and installation of treated water reutilization systems. Various financial agreements, both international and bilateral institutions are heavily investing in the development of reclaimed land. The utilization of treated water serves two purposes: the handover of technology or concept to others and capital investments. The financial situation has improved as a result of these new measures. In the area of wastewater management, a large number of PPP models are running successfully in India (Kumar & Indira Khurana, 2014; Kumar & Kirti, 2020).

4.6.2 Legal initiative

The overall success of such programs is dependent on the effective implementation of policies, infrastructures, and legislation. Ecological policy aids the implementation of recovered water reutilization as a division of water management. The durability of such a project necessitates sustainable principles and a stringent regulatory framework. To manage such projects in India, policies are framed by the local administration or government. The federal and state governments have taken several steps to support the reutilization of treated water. To promote such projects several improvements in the policy have been incorporated, such as cost recovery, implementation of the PPP model, excise duty on water use, etc., paving the path for integrated reliable development, including privatization (Shah, 2016; Thawale, Juwarkar, & Singh., 2006).

4.6.3 Market viability

It is only possible to reutilize treated wastewater successfully if such water has a market value. Market viability refers to the project's ability to find end-users for reclaimed water as well as its long-term financial viability. It is dependent on several factors. As evidenced by the following examples, treated wastewater has broad applications in the Indian market. Many cities currently get their water from far away sources, resulting in pollution and extra conveying cost. Some commons examples in India are Chennai, which obtains water from Lake Victoria resulting in high conveyance costs. Similarly, the Cauvery River, 95 km away from Bangalore, supplies water to Bangalore city, and is pumped at a height of 1000 m; likewise, the Krishna River is the source of Hyderabad's water, which is 130 km away and requires expensive multi-stage pumping, and so on (Recycling and reuse of treated wastewater in urban India: a proposed advisory and guidance document. CGIAR Research Program on Water, Land and Ecosystems, 2016). Using treated wastewater will save money on water.

The usage of treated wastewater will help to lower greenhouse gas emissions. Pollution controlling or regulating agencies in India like Central Pollution Control Board recommended zero liquid discharge. Water is now needed to be recycled and reused in industries. According to studies, reutilization of treated water for cultivation has a lot of financial advantages and such water acts like cash for several government organizations because this water has created revenue for the businesses to help offset costs. As a result, it is a desired source of water. The Chandigarh municipality, for example, charges 500 rupees per acre to supply such treated water to farmers for crops cultivation, rupees 50 for 500 yards for 2 months for irrigation of green spaces and rupees 8 to 11/m^3 to industries (Kumar & Kirti, 2020).

4.7 Conclusions

Population expansion and economic development in India have resulted in a rising need for water. The inadequate sources of existing water supplies have compounded the problem. Attempts to bridge the country's expanding water supply–demand gap had to be started a long time ago, but they started comparatively late. Therefore, some possible substitutes for augmentation of water were identified and utilization of effluent or wastewater after a proper treatment had been established as a possible solution to this issue. Reusing treated wastewater, according to research, is the most practical alternative because it provides multiple environmental and economic benefits while using the fewest resources. The notion of treated wastewater reuse is still being developed in India. Although there are several regulating rules and regulatory frameworks at the federal level, implementation and adjustment at different levels still necessitate a noteworthy effort to preparation of some simple rules and awareness programs to utilize them. As a result of improvements in environmental legislation, the capacity of the industrial reuse market has risen dramatically in recent years. Even though there are various chances to stimulate the reutilization of treated wastewater across the nation, generally it is not so acceptable till now. Various technical and commercial models, infrastructure for their implementation, and possible reclaimed water for users must be established and investigated for finance and revenue to achieve long-term reutilization of such treated water. For such projects, design, operation, and maintenance concerns, as well as the business plan to sell such treated water for different purposes, are required. There is an immediate need for a national-level policy that covers the aims, regulations, and financial measures for the reutilization of such water.

Current regulations related to water must also be updated and expanded to reflect the diminishing availability, quality, and accessibility of freshwater per capita. For recovered water reuse to thrive in India, sociological variables such as education levels, public attitudes toward reclaimed water, and environmental awareness must also be considered. The design of treated wastewater reuse systems is influenced by their willingness and ability to pay, as well as their desire to safeguard the environment. It is also necessary to have the political will and institutional support. Treated wastewater reuse, as well as agricultural and industrial reuse, has a lot of potential in the country. Reclaimed water reutilization must be integrated into the present and future water supply and treated water management programs through a viable strategy, proper governing structure, and efficient organizations, particularly in water-stressed areas, if it is to receive mainstream attention. A broad strategy is necessary at country-wide levels. Decision makers, contributors, public society organizations, and other shareholders, particularly in cities drafting sanitation plans, must promote continuing treated wastewater reuse activities and stimulate future endeavors. A shift in policy from the perception of water as community property to identifying the marketable commercial value for water. The government's role might be transformed by moving some of the government's existing privileges to consumers and the nongovernment sector, allowing reuse initiatives to be more financially viable. If the priority was given to demand management instead of supply management, water planning and management would improve. The creation of a holistic approach would aid towns in fulfilling future demographic, industrialization, and economic growth demands.

References

Alley, Kelly, D., Maurya, Nutan, & Sukanya, Das (2018). Parameters of successful wastewater reuse in urban India. *Indian Politics & Policy, 1*(2).

Angelakis, A. N. (2014). Water reuse: Overview of current practices and trends in the world with emphasis on EU states. *Water Utility Journal, 8*(67), e78.

Angelakis, A. N., & Snyder, S. A. (2015). Wastewater treatment and reuse: Past, present, and future. *Water, 7*(9), 4887–4895.

Arceivala, Sol, J., & Shyam, R. Asolekar (2006). Wastewater treatment for pollution control and reuse. Tata McGraw-Hill Education.

Ashton, John, & Janet, Ubido (1991). The healthy city and the ecological idea. *Social history of medicine, 4*(1), 173–180.

Barber, W. P. F., & Lancaster, R. (2012). Thermal hydrolysis: the missing ingredient for better biosolids p. 4. West Chester, PA: WaterWorld.

Barringer, J. (2014). Urban domestic and commercial water reuse in Pune and its influence on the present water crisis. *Water Quality, Exposure and Health, 6*(1), 35–38.

Bazza, M. (2003). Wastewater recycling and reuse in the Near East Region: experience and issues. *Water Science and Technology: Water Supply, 3*(4), 33–50.

Byrd, J. F. (1957). Combined Treatment of Industrial and Municipal Wastes: An Industrial Viewpoint. *Sewage and Industrial Wastes, 29*(4), 414–421.

Cairncross, S. (2018). Environmental health engineering in the tropics: Water, sanitation and disease control. Routledge.

Central water commission. Water-its conservation, management and governance. Government of India, 2017. (finalcwcar2017-18_1.pdf)

Chauhan, S., Jensen, O., Sengaiah, G., and Screenivas, N. (2016). "Closing the Water Loop: Reuse of Treated Wastewater in Urban India, (September)."

Composite water management index : a tool for water management (2018). Government of India.

Cosgrove, William, J., & Frank, R. Rijsberman (2000). Challenge for the 21st century: Making water everybody's business. *Sustainable Development International, 2*, 149–156.

Crites, Ronald, W., Joe Middlebrooks, E., & Sherwood, C. Reed (2010). *Natural wastewater treatment systems*. CRC press.

Devi, Mekala Gayathri, & Samad, Madar (2008). Wastewater treatment and reuse: an institutional analysis for Hyderabad, India. In *International Water Management Institute Conference Papers* no. h041888.

Dutta, S., & Bhaskar, S. (2018). "Surface and groundwater pollution are pushing India towards a water crisis."

Escher, B., Leusch, F., Chapman, H., & Poulsen, A. (2012). *Bioanalysitical tools in water quality assessment*. London, UK: IWA Publishing.

Escher, B. I., Allinson, M., Altenburger, R., Bain, P. A., Balaguer, P., Busch, W., … Hilscherova, K., et al. (2014). Benchmarking organic micropollutants in wastewater, recycled water and drinking Water with *in Vitro* Bioassays. *Environmental Science & Technology, 48*(3), 1940–1956.

Goyal, Kirti, & Kumar, Arun (2021). Development of water reuse: a global review with the focus on India. *Water Science and Technology, 84*(10-11), 3172–3190.

Grafton, Q. R., & Hussey, K. (2011). *Water resources*. Newyork: Cambridge University Press.

Gutierrez-Mata, A. G., Velazquez-Martínez, S., Álvarez-Gallegos, Alberto, Ahmadi, Mehdi, Hernández-Pérez, José Alfredo, & Ghanbari, F. (2017). Recent overview of solar photocatalysis and solar photo-Fenton processes for wastewater treatment. *International Journal of Photoenergy*.

Hedberg, Annika, Romain Pardo, Andrea Frontini, and Toutia Daryoush. (2015). "REACHING FOR BLUE GOLD-How the EU can rise to the water challenge while reaping the rewards. EPC Issue Paper No. 80, November 2015."

Henze, M., van Loosdrecht, M. C., Ekama, G. A., & Brdjanovic, D. (2008). *Biological wastewater treatment: Principles. Modelling and Design, **2008***, 493–494.

Kog, Y. C. (2015). *Springer Science and Business Media LLC*, (Vol. *15*, pp. 575–591). doi:10.1007/978-94-017-9801-3_26.

Khopkar, S. M. (2007). *Environmental pollution monitoring and control*. New Age International.

Kumar, Shiv, Indira Khurana, and P.R.K. Sobhanbabu. (2014). "Scoping Study: India's Global Resource Footprint in Food, Energy and Water (FEW)".

Kumar, Arun, & Kirti, Goyal (2020). Water reuse in India: Current perspective and future potential. *Advances in chemical pollution, environmental management and protection* (pp. 33–63). Elsevier 6.

Lautze, J., Stander, E., Drechsel, P., da Silva, A. K., & Keraita, B. (2014). Global experiences in water reuse. In CGIAR Research Program on Water, Land and Ecosystems (WLE) (p. 31). International Water Management Institute (IWMI).

Levine, A. D., Tchobanoglous, G., & Asano, T. (1985). Characterization of the size distribution of contaminants in wastewater: treatment and reuse implications. *Journal of the Water Pollution Control Federation, 57*, 805–816.

Logan, B. E., Hamelers, B., Rozendal, R., Schröder, U., Keller, J., Freguia, S., … Rabaey, K. (2006). *Environmental Science and Technology, 40*(17), 5181–5192. doi:10.1021/es0605016.

Logan, B. E., & Regan, J. M. (2006). Environmental Science and Technology, 40(17), 5172–5180. doi:10.1021/es0627592.

Mahmood, Qaisar, Pervez, Arshid, Zeb, Bibi Saima, Zaffar, Habiba, Yaqoob, Hajra, Waseem, Muhammad, & Afsheen, Sumera (2013). Natural treatment systems as sustainable ecotechnologies for the developing countries. *BioMed Research International, 2013*.

McMillan, M. F., Hampton, J., & Lennert, L. (2019). Control your water, control your results: improving irrigation audits and reducing soil hydrophobicity. Irrigation Association.

McNabb, D. E. (2019). The population growth barrier. In: Global pathways to water sustainability (pp. 67–81). Cham: Palgrave Macmillan. https://doi.org/10.1007/978-3-030-04085-7.

Mekala, Gayathri Devi, Brian Davidson, Madar Samad, and Anne-Maree Boland. (2007). "Wastewater reuse and recycling systems: a perspective into India and Australia."

Moefc, G. (2018). "India: second biennial update report to the United Nations framework convention on climate change."

Mudgal, S., Van Long, L., Saidi, N., Haines, R., McNeil, D., Jeffrey, P., Smith, H., & Knox, J. (2015). Optimization Water Reuse in EU: Final Report. *BIO by Deloitte: Brussels, Belgium*, (2015), 199.

Neyens, E., & Baeyens, J. (2003). *Journal of Hazardous Materials, 98*(1–3), 51–67. doi:10.1016/S0304-3894(02)00320-5.

Notification for not using treated wastewater for agriculture dated April 15, 2015. India: Uttar Pradesh Jal Nigam, 2015, 63–75.

Paranychianakis, N. V., Salgot, M., Snyder, S. A., & Angelakis, A. N. (2015). *Critical reviews in environmental science and technology, 45*(13), 1409–1468. doi:10.1080/10643389.2014.955629.

Rai, Raveendra Kumar, Singh, Vijay P., & Upadhyay, Alka (2017). *Planning and evaluation of irrigation projects: methods and implementation*. Academic press.

Ravikumar, Joseph, Raman, Rajiv, & Shubhra, Jain (2014). Creating an Economic Value for Wastewater through Industrial and Agricultural Reuse. *THE ASIAN JOURNAL*, 38–49.

Recycling and reuse of treated wastewater in urban India. (2016). a proposed advisory and guidance document. *CGIAR Research Program on Water, Land and Ecosystems*. International Water Management Institute (IWMI).

Rode, A. R., Shirbhate, S. S., & Shewale, S. R. (2013). Planning and hydraulic design of sewage treatment plant for Yavatmal city. *International J Research, 1*(4), 159–163.

Shah, Mihir. (2016). *Urban water systems in India: a way forward*. No. 323. In *Working Paper*.

Shiklomanov, I. (2000). World water resources and water use: Present assessment and outlook for 2025 (supplemented by CD-ROM: Shiklomanov, I., World freshwater resources, available from: International Hydrological Programme, UNESCO, Paris). *World Water Scenarios: Analysing Global Water Resources and Use. London: Earthscan Publications*.

Snyder, S. A. (2014). *Journal – American Water Works Association, 106*(8), 38–52. doi:10.5942/jawwa.2014.106.0126.

Strande, Linda (2014). *Faecal sludge management: Systems approach for implementation and operation*. Damir Brdjanovic, eds. IWA publishing.

Tare, V. (2011). Review of Wastewater Reuse Projects Worldwide. *Collation of Selected International Case Studies and Experiences, Report code: 012_GBP_IIT_SOA_01_Ver 1_Dec, 2011*, 11–38.

Tchobanoglous, G, Burton, FL, & Stensel, HD (2003). *Wastewater engineering: treatment and reuse* (4th edn.). New York, USA: McGraw-Hill.

Thawale, P. R., Juwarkar, A. A., & Singh, S. K. (2006). Resource conservation through land treatment of municipal wastewater. *Current Science*, 704–711.

Tzanakakis, V., Koo-Oshima, S., Haddad, M., Apostolidis, N., Angelakis, A., & Angelakis, A. (2014). The history of land application and hydroponic systems for wastewater treatment and reuse. *Evolution of Sanitation and Wastewater Technologies through the Centuries* (1, p. 457). London, UK: IWA Publishing.

Tzanakakis, V. E., Paranychianaki, N. V., & Angelakis, A. N. (2007). *Water Supply, 7*(1), 67–75. doi:10.2166/ws.2007.008.

Walmsley, N., & Pearce, G. (2010). Irrigation and drainage systems, *24*(3–4), 191–203. doi:10.1007/s10795-010-9100-z.

Chapter 5

Wastewater treatment plant's tracking of resistant bacteria and gene in wastewater

Rachana Tiwari[a], Shahina Bano[a] and Manisha Agrawal[b]

[a]*Rungta College of Sciences & Technology, Chhattisgarh, India,* [b]*Rungta College of Engineering & Technology, Bhilai, Chhattisgarh, India*

5.1 Introduction

Water after use for domestic, industrial or commercial purposes gets contaminated, such water is known as wastewater. The compositions of wastewater are in varieties but roughly in proportion. It contains almost 99.9% part water and around 0.1% of other materials which needs to be removed. In total, 0.1% contains generally the organic matter, microorganisms and inorganic compounds. These contaminated water moves from one place to other through the sewer system and finally gets discharged into water sources generally lakes, ponds, streams, rivers, estuaries, and oceans.

Wastewater treatment plants (WWTPs) are the industries that reclaim such kinds of wastewater. WWTP is a facility in which a combination of various processes (e.g., physical, chemical, and biological) used to treat industrial wastewater and remove pollutants. After the removal of all the contaminants in permissible limit, such recycled water is further used in industries and cleaning processes. The major drawback factor which is connected with recycled water is the presence of antibiotic resistant bacteria (ARB) and antibiotic resistant gene (ARG). Problem arises because of the industry that deals with antibiotics, like various pharmaceutical industries, and through insufficiency of the human toward metabolizing of these antibiotics. All this ends up in wastewater thus increasing ARBs and ARGs. WWTPs track the count of ARBs and ARGs in the reclaimed water before releasing it for further use.

In this swot up, the ARBs are mainly tested against tetracycline and sulfonamides. Tetracycline is a broad-spectrum antibiotic. Infections that are generally caused by bacteria are treated with tetracycline. Tetracycline is widely used for the healing of bacterial infections such as—skin, intestines, respiratory tract,

Antimicrobial Resistance in Wastewater and Human Health.
DOI: https://doi.org/10.1016/B978-0-323-96124-0.00005-2

urinary tract, genitals, and other body systems. It is also used for the treatment of severe acne, or STDs such as syphilis, and gonorrhea. Sulfonamides are the firstly adapted chemotherapeutical substance that was used for the treatment of bacterial infections in living things. Amides or sulfa drugs contain the sulfonamide group and in the early days, it was used as an antimicrobial agent. They serve as the foundation of several groups of drugs. In the present study, the effect of tetracycline finds more significant than sulfonamides.

5.1.1 Necessity of wastewater treatment plants

More than 80% amount of reclaimed water is being used globally for daily requirements of water. Such wastewater has potential to transfer or generate many kinds of diseases to living animals and the environment. We may summarize its necessity of it as:

1. Decaying material utilizes the oxygen present in water which eventually leads to the death of the entire biotic components of water.
2. Some nutrition like phosphorus and nitrogen gives rise to eutrophication, thus increases the toxicity level of water.
3. The aquatic animals' invertebrates, algae, and fish are affected by the large quantity of chloride.
4. Microorganisms like bacteria, viruses, and disease-causing pathogens alters the marine life like shellfish which ultimately leads to human toxicity upon consumption.
5. Heavy metals: mercury, lead, cadmium, and arsenic causes acute and chronic toxic effects on many different species.
6. Industries like pharmaceutical or those concerned with care products also pose a very harmful effect on environment and it also leads in the increase in ARBs and ARGs.

Hence to control all these threats, WWTPs are required.

5.2 Methodology required for the treatment of reclaimed water

Basic WWTPs have been applied for a very long time using certain processes. These processes are as follows:

1. Physical process.
2. Biological process.
3. Chemical process.

5.2.1 Physical process

This process is often called the primary treatment. It is mostly based on the principle of precipitation. In this process, wastewater is collected in large tanks where solid material gets deposited or settled, which later gets extracted for further processing. Sedimentation is achieved by adding chemical substances like alum or potassium permanganate.

For colloidal units, flocculation, floatation, or precipitation method is generally used.

5.2.2 Biological process

Biological process is known as the secondary treatment method of wastewater plants. In this process, organic matters such as oil, care product, medicinal waste, industrial waste, human wastes get decomposed with the help of bacteria. It may be further divided in two categories:

a. Aerobic treatment.
b. Anaerobic treatment.

Aerobic treatment is a biological process that utilizes oxygen for the breakdown of the organic matter as well as certain gases like nitrogen and phosphorus. Organic matters get broken into carbon dioxide and biomass by aerobic microorganism metabolizes.

Example of an aerobic treatment: Well-known example of it is a trickling filter. **Trickling filter**—after the primary treatment of wastewater, the recycled (treated water) is passed over the trickling filter. The image of a trickling filter is shown in Fig. 5.1. Trickling filter consists either of plastic beads, stone, gravel, rock, or slag, through which the primary treated wastewater is passed. As the wastewater contains bacteria in it, these bacteria start coagulating over the culture bed and simultaneously degraded or decompose d the organic matter present in it into simpler form.

As a result, the water gets treated or eradicated of the organic matter thus leaving filtered water which is further collected as an effluent and sent for tertiary treatment.

5.2.2.1 Anaerobic treatment

The biological process which takes place in the absence of oxygen is known as an anaerobic treatment. It takes place in presence of anaerobic bacteria and other microorganisms. Anaerobic microorganism degrades materials which found in industrial effluents of agricultural, food, dairy, paper, textile, as well as municipal wastes. A basic diagram of the anaerobic system is shown in Fig. 5.2.

In UASB, the influent is pumped through a peristatic pump from the bottom toward upward. The influent is then passed through an evenly distributed sludge bed that contains high concentration of anaerobic biomass where influent's organic matter gets decomposed. After that, it is conceded through a less concerted biomass level called as the sludge blanket that treats the remaining crude matter. As the influent gets treated, it produces a large amount of biomass which is collected via a gas collection dome, which gets further collected through gas

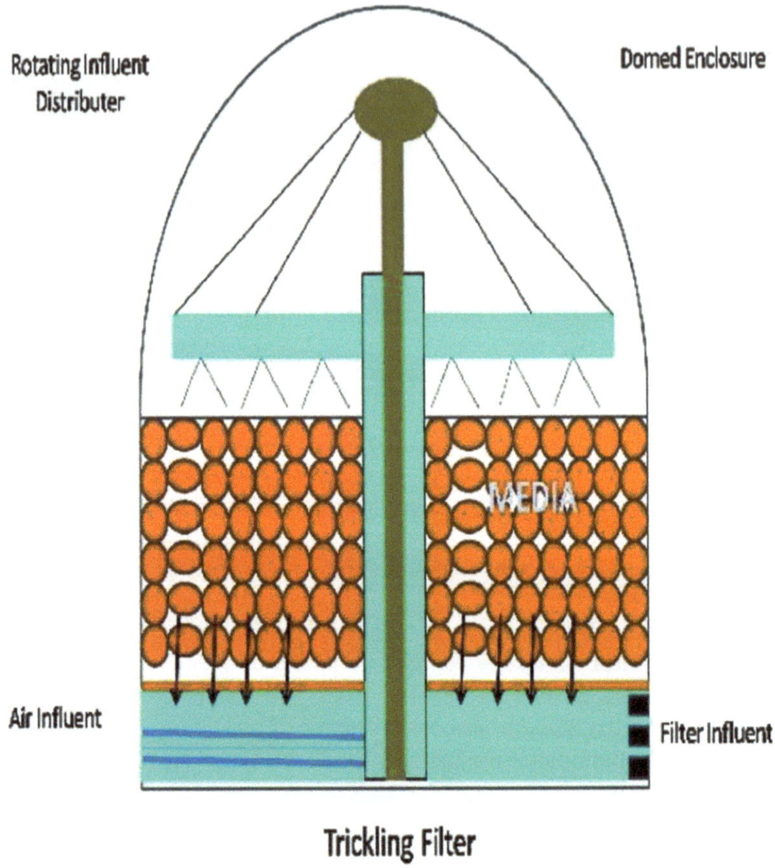

FIGURE 5.1 Trickling filter.

collecting tube. The treated water is then collected through an effluent collecting tube.

In the next stage, the treated water is further sent for the next process, that is, tertiary treatment.

5.3 Chemical process

The chemical process is being used for the last stage for the treatment of wastewater. The chemical process or tertiary treatment deals with the final cleaning of effluent before it gets released for the reuse by removing the remaining inorganic waste like nitrogen and phosphorus. At this stage, only all the microorganisms like bacteria, virus, or parasite are removed too.

For the removal of phosphorus alum is added, while the other solid material gets clustered together to form floc. It is then gravity fed through the sand where the cluster gets blocked while leaving the remaining effluent for the further

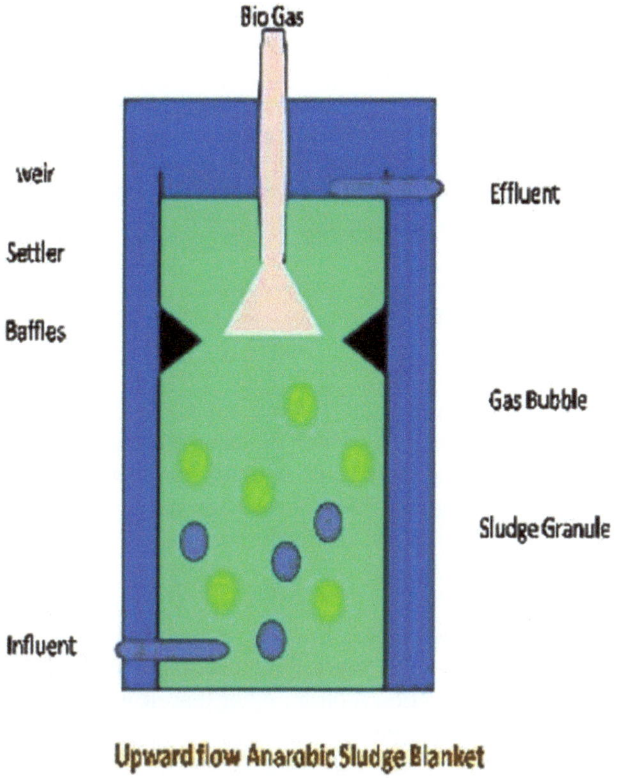

FIGURE 5.2 Up-flow anaerobic sludge blanket reactor.

treatment through chlorine. Chlorine gets rid of the left-over microorganisms—bacteria, virus, or parasite like giardia and cryptosporidium.

The recycled water is tested for ARBs and ARGs at the end of the process.

The chapter deals with the study of the coliform bacteria tested against tetracycline using disc diffusion assay for recognition of ARGs, polymerase chain reaction was applied followed by agarose gel electrophoresis.

5.4 Materials and methods

The study may be divided into several steps for ease of understanding.

5.4.1 Sample collection

Area selected for the study is located in Bhilai, Chhattisgarh. Samples were collected from three different sites. These are as follows:

- Wastewater outlet of an industry in sector area, Bhilai.
- Hospital area, Nehru Nagar, Bhilai.
- Residential area, Kohka, Bhilai.

Recycled waters were collected from these areas and were treated with alum and chlorine. Then, it was tested against tetracycline and sulfonamide.

5.4.2 Media preparation

Laboratory grade chemicals are used for the preparation of culture media. Standard nutrient agar medium was utilized for the testing process with 3 g beef/meat extract, 5 g each peptone and NaCl, and 15 g agar–agar powder for 1000 mL.

5.4.3 Instruments used

5.4.3.1 Autoclave

For the wet sterilization of media and glassware, autoclave is the instrument that is used. The nutrient agar media that was prepared for the experiment was first homogenized and then it was autoclaved at 121°C for an hour.

After the autoclave process which is used for sterilization, the media was shifted into the laminar air flow (LAF) for further.

5.4.3.2 Laminar air flow

LAF is a covered workspace that is utilized to generate a contamination-free atmosphere with the use of HEPA filters. For the work, firstly the LAF was wiped completely using single direction motion. After that, it was left for 15 minutes in the presence of ultraviolet lights.

After that, rest of the process of plating, spreading of wastewater, and disc diffusion method was done. The properly taped plates are kept in an incubator at 37°C. After 18–24 hours, plates were analyzed for inhibition zone bacterial growth.

5.4.3.3 Hot air oven

A hot air oven is an equipment that is utilized for dry sterilization.

Hot air oven is generally used for sterilization of glassware. Firstly, the glassware were wiped using 70% ethanol, and then it was wrapped and kept in oven at 75°C for 24 hours.

5.5 Antibiotic resistivity test

The recycled water is tested for antibiotic resistance activity which is tested against two antibiotics using disc diffusion assay method. In this method, NAM plates were prepared and were marked as S_1, S_2, and S_3, respectively, for industrial, hospital, and residential water sample.

A. Three replicates of each test were prepared for experiments. Each set contains six plates including all sites of water sample for tetracycline and sulfonamide antibiotics.

TABLE 5.1 Abbreviations of sampling sites.

S. no.	Sites	Abbreviation	Tetracycline antibiotic	Sulfonamide antibiotic
1	Industrial sector area	S_1	TS_1	SS_1
2	Hospital Area Nehru Nagar	S_2	TS_2	SS_2
3	Residential area	S_3	TS_3	SS_3

B. The water sample was inoculated using the spread plate technique and three discs were applied per plate.

C. At 37°C for 24 hours plates were incubated. Observation for the antibiotic resistive activity was done after via measuring the resistivity zone created by the bacteria present in the respective sample. Table 5.1 summarizes the abbreviations of sampling sites

5.6 Mechanism of antibiotics during recycling of water

While recycling process of wastewater, five major steps get followed (Functions of Antimicrobial Drugs, 2021):

A. Inhibition of cell wall synthesis.
B. Disruption of cell membrane function.
C. Inhibition of protein synthesis.
D. Inhibition of nucleic acid synthesis.
E. Action of antimetabolites.

5.7 Inhibition of cell wall synthesis

The work of many antibiotics is attacking the cell wall of bacteria. Specifically, the drugs prevent the bacteria from synthesizing a molecule in the cell wall called peptidoglycan, which provides the wall with the strength it needs to survive in the human body. Antibiotic that is beta-lactam and glycopeptide inhibits the cell wall synthesis of bacteria. They bind to the 30S subunit of ribosome and inhibit the protein synthesis. These target the proteins which catalyze the synthesis of the peptidoglycan that forms the cell wall of bacteria.

Examples of such antibiotics are penicillin and cephalosporin.

5.8 Disruption of cell membrane function

The cell wall or membranes are the wall that surrounds the bacterial cell. In principle, there is three main antibiotic which targets the bacteria. Disruption of the cell membrane or plasma membrane causes depolarization due to depolarization

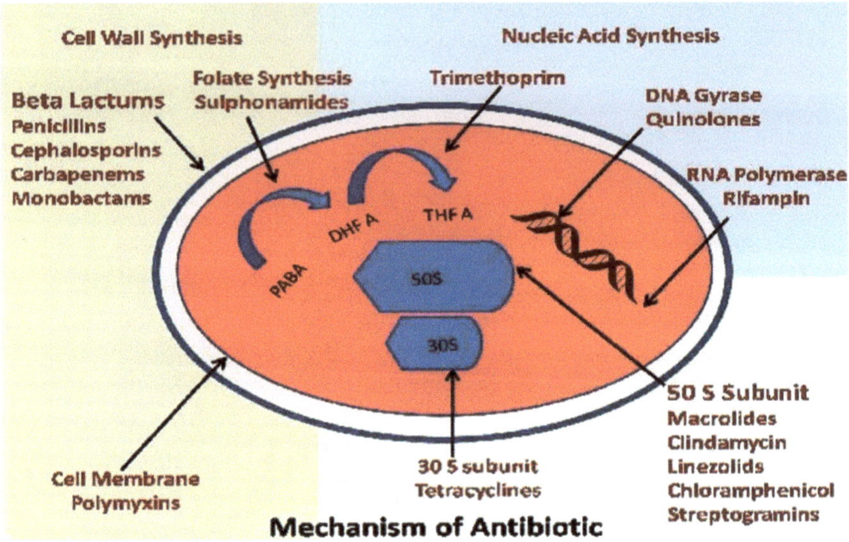

FIGURE 5.3 Mechanism of antibiotic.

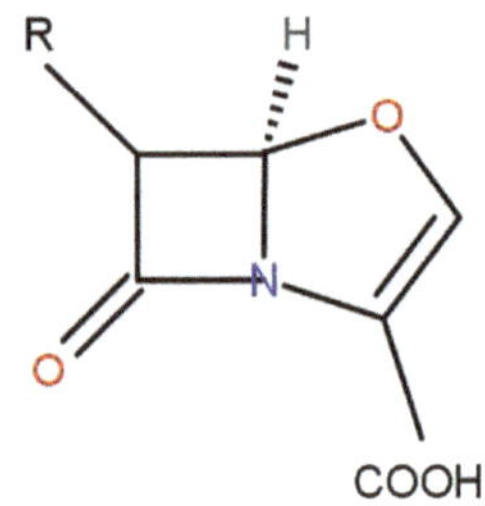

FIGURE 5.4 Beta-lactam.

of membrane which is the rapid loss in the potential of the cell membrane occur because of cell wall potential decrement bacterial cell becomes dead or inactive, as shown in Fig. 5.3.

An example of such antibiotic is daptomycin.

5.9 Inhibition of protein synthesis

The bacteria and eukaryotes have ribosomes that are structurally different. Bacteria have 70S ribosomes (Munita & Arias, 2016) There is certain antibiotic which can either stop or slow down the translation process by inhibiting the ribosomal 30S subunits (Figs. 5.4–7).

Example of such antibiotic are 50S inhibitors erythromycin chloramphenicol. An example of such antibiotic is 30S inhibitor tetramycin streptomycin.

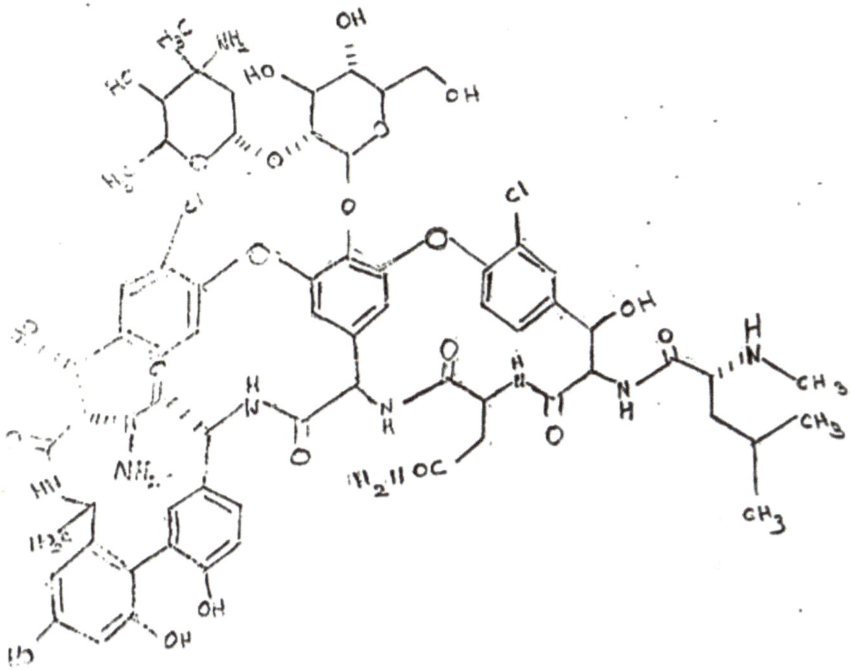

FIGURE 5.5 Vanomycin.

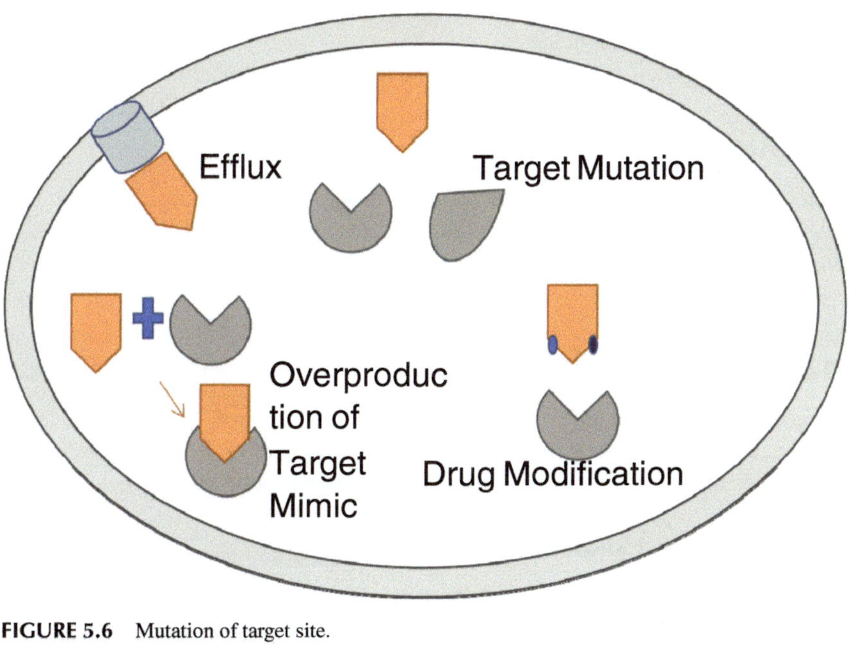

FIGURE 5.6 Mutation of target site.

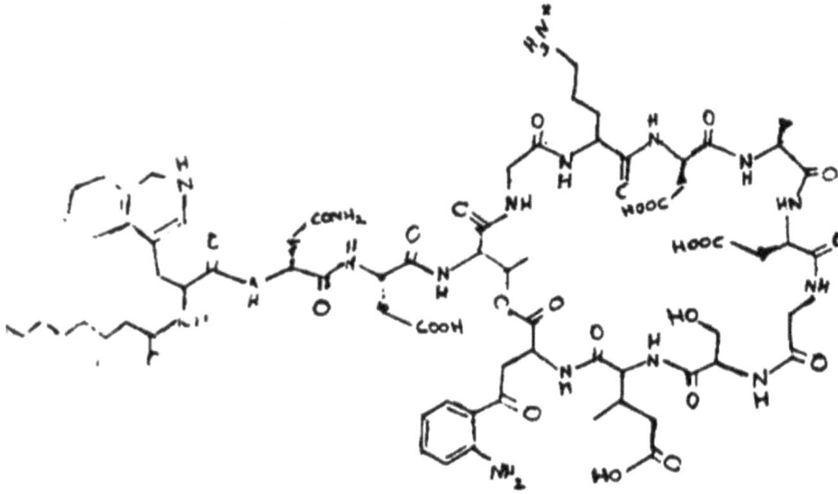

FIGURE 5.7 Daptomycin.

5.10 Inhibition of nucleic acid synthesis

Certain antibiotic inhibits the nucleic acid activities by inhibiting topoisomerase. Topoisomerase is also known as gyrase. This allows the DNA strand to be replicated by DNA or RNA polymerases. They block the initiation of RNA synthesis. A nucleic acid inhibitor is a type of antibacterial that acts by inhibiting the production of nucleic acids.

Examples of such antibiotics are ofloxacin and ciprofloxacin.

5.11 Action of antimetabolites

Nucleotides are essential for nucleic acid synthesis. For this process, antimetabolites antibiotic is used. The nucleotide that gets inhibited is dTMP which is a pyridine.

Examples of such antibiotics are sulfonamide and trimethoprim.

5.12 Antibiotic resistance

The common bacterial mechanism to deal with the action of antibiotic is to produce some substances that can stop the drug action by adding specific chemical or that destroy the molecule, it can be two types.

5.12.1 Chemical alteration

The change in molecules chemical alteration is occur due to production of enzyme. This production of enzyme which alters the molecules' chemical structure inhibits the process of protein synthesis.

Some of these types of enzymatic activities are acetylation phosphorylation. This enzyme modifies the functional group present chemical.

5.12.2 Destruction of antibiotic molecule

By action of beta-lactamases, the destruction of antibiotic molecule will occur. In the lactam ring, beta-lactamases enzyme destroys the bonding present in the ring (Arabi, Pakzad, Nasrollahi, Hosainzadegan, & Azizi Jalilian, 2015).

- ➢ This beta-lactam is basically used to stop penicillin resistance.
- ➢ These are effective on Gram-negative bacteria.
- ➢ This beta-lactam further divided into two categories by bush Jacoby.
- ➢ Chelating ligand are inhibiting the beta-lactamases. This kind of chelating ligand contains ion like EDTA.

5.13 Decreased antibiotic penetration

- ➢ Some antibiotic attacks the inner layer of membrane.
- ➢ For example, in Gram-negative bacteria, which is present in the inner membrane antibiotic attacks in inner layer.
- ➢ These kind of bacteria or microbes inhibit the excess of antibiotic in the cytoplasmic part.
- ➢ The entry of antibiotic in the part of cell is defend by outer layer.
- ➢ The main example for this type of resistance is action of vancomycin.
- ➢ Which is a glycopeptide antibiotic being not effective on Gram-negative bacteria.
- ➢ There are many portions by which absorption can reach the attacking part.

5.14 Efflux pump

Efflux pumps allow the microorganism to regulate their internal environment by removing toxic substances, including antimicrobial agents and metabolites.

- ➢ This type of machineries or pump extrude the chemical out of the membrane.
- ➢ This type of pump stops the exchanges of protons.
- ➢ This kind of stoppage of exchange of protons is generally found in Gram-negative bacteria.
- ➢ The best-defined efflux pump is MEF ARGs that excludes the macrolides of antibiotic.

5.15 Change of target sites

Target modification is a self-resistance mechanism that act against several kinds of antibiotic. Change of target site is the most effective way to stop initiating action of antibiotic. Microbes can protect their target site or can modify their target site to avoid antibiotic action.

5.16 Modifications of target site

- These modifications can be enzymatic alteration of binding site. Or it can be occurred by small alteration in ARGs.
- In enzymatic alteration some subgroups are added on the target site like methyl or ethyl groups.
- In this type of alteration, methylation is catalyzed by enzyme that results in macrolides resistance.
- They can also change their actual target site.
- The best example of this resistance is methicillin resistance in S. Aureus7.
- In the Vanomycin resistance peptidoglycan structure altered by van gene cluster.
- Mutation of target site inhibits the DNA-dependent RNA polymerase
- Inhibited by the target site mutation because of this process bacterial transcription stops.

5.16.1 Resistance due to global adoption

In the era of evaluation bacteria have developed several methods (Zhang, Huang, Zhao, Cao, & Li, 2020) to deal with environmental compressor. Bacteria needs to fight against several substances and the immune system of human or living things.

Examples of such antibiotics are daptomycin and vancomycin.

5.17 Result and discussion

Tetracycline and sulfonamide consist of disc diffusion study and polymerase chain reaction (Nahar et al., 2019; Ashrafun, 2019) test against three different water samples sourced from residential, industrial, and hospital sectors.

In the primary test of detecting ARB using disc diffusion method, six plates were prepared from sample in duplets for both tetracycline shown in Fig. 5.8 and sulfonamide shown in Fig. 5.9, respectively.

The zone of inhibition for tetracycline for industrial and residential wastewater was found to be 0.5 and 0.4 mm. Whereas it did not show any zone of inhibition against hospital wastewater. On the other hand, sulfonamide's zone of inhibition was found to be 0.4 and 0.3 mm for industrial and residential wastewater sample, respectively, are shown in Table 5.2. Apparently, sulfonamide also did not show any zone of inhibition against microorganisms present in hospital wastewater (Figs. 5.10–12).

In the secondary test, the plates that were found to have resistant bacterial colonies against tetracycline and sulfonamide were tested again using polymerase chain reaction.

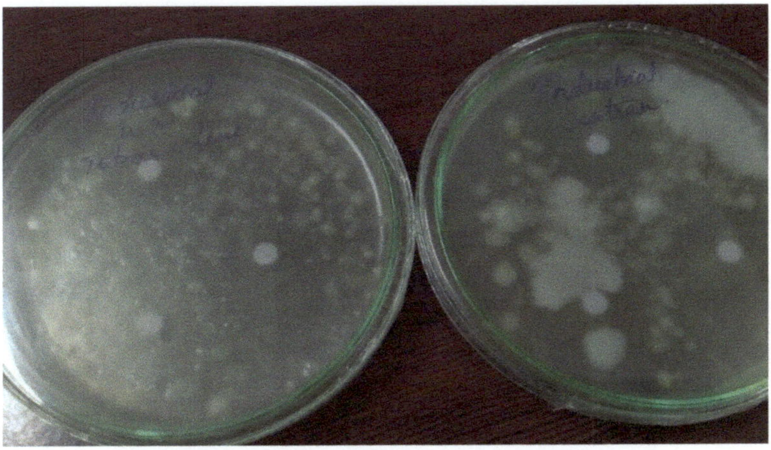

FIGURE 5.8 Inhibition zone in the water sample.

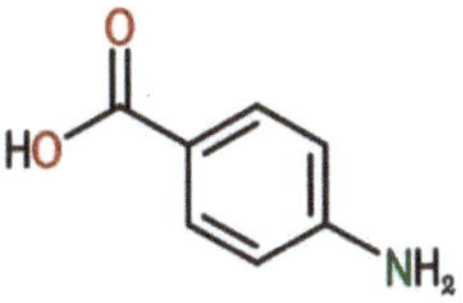

FIGURE 5.9 Sulfonamide.

TABLE 5.2 Inhibition zone developed in various wastewater samples.

Drug	Industrial area WW	Hospital area WW	Residential area WW
Sulfonamide	0.5	Nil	0.4
Tetracyclines	0.4	Nil	0.3

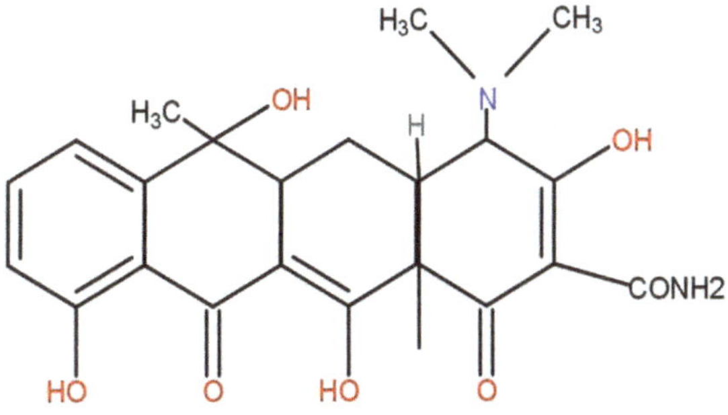

FIGURE 5.10 Tetracycline.

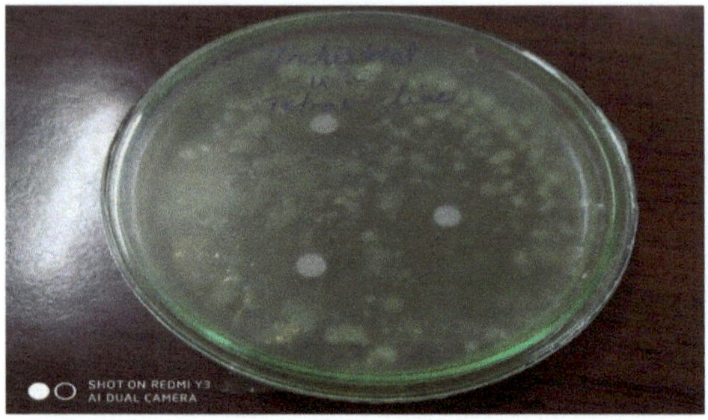

FIGURE 5.11 Inhibition zone against tetracycline plate TS[1].

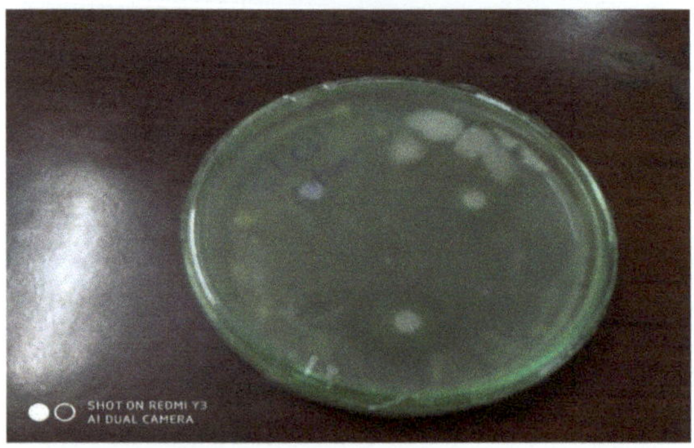

FIGURE 5.12 Bacterial colony growth in sample.

TABLE 5.3 ARGs found in bacterial sample.

Antibiotic	Industrial water	Hospital water	Residential water
Tetracycline	Nil	1	Nil
Sulfonamides	Nil	2	Nil

A total of three ARGs were found, as shown in Table 5.3. One against tetracycline was found to be "Tet A" and the one against sulfonamide was found to be Sul1 and Sul 2.

5.18 Discussion

5.18.1 Chapter deals with wastewater treatment plant's tracking of antibiotic resistant

Bacteria and antibiotic-resistant gene: The zone of inhibition found from the disc diffusion method tested with tetracycline and sulfonamide against wastewater (residential, industrial, and hospital) showed that industrial and residential water did not have any resistant bacteria whereas hospital water showed the presence

TABLE 5.4 Bacterial antibiotic colonies count.

Antibiotic	Industrial water	Hospital water	Residential water
Tetracycline	13	35	17
Sulfonamides	16	42	21

of ARB and gene as well. Bouki, Venieri, and Diamadopoulos (2013), in their paper Detection and fate of ARB in WWTPs: A review, found that ARB and ARGs were prominently found more in wastewater whether its raw or treated as compared to the surface water (Table 5.4).

On the other hand, Marily (1996) stated that tetracycline resistance is due to accession of new plasmid or transposable elements or energy-dependent efflux. It also showed that tetracycline resistance is species-specific like Gram-positive or Gram-negative bacteria.

Sköld (2000) in their paper studied sulfonamide against both ARB and gene and two isolates of ARGs were derived. Edmund, Wise, and Martha (1975) found that *E. coli* isolates showed resistance against sulfonamide and is determined with the help of R plasmid. Apart from *E. coli* Klebsiella and pneumonia strain also showed resistance against sulfonamide.

5.19 Conclusion

With an increase rate of antibiotic usage, the rate of ARB has been elevated. The reason behind antibiotic resistance is the ability of microorganism to grow and multiply in the presence of antibiotics. This ability occurs when a certain type of microorganisms grows in an environment that contains antibiotic in it. As a result, these microorganisms start using the same antibiotic as the nutrient source or growth factor. This scenario does not occur at once. It occurs when a type of microorganism's strain keeps on recurring in the same environment again and again and because of that, it develops a resistant gene that enables it to grow and resist antibiotics. This resistivity is generally seen in pharmaceutical wastewater, hospital wastewater, laboratory wastewater, textile wastewater, etc.

For testing, the water sample is accumulated from all the sources and is tested against antibiotics (tetracycline and sulfonamide). As the bacteria grew, one that carried antibiotic resistant gene showed resistance against those antibiotics. Rest showed a zone of inhibition. The zone of inhibition found in both residential and industrial plates showed that the microorganisms found in those examples were not able to resist against tetracycline and sulfonamide. Whereas the remaining one, that is, hospital wastewater did showed resistivity.

Guillaume, Verbrugge, Chasseur-Libotte, Moens, and Collard (2000) studied tetracycline determinant Tet A–E against 40 salmonella enterica and sludge from household and hospital wastewater and found that Tet A was present in around activated sludge and Tet C was found to be in only one of the 40 clinical isolates and in seven samples.

References

Iwapublishing up flow - anaerobic sludge blanket reactor (uasb) Up Flow Anaerobic Sludge Blanket Reactor (UASB) about anaerobic wastewater.

Eponline/Articles/2018/02/08/Four-Effective-Processes-to-Treat-Wastewater/Page 2.

Published in 2017 by the United Nations Educational, Scientific and Cultural Organization, 7, place de Fontenoy, 75352 Paris 07 SP, France.

MASTERFLEX tech-article/importance-of-wastewater-treatment/05/26/21.

The Benefits of a Modern Wastewater Treatment System, Carlow Tanks (2014).

A Visit to a Wastewater Treatment Plant by USGS/primary treatment process//August 30 (2018).

SAMCO What Is Anaerobic Wastewater Treatment and How Does It Work JULY 9, (2019).

Arabi, H., Pakzad, I., Nasrollahi, A., Hosainzadegan, H., & Azizi Jalilian, F. (2015). Sulfonamide resistance genes (*sul*) M in extended spectrum beta lactamase (ESBL) and Non-ESBL producing *Escherichia coli* isolated from Iranian hospitals. *Jundishapur Journal of Microbiology, 8*(7), 1–6.

Ashrafun, N. (2019). Detection of tetracycline resistant E. coli and Salmonella spp. in sewage, river, pond and swimming pool in Mymensingh, Bangladesh/Article Number - 4E8CC3161497 *Vol. 13*(25), pp. 382–387,

Bouki, C., Venieri, D., & Diamadopoulos, E. (2013). Detection and fate of antibiotic resistant bacteria in wastewater treatment plants: A review. *Ecotoxicology and Environmental Safety, 91*, 1–9. doi:10.1016/j.ecoenv.2013.01.016.

Edmund, M., Wise, J., & Martha, M. (1975, July). Sulfonamide resistance mechanism in Escherichia coli: R plasmids can determine sulphonamide-resistant dihydropteroate synthases. *Proceedings of the National Academy of Sciences of the United States of America, 72*(7), 2621–2625 Biochemistry.

Functions of Antimicrobial Drugs, Functions of antimicrobial drugs. (2021). https://bio.libretexts.org/go/page/8707. (Accessed on 4 January 2021).

Guillaume, G., Verbrugge, D., Chasseur-Libotte, M., Moens, W., & Collard, J. (2000). PCR typing of tetracycline resistance determinants (Tet A-E) in Salmonella enterica serotype Hadar and in the microbial community of activated sludges from hospital and urban wastewater treatment facilities in Belgium. *Fems Microbiology Ecology, 32*(1), 77–85 2000.tb00701. x. PMID: 10779622. doi:10.1111/j.1574- 6941.

Marily, C. R. (1996). Tetracycline resistance determinants: Mechanisms of action, regulation of expression, genetic mobility, and distribution FEMS. *Microbiology Reviews, 19*(Issue 1), 1–24.

Munita, J. M., & Arias, C. A. (2016). Mechanisms of antibiotic resistance. *Microbiology Spectrum, 4*(2), 1–24 10.1128/microbiolspec.VMBF-0016-2015PMID: 27227291; PMCID: PMC4888801. doi:10.1128/microbiolspec.VMBF- 0016-2015.

Nahar, A., Islam, Md. A., Sobur, Md. A., Hossain, Md. J., Zaman, S. B., Rahman, Md. B., … Rahman, Md. T., et al. (2019). Detection of tetracycline resistant E. Coli and Salmonella spp. In sewage, river, pond and swimming pool in Mymensingh, Bangladesh. *African Journal of Microbiology Research, 13*(25), 382–387.

Sköld, O. (2000). Sulfonamide resistance: Mechanisms and trends. *Drug Resist Update, 3*(3), 155–160 PMID: 11498380. doi:10.1054/drup.2000.0146.

Zhang, S., Huang, J., Zhao, Z., Cao, Y., & Li, B. (2020). Hospital wastewater as a reservoir for antibiotic resistance genes: A meta-analysis. *Frontiers in Public Health, 8*, 1–12, 574968. Published 2020 Oct 28. doi:10.3389/fpubh.2020.574968.

Chapter 6

Techniques to stop spread and removal of resistance from wastewater

Dhruti Sundar Pattanayak[a], Dharm Pal[b], Chandrakant Thakur[a] and Awanish Kumar[c]

[a] *Department of Chemical Engineering, National Institute of Technology Raipur, Raipur, Chhattisgarh, India,* [b] *Associate Professor, Department of Chemical Engineering, National Institute of Technology Raipur, Raipur, Chhattisgarh, India,* [c] *Associate Professor, Department of Biotechnology, National Institute of Technology Raipur, Raipur, Chhattisgarh, India*

6.1 Introduction

Water is a necessary element of living organisms that play an important part in the lives of both humans and animals (Mohanty, Pattanayak, & Dash, 2021). It is defined as a substantial constituent of the blood that transports nutrients and life-saving oxygen to all cells, making it essential. As long as these vital functions are in existence, access to both quality and quantity of water should be prioritized. Pollution, on the other hand, hinders water quality and the quantity released into wastewater management facilities from many sources (Jia et al., 2019). Water recycling has been proposed as a way out of water scarcity, and effective handling procedures will aid in achieving good quality water to ensure the survival of all living species (Mohanty, Sundar Pattanayak, Singhal, Pradhan, & Kumar Dash, 2022). Enhancing water excellence necessitates core measures such as the obstruction of pollutants such as pesticides, medicines, heavy metals, dyes, treatment of polluted water, and ecosystem restoration and protection (Ezeuko, Ojemaye, Okoh, & Okoh, 2021). Antibiotics are pharmacological medications that kill microorganisms in order to prevent bacterial infections in living creatures. Since the 1920s (Hutchings, Truman, & Wilkinson, 2019), the main issue has been the huge use of antibacterial medication, with the possible threat of antibiotic resistance gene (ARG) amplification, which affects treatment strategies effectively (E. Implementation WHO Report on Surveillance of Antibiotic Consumption, WHO Report on Surveillance of Antibiotic Consumption, 2018). As a result, there are global attempts to minimize antibiotic

Antimicrobial Resistance in Wastewater and Human Health.
DOI: https://doi.org/10.1016/B978-0-323-96124-0.00009-X

usage, yet they are still far from achieving this aim. Data on antibiotic usage are sparse and frequently unavailable (Pattanayak, Mallick, Thakur, & Pal, 2020). They also predicted a 15% rise in worldwide antibiotic consumption if present antibiotic use levels are not altered. It is significant to note that the current worldwide epidemic of SARS-Cov-2 may lead to an increase in antibiotic usage worldwide. Despite the fact that COVID-19 is a viral illness, antibiotics are generally administered to take care of bacterial coinfections (Rawson et al., 2020). Antibiotics, on the other hand, are frequently administered to patients without evidence of bacterial infection and may be used as self-prescribed. As a result, the SARS-Cov-2 deadly disease may add to the presence of antibiotics in water bodies and put selection pressure on the spread of ARG (Langbehn, Michels, & Soares, 2021). As previously stated, inadequate treatment technology causes wastewater treatment plants (WWTPs) and their effluents to retain and transmit antibiotic-resistant bacterias (ARBs) or ARGs. ARB is defined by the World Health Organization (WHO) as bacteria that acquire resistance mechanisms to antibiotics that are used to treat them. On the other hand, ARGs are acquired by random mutation or genetic transmission with bacteria that can resist one or more antibiotics (Ezeuko et al., 2021). As a result, it is unsafe to expose human beings to wastewater containing ARB or ARG. As a result, wastewater from manufacturers, hospitals, and the agricultural sector should be thoroughly cleaned to avoid the development of ARB and ARGs in a variety of settings (Zhuang et al., 2015). Antibiotic resistance (AR) arises as a result of widespread or excessive antibiotic usage, which leads to the development of resistance mechanisms in organisms. AR is the ability of bacteria to surpass the efficacy of antibiotics (Manaia et al., 2018). AR in the natural environment is caused by antibiotic overuse, patients failing to complete their antibiotic course, overuse of these antibiotics in domestic animals and fish farming, cleanliness in hospital environments, and a lack of newer antibiotics. ARB and ARGs are considered emergent or uncontrolled contaminants in water bodies due to their poor knowledge, destiny, and threat to living beings' health in the ecosystem. Because of the potential hazards to human health, the appearance of ARB has now become a worldwide concern (Sano et al., 2020). ARBs are bacteria that can stay alive and proliferate in the presence of parent antibiotics that are expressly meant to destroy them. ARB generation is a complicated process in which a microbe might be innately resistant to an antibiotic or gain resistance through a different mechanism (Kumar & Pal, 2018). Multiple paths exist for the interchange of inherent matter inside microbes. Particularly, medication inactivation, target modification, reduced antibiotic permeability, and antibiotic efflux pumps are the four major processes underlying AR generation. As a result, antibiotics are ineffective in treating illnesses caused by ARB, and they are responsible for about 7,00,000 human fatalities per year (Cassini et al., 2019). Furthermore, horizontal gene transfer activities, as well as vertical gene transmission, add to the spread of AR. During conjugation, when the active donor cell and the receiving cell are proficient to initiate and maintain physical contact, genetic elements are exchanged. Following cell lysis, however, the transmission of foreign genetic elements requires just a metabolically active receiver capable of being infected (transduction) or actively absorbing these genetic elements into its own chromosome (natural transformation) (Dodd, 2012). Furthermore, because of their significance in the spread of AR, mobile genetic elements such as integrons and transposons have been widely studied in recent years. Integrons are genetic structures that can trap gene cassettes that encode AR. Integron mobility is linked to transposons, jumping gene systems with integrated resistance genes, and plasmids (Ghaly, Geoghegan, Tetu, & Gillings, 2020). AR spreads as a result of these moving components containing genetic information being passed to the bacterial genome (Herraiz-Carboné et al., 2021; Sabbagh, Rajabnia, Maali, & Ferdosi-Shahandashti, 2021). Furthermore, regulating the spread of ARB/ARGs and avoiding their detrimental health consequences necessitates the use of numerous management technologies to eliminate these contaminants from wastewater inlets and outlets. Physical, chemical, and biological treatment techniques are used to remove resistance. These techniques have been shown to be effective in eliminating potential contaminants. Nonetheless, they are incompetent for eliminating ARB or ARGs from wastewater because they cannot be held accountable for all of the pollutants removed. The majority of them have been shown to accelerate the growth of ARBs and ARGs in wastewater (Ezeuko et al., 2021).

Even though microbial contamination is still a serious hazard to human being health and environmental sustainability, this chapter focuses on the prospective treatment method for eradicating ARB and ARGs in the aquatic milieu.

6.2 Antibiotic resistance bacteria/superbugs

The spread of antibiotics and the growth of resistance are inextricably linked. AR has become a prerequisite for bacterial survival, with resistant bacteria being favored for growth and spread. Even at modest doses, their occurrence can increase bacterial resistance, allowing ARB to proliferate (Wang et al., 2020). These bacteria were dubbed "multidrug-resistant" or, more colloquially, "superbugs." ARB can be made by expressing ARGs in cells and producing the associated antibiotic-resistant protein (Zhang, Li, Chen, Zhou, & Chen, 2020). The *Salmonella* strain, which transmits a disease to around 21.6 million individuals globally, is an example of a superbug (Zhang et al., 2019). It is resistant to antibiotics like ampicillin, aztreonam, cefoxitin, ceftriaxone, etc. (Wang et al., 2020). Bacteria evolve faster than new drugs are developed. This occurrence resulted in a significant environmental alteration, as well as a harmful alarm to the bacterial community. In the case of sickness, a combination of antibiotics is used to control ARB, which is the most serious challenge to contemporary medicine (Sagar, Kaistha, Das, & Kumar, 2019). Bacteria and viruses have both been linked to some of history's most heinous diseases. ARB and its genes are one of the 21st century's most severe and complex concerns

(Chattopadhyay, Chakraborty, Grossart, Reddy, & Jagannadham, 2015). They are easily spread to human beings and animals by horizontal gene transfer in aquatic habitats, providing a significant risk (Cheng et al., 2020). According to the Centers for Disease Control and Prevention, AR sickens more than two million people in the United States each year and kills 7,00,000 people globally. According to one study, ARs contagions will be the main cause of fatality worldwide by 2050. Transformation or the acquisition of resistance genes from other bacteria may potentially play a role in superbug drug resistance (Baaloudj et al., 2021).

6.3 Antibiotic resistance genes (ARGs)

6.3.1 Origin of antibiotic resistance genes (ARGs)

The detection and spread of ARB/ARGs in various water environments, including wastewater handling plants, is an important issue for the general public. Knowing about these harmful contaminants necessitates lifestyle modifications and extreme remedies. Human sources of ARGs in the environment are ARGs, whereas animal sources are manure applications on land (Barancheshme & Munir, 2018). Adefisoye and Okoh studied the occurrence of ARGs in municipal WWTPs in the Eastern Cape and South Africa. From the study, it was found that WWTPs are ARB and ARG point sources. It might be related to ARGs linked to clinical pathogens in sewage and hospital waste. AR is common in WWTPs, and the harm posed by this menace cannot be understated (Adefisoye & Okoh, 2016). Antibiotics, ARBs, and ARGs are found at high levels in sediment samples from several rivers. Although their natural origins are devoid of antibiotics and AR contamination, the majority of these rivers are impacted by urban runoff. ARGs are known to be abundant in agriculturally impacted areas, such as poultry and fish ponds (Yu, Zhan, Shen, Zhou, & Yang, 2017). As being among the primary contributors of ARB and ARGs, leachates from municipal solid waste dumps should be given more attention. ARG persistence in landfill leachate and municipal solid waste is exacerbated by the availability of metals and organic pollutants, as well as antibiotics (Wu, Huang, Yang, Graham, & Xie, 2015). This is confirmed by the papers published on ARB or ARGs in wastewater from across the globe. Furthermore, due to a lack of treatment facilities, the WWTPs receive a large quantity of antibiotic residue, ARB/ARGs, from pharmaceutical firms (Anthony A, Adekunle C, & Thor A, 2018). A schematic depiction illustrating antibiotic sources and exposure pathways, as well as the distribution of ARB or ARGs in aquatic environments, is depicted in Fig. 6.1.

There is a possibility that stable antibiotics that are released into the environment get carried into water bodies by rain, particularly groundwater near dumpsites. Furthermore, all the by-products of water disinfectant (DBPs), which include surface water and other bodies of water, might cause AR (Antunes, Machado, Sousa, & Peixe, 2005).

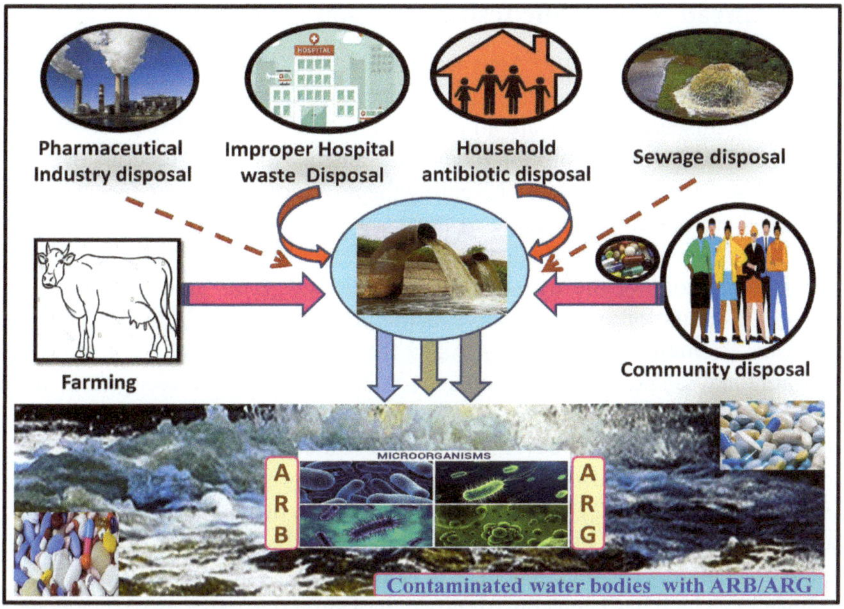

FIGURE 6.1 Schematic diagram of antibiotic sources and exposure routes, and the spread of ARB/ARGs in aquatic system.

6.3.2 Evolution of antibiotic resistance genes (ARGs)

Natural selection can cause low-level AR. Alternatively, the high number of ARB and ARGs in the surroundings is owing to human activity. ARGs can infiltrate the environment through a variety of routes, the most common of which are human and animal. ARG incidence and dissemination are major anthropogenic sources of ARG incidence and dissemination (Rizzo et al., 2013). ARGs come from a variety of sources and are said to be hotspots for ARGs connected to clinical illnesses since they receive waste from a range of sources. ARGs, which have been associated with clinical illnesses, have been found in wastewater in a number of studies (Devarajan et al., 2015). These ARGs propagate all through the water cycle via wastewater discharged into nearby water bodies. Aquatic habitats are suitable venues for the emergence and spread of ARGs because antimicrobial compounds are typically present in them (Rodriguez-Mozaz et al., 2015). The phrase "genetic reactor" refers to areas where genetic evolution happens regularly, perhaps leading to AR. As a result, AR develops in four major sites where genetic change occurs regularly. These are hospitals and health care facilities, wastewater, all kinds of biological waste, and groundwater ecosystems. The receiving habitats are affected by wastewater discharge and antimicrobial chemicals present at measurable concentrations since urban wastewater may not be treated properly (Barancheshme & Munir, 2018).

6.3.3 Spread of antibiotic resistance genes (ARGs)

Humans are often exposed to ARB from other people, whether in hospitals or in the public. Physical contact or indirect contact transfer, aerosols, and food produced by diseased people are all common methods for the disease to spread. These are also the most common means of transmission for infectious bacteria generally, and the therapies for preventing the development of resistant illnesses in humans are basically the same as those for preventing the extension of any bacterial infection (Levin, Baquero, & Johnsen, 2014). Importantly, sanitation is critical in avoiding the transmission of resistant germs among humans, as appropriate sanitation practices serve as the primary dispersal barrier for resistant diseases(Mattner et al., 2012). Aside from human-to-human transmission, environmental dispersion mechanisms for ARB have been recognized as a major element in the emergence of AR (Huijbers et al., 2015). Conditions that help resistant bacteria develop also help nonresistant human illnesses and, more broadly, opportunistic pathogens spread. As a result, sewage, wastewater treatment facilities, water bodies, airborne aerosols, dust, and bacteria-infested food serve as important environmental channels for bacterial transmission (Pal, Bengtsson-Palme, Kristiansson, & Larsson, 2016). STPs often release their wastewater (which has been consistently demonstrated to carry resistance genes) into bodies of water. Polluted water from STP effluent is generally used for crop irrigation, recreational swimming, and drinking purpose following additional treatment. Domestic animals drink untreated surface water on a regular basis, potentially exposing humans to bacteria that are resistant to antibiotics. Untreated wastewater released into waterways, on either hand, provides a far higher risk for the propagation of resistant bacteria than STP effluents since STPs typically lower the relative abundance of the vast majority of resistance genes, as well as overall bacterial richness, by tens to thousands of times (Karkman et al., 2016). Understanding the existing environmental dispersion barriers is critical for restricting the dissemination of human-associated bacteria (HAB), both resistant and nonresistant. In comparison to clinically and community-transmitted germs, establishing significant constraints to dispersion in the environment is far more difficult. We may use a meta-community ecology approach here and consider the human beings and animal hosts of primary diseases to be livable patches, while the majority of other external habitats would act as a dispersion matrix (Leibold et al., 2004). Importantly, the dispersion barriers are usually species-specific, since various bacteria require quite different conditions to survive outside of their natural home. Many gut bacteria, for example, are obligate anaerobes and hence have poor survivability and, as a result, restricted dispersion potential outer of the human body. Because survival instead of development is more important in the spread of HAB (resistant or nonresistant) through the surroundings. The benefit of carrying resistance genes in the presence of low antibiotic concentrations is likely to be negligent for species using the environment for various ways of dispersal. Conversely, for microbes that use the environment as an alternate or primary habitat and hence survive there, subsequent antibiotic contact is likely to contribute to the selection of resistance determinants during environmental dissemination. Recently, the air has received a lot of attention as an unexplored route for resistant microbe transfer (Gat, Mazar, Cytryn, & Rudich, 2017). Due to a shortage of nutrients in the air environment; however, the ability to develop in the presence of antibiotics takes on a marginal role in this dispersal scenario when compared to survival and persistence. As a result, both resistant and nonresistant bacteria could spread by air. While horizontal resistance transmission between bacteria in aerosols and on dust particles is still possible, such activities are less prone to being permanently implanted in their genomes unless the recipients are exposed to antibiotics in a more favorable environment for growth. Bacteria-carrying resistant genes have been found in wild birds and animals that reside close to humans, perhaps contributing to the transmission of such genes across large areas (Stedt et al., 2015). There is very little information on nonpathogenic environmental bacteria that carry resistance to spread and move on to interact with HAB. Pathogens play a critical role in the process of mediating resistance from environmental bacteria toward the human microbiome. The transmission paths from conditions that provide selection pressure for the spread, mobilization, and maintenance of resistance genes all the way to people and animals remain unknown. These mechanisms, as well as the factors that influence the survival of environmental bacteria in various dispersion matrices, must be characterized. There are attempts to track infection and resistance gene prevalence in a variety of settings (Bengtsson-Palme, Kristiansson, & Larsson, 2018).

6.3.4 Consequences of antibiotic resistance genes (ARGs)

AR is a global health issue in a number of water bodies. Therefore, recognizing and eliminating AR determinants is crucial. AR has been related to global health and food security, according to the WHO (Safdari et al., 2017). It is also believed to be a future disease that would wreak havoc on people's lives. According to the WHO, *methicillin-resistant staphylococcus aureus (MRSA)*, which causes skin infections, pneumonia, and toxic shock syndrome, as well as *Enterococcus,* becoming resistant to Vancomycin, *Mycobacterium Tuberculosis* becoming resistant to multi drugs, and other resistant bacteria such as *Enterobacteriaceae* gut bacteria that cause severe health issues, according to the WHO. In July 2017, according to WHO scientists, AR *gonorrhea* is a life-threatening disease that is difficult to treat. ARBs and ARGs also pose an ecotoxicological threat to the aquatic environment. It also promotes the structure of the algal population, causing a change in all food chains (Singh, Singh, Kumar, Giri, & Kim, 2019). If these contaminants were discharged into the soil on a regular basis, they could pose a threat to humans and animals (Karkman, Do, Walsh, & Virta, 2018). All of these factors contribute to resistance evolution, which requires immediate action in a number of countries to mitigate its negative effects on environmental conditions. The access of antibiotics to wastewater treatment plants might help to avoid the health risks associated with AR. In the meantime, the WHO revealed in 2015 that AR develops spontaneously as a result of random mutations. According to one investigation, eliminating resistant bacteria without addressing the genetic alteration that interacts with the DNA sequence posed a risk to treatment efforts (Windels et al., 2019). These genetic alterations serve as defense mechanisms for bacteria, allowing them to withstand the effects of contemporary antibiotics designed to kill them. They are perhaps the most important factors that foster biological variety. Genetic alterations happen in a cell that makes the next generation and alters the microorganisms' hereditary parts by changing the pre-existing genetic composition of a cell's DNA (Sharma, Johnson, Cizmas, McDonald, & Kim, 2016). As a result, the genetic composition of DNA may induce changes in every part of living organisms. Adequate steps need to be taken to prevent the spread of ARB or ARGs in water bodies. These actions may prevent the 10 million deaths predicted by ARB/ARGs for the year 2050 (Tadesse et al., 2017).

6.4 Mechanism of antibiotic resistance

AR is either genetically inherited or acquired. Some ARBs and ARGs are found in the surroundings naturally such as an assessment of Beringian permafrost deposits, where resistance to tetracycline, lactam, and glycopeptide antibiotics existed 30,000 years ago, significantly earlier than Fleming found the first antibiotic in 1928 (Levin, Baquero, & Johnsen, 2014). AR can be observed in pristine ecosystems that have not been influenced by anthropogenic activity. While some resistances are innate, many are acquired. Horizontal gene transfer through conjugation, transformation, transduction, or gene transfer agents (GTAs) can occur as a result of a bacterial DNA mutation.

6.4.1 Mutation and coselection

A mutated gene is a permanent change in the DNA sequence of bacteria that can be introduced by stimuli that trigger the SOS DNA stress response or the oxidative stress response. The oxidative stress, stress tolerance, and antioxidant stress response are global responses to DNA breakage in the cell. An increase in reactive oxygen species (ROS) like superoxide ($O_2^{-\cdot}$), hydroxyl ($OH^{\cdot}$), and single oxygen triggers the oxidative stress response (Mittler, 2002). Spontaneous mutation occurs at a very low rate, ranging from 1 per million to 10 per million. Contaminants in metropolitan water systems, on the other hand, can hasten the process of mutation. Major antibiotic families, such as quinolones, lactams, and aminoglycosides, can enhance ROS generation for both gram-positive and negative bacteria. Consequently, SOS response and genetic mutation are triggered (Kohanski, DePristo, & Collins, 2010). Toxins in the environment, such as heavy metals, disinfectants, pharmaceutical waste, and personal hygiene products, may even increase the frequency of mutations. Coselection is a sort of indirect selection that occurs both physically (coresistance) and genetically (cross-resistance) (Baker-Austin, Wright, Stepanauskas, & McArthur, 2006). Cross-resistance occurs when one resistance gene imparts resistance to a different antibiotic. When bacteria are exposed to metals, for example, multidrug efflux pumps release intercellular toxins, and multidrug efflux pumps were found to lessen sensitivity to both metals and antibiotics. Coresistance is a genetic technique that involves the discovery of multiple resistance genes on the same mobile genetic material. Because it is frequently located on an integron, the *sulfonamide* resistance gene sul1 is frequently coselected via coresistance (varied portions of which typically include ARGs) (Antunes, Machado, Sousa, & Peixe, 2005). When an ARG is chosen in the context of antibiotics or other environmental pollutants with which it develops resistance, this is known as coselection.

6.4.2 Horizontal gene transfer

It is a process by which DNA is carried to a recipient cell (either from a donor cell or the external environment) and becomes a part of the recipient genome (Von Wintersdorff et al., 2016). This can occur in a variety of bacterial strains and species, and it is a major factor in the extent of AR in water bodies. Horizontal gene transfer is accomplished by four key mechanisms: conjugation, transduction, transformation, and GTAs.

The four major transfer mechanisms are briefly outlined here:

Transformation: In order to achieve a transformational state, bare double-stranded DNA is taken up by a competent bacterial cell. It also has the capacity to deliver DNA to bacteria of different species. Bacterial population diversity and mixing are among the evolutionary benefits (Carvalho et al., 2020). By allowing bacteria to interact favorably through mutations, genetic mixing during transformation might confer selection benefits. After that, a cell can absorb and merge its chromosome and extracellular DNA (eDNA); the eDNA that is absorbed serves as a repository of information and food (Takeuchi, Kaneko, & Koonin, 2014).

Conjugation: Cell-to-cell interaction is needed when DNA is transferred from the donor cell to the receiving cell via sexual pilus and adhesins. Freshly acquired resistance genes cause donor cells to become resistant (Somensi et al., 2015). A conjugative pilus is defined as the physical link that allows DNA to be transferred from donors to recipients. It plasmids and plays an important role in the spread of plasmid-borne ARGs.

Transduction: Transduction is the transfer of DNA from one bacteria to another using bacteriophage. It facilitates the propagation of plasmid-borne ARGs by transferring plasmids. This is the most important gene transfer route. Bacteriophages have a wide range of hosts and, as carriers of other bacteria, can infect other bacteria hosts (Berglund, 2015). It can also transmit its genes to the new bacterial host's DNA. Transduction phage particles are tiny phage particles that facilitate DNA transfer in the environment and are resistant to environmental deterioration. They have a small size that aids in fast dissemination(Bengtsson-Palme et al., 2018). Because of the special qualities of transduction, genes can be transferred from the environment of bacterial colonies to human microbiomes.

Gene transfer agent (GTA): GTAs are virus-like particles that transport host DNA and convert it to a target cell. It provides for random gene transfer among different species, and particles can be discharged from cells, and the particles are discharged and move to target cells (Fogg, 2019).

Bacteria, MGEs, and bacteriophages cohabit in urban water systems (including WWTPs, and receiving waters), leading to horizontal gene transfer with a high frequency and extensive range, culminating in the spread of AR (Karkman et al., 2018). Furthermore, it has been demonstrated that aquatic pollutants might enhance the horizontal spread of AR, subinhibitory antibiotic doses, such as-lactams, amino-glycosides, fluoroquinolone antibiotics, and mitomycin C, might, for example, promote the transfer of tetracycline resistance plasmids in *Staphylococcus aureus* or induce transformability in *Streptococcus pneumonia* (Prudhomme, Attaiech, Sanchez, Martin, & Claverys, 2006). Other commonly observed environmental pollutants, in addition to antibiotics, have been demonstrated to enhance the horizontal spread of AR. Pollutants include nonantibiotic drugs, nanomaterials, ionic liquids, and even disinfectants, and disinfection by-products (Wang et al., 2019).

6.5 Approaches to abate antibiotic resistance

WWTPs are always considered repositories of antibiotic genes, hence removing these ARGs during WWTPs operation is critical. In general, commonly used WWTPs techniques, like physical, chemical, and biological processes may have a significant impact on ARB and ARG outcome and modification until they are released into an aquatic system. To better understand ARB and ARG elimination patterns in WWTPs, the elimination efficacy of ARGs throughout normal operations was studied and discussed.

6.5.1 Physical approach for removal of resistance

6.5.1.1 *Coagulation process*

Coagulation is a process used to improve water quality and eliminate dispersed particulates and impurities in WTPP (Li, Sheng, Lu, Zeng, & Yu, 2017). During the coagulation process, coagulants reduced the repelling tendency of the colloidal dual electrical layer, resulting in colloidal aggregation and instability, along with colloidal solids in the water sample. Huge flocs were subsequently divided by gravity into solids and liquids and finally separated during the sedimentation process. Chemical coagulants, such as ferric and aluminum-based coagulants, are widely used in the coagulation process due to their wide pH and temperature range. The exclusion of ARGs from wastewater is depicted in Table 6.1 using the coagulation technique. Several factors influence removal efficiency, including turbidity, pH, water temperature, coagulant variety and dose, and hydrodynamic state, among others (Sillanpää, Ncibi, Matilainen, & Vepsäläinen, 2018). It was revealed that poly aluminum chloride (PAC) had an effect on ARG elimination and that a higher PAC dosage resulted in superior total ARG removal efficiency (Lee et al., 2017). The type and dosage of coagulants influenced the removal efficiency of $FeCl_3$ and poly ferric chloride (PFC) in reducing the amount of ARGs (sul1, sul2, tetO, tetW, tetQ) and IntI1genes (Li et al., 2017). The elimination effectiveness of ARGs by PFC increased steadily as the PFC dose increased, eventually reaching its maximum. The presence of iron particles in the mixture may affect the effectiveness of ARG removal. $FeCl_3$ solution included around 75% Fe monomeric species (Fea) and 25% Fe polymeric species (Feb), whereas PFC solution with 0.8 basicity comprised around 33% Fea, 57% Feb, and 10% amorphous hydroxide precipitate (Fec). In the case of $FeCl_3$, following a series of hydrolysis processes, the unstable Fea can be transformed into a positively charged colloid or Fec. Thus, the eradication of ARGs might be ascribed to the compression of the electric double layer and charge neutralization. Due to the preponderance of Feb in the PFC solution, adsorption, charge neutralization, and capture may be the key processes for ARG elimination by coagulation with PFC. In addition, after 3 days of storage, RTan was shown to be reactivated, and the incidence of qnrS in RTan was higher than in wastewater (Grehs et al., 2019). As a result, the storage's reactivity after coagulation must be assessed. Though, the processes of microbes elimination by diatomite coagulation require additional investigation. Coagulation with $Al_2[SO_4]_3$ (Alu) and tannin (Tan), a green polymeric coagulant, was investigated for ARGs elimination (Grehs et al., 2019).

Alu was observed to dramatically diminish the abundance of blaTEM and qnrS. Coagulant action and the mechanism of ARG elimination deserve additional investigation. It was vital to notice that the increasingly sludge-generating treatment, which comprised ARBs and ARGs, should be administered afterward to disrupt ARBs and remove ARGs.

6.5.1.2 *Membrane separation*

Membrane separation technologies associated with physical segregation, such as microfiltration (MF), ultrafiltration (UF), nanofiltration (NF), and reverse osmosis (RO), have long been recognized as a viable approach for treating wastewater and reclamation (Arola, Van der Bruggen, Mänttäri, & Kallioinen, 2019). Because of their enormous molecular weight, ARGs might theoretically be caught by membranes. In recent years, researchers have discovered that the membrane filtration process, which is primarily based on the size exemption system, could be used as a physical wall to effectively eliminate bacteria from water bodies, reducing the abundance of ARBs and ARGs and, subsequently,

TABLE 6.1 Removal of ARB or ARG by coagulation and membrane separation process.

S.no.		Coagulant/ membrane separation methods	Matrix	Target ARGs	Removal of ARGs	References
1.	COAGULANT	$FeCl_3$, polyferric-chloride	Wastewater effluent	sul1, sul2, tetO,tetW, tetQ	0.5–3.1 log reduction	Li et al. (2017)
2.		PACl	Municipal wastewater	tetX, tetM, tetA,sul1, sul2, ermB,qnrD, blaTEM	In WWTP2, total ARGs declined by 58.1%, while total ARGs grew by 59.6% in WWTP1	Lee et al. (2017)
3.		Alu, Tan	Secondary effluent	blaTEM, qnrs	After coagulation with Alu, blaTEM and qnrs were reduced by roughly 1 log, but after coagulation with Tan, they were reduced by 0.4–0.7 log	Grehs et al. (2019)
4.	MEMBRANE SEPARATION METHODS	MF, UF, NF, RO	Distilled water, wastewater effluent	blaTEM, sul1, qnrS1,ctx-m-32, vanA	Membranes of 2500 Da MWCO eliminated more than 99.0% of total free DNA	Slipko et al., 2019
5.		NF, RO	Swine wastewater	sul1, sul2, tetA,tetM, tetW	The absolute abundance of ARGs fell by 4.98–9.52 logs, whereas the relative abundance of sul genes remained nearly stable and the absolute abundance of tet genes declined by 0.88–3.47 logs	Lan et al. (2019)
6.		PAC/BPAC–UF	Secondary effluent	sul1, sul2, tetA, tetW	UF alone reduced by 0.86–1.56 logs; UF with 60 mg/L PAC reduced by 1.35–3.35 logs; UF with 80 mg/L BPAC reduced by 1.81–3.73 logs	Sun et al. (2019)
7.		UF, NF, RO	Ultrapure water	pCRVR II-TOPO plasmid DNA	With a molecular weight cutoff of 1 kDa or below 1 kDa, > 99% of ARGs could be removed through membrane	Krzeminski et al. (2020)

preventing the spread of AR into the water habitats (Krzeminski et al., 2020; Slipko et al., 2019). Lan, Kong, Sun, Li, and Liu (2019) discovered that NF and RO procedures may reduce the large quantity of ARGs (sul1, sul2, tetA, tetM, tetW) in the effluent of a swine firm. The elimination of ARGs by membrane partition techniques was presented in Table 6.1. ARG type, DNA type, filtration technique, quality of water, and operating circumstances all influenced the efficacy of ARG elimination by membrane filtering (Krzeminski et al., 2020; Slipko et al., 2019). The removal rate of ARGs by polyethersulfone (PES)-UF membrane was higher than that of the PVDF-UF membrane. Krzeminski et al. (2020) obtained comparable findings while utilizing cross-flow filtering mode. It was revealed that when the membrane pore sizes shrank, the retention effectiveness of ARGs in deionized water through UF, NF, and RO increased. The resistance of eARGs dissolved in water was higher than that of iARGs, which could be easily removed by removing ARBs (Lan et al., 2019). Furthermore, the incidence of dissolved organic carbon, the intI1 gene, and 16S rRNA may alter the effectiveness of ARG removal (Sun, Shi, He, Zhang, & Duan, 2019). After the MF method, the quantitative density of ARGs was higher than that of the 16S rRNA gene, indicating that iARGs persisted in the solution even after the microbes were eliminated (Lu, Zhang, Wu, Wang, & Cai, 2020). The effects of membrane type, free DNA type, and aqueous environments on the elimination of ARGs were investigated (Slipko et al., 2019). NF and RO (0.88–3.47 log in winter and 0.02–2.28 log in summer) significantly reduced the quantity of tetracycline resistance genes, while sulfonamide resistance genes did not change (Lan et al., 2019). As a result, alternative measures along with the membrane filtering procedure should be implemented to increase the elimination effectiveness of ARGs and relieve the foulant layer. Furthermore, unbound DNA was able to permeate membrane pores much smaller than their size due to its structure and elasticity (Slipko et al., 2019). The use of membrane filtration in the removal of ARGs may be hampered by poor eARG removal and substantial membrane fouling. More research is needed to ascertain how long ARGs last on membranes as well as the viability of the integrated ARG removal procedure.

6.5.2 Chemical approach for removal of resistance

Chemical techniques, such as chlorine disinfection and advanced oxidation processes (AOPs), were widely used to remediate ARG-polluted water samples.

6.5.2.1 Disinfection

Chlorine is an oxidant that acts by destroying microorganisms' nucleic acids and cell membranes. It is the most widely used disinfectant in many countries. The removal efficiency of sul1, tetX, tetG, and intI1 (0.1 log to 1.30–1.49 logs) increases significantly when the FC dosage increases from 5 to 30 mg/L (Tandukar, Oh, Tezel, Konstantinidis, & Pavlostathis, 2013). Likewise, Oh et al.

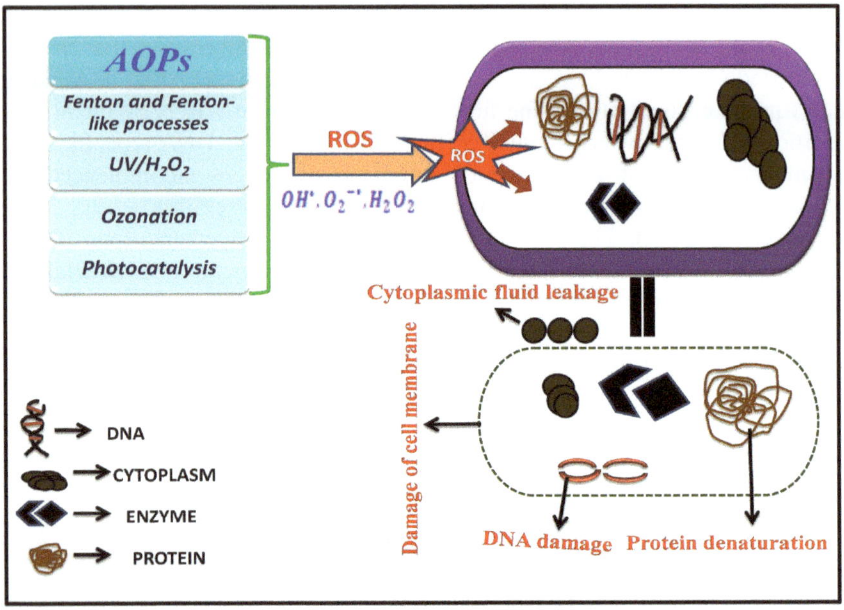

FIGURE 6.2 Schematic diagram of different AOPs for ARB or ARG inactivation.

reported that an FC dose of 30 mg/L could remove up to 90% of ARGs from the effluent of a biological aerated filter (BAF) process (Oh, Salcedo, Medriano, & Kim, 2014). It's worth noting that the presence of NH_3-N in a water sample, which competes with free chlorine, lowers ARGs' bulk removal effectiveness. According to Zhang et al., when the NH_3-N dosage is increased from 2.56 to 5 mg/L (30 mg/L FC concentration), the inactivation of ARG reduces from 1.20 to 1.49 logs to 0.63 to 0.79 logs when the NH_3-N dosage is increased from 10 mg/L (Y. Zhang et al., 2015). This finding might explain why 40% of erythromycin resistance genes and 80% of tetracycline resistance genes survived after chlorination (to 60 mg Cl_2/L) with a loading of 0.35 kg NH_3-N/L) (m^3d) (Jennings, Minbiole, & Wuest, 2016).

6.5.2.2 *Advanced oxidation processes*

As previously stated, traditional disinfection has several drawbacks. As a result, sophisticated technologies such as AOPs have been considered since 1980 (Ghernaout & Elboughdiri, 2020). Fig. 6.2 depicts the basic mechanism of many AOPs for ARB and ARG deactivation via ROS.

The ability of several AOPs to damage the cell membrane and DNA sequence via ROS reactions has been discovered, hinting that AOPs might be a viable practice for damaging microbial cells and removing ARGs (Zhang et al., 2016). The technique employed will be determined by the kind of waste and its particular water quality features. Not only should operation expenses be considered, but also ARG's highest removal and possible environmental consequences, as well as wastewater type and pollutant concentration, in varied scenarios. The following is a detailed breakdown of how each AOP approach removes ARGs.

6.5.2.2.1 Fenton and Fenton-like processes

Fenton is a very effective water purification method that uses the OH˙ produced in an acidic environment by H_2O_2 and Fe^{2+}. According to Wang et al. (2020), the Fenton reaction lowered the quantity of ARGs and intI1. Furthermore, the researchers revealed that using the Fenton approach in conjunction with alkaline or microwave irradiation may nearly totally eradicate the particular ARGs (Wang et al., 2020). Additionally, when using *E. coli* DH5 as a model ARB with dual plasmid-encoded ARGs (tetA and blaTEM), the photo-Fenton approach revealed a higher microbial inhibition efficacy, eliminating more than 6.17 log ARB after 30 minutes of visible irradiation at a neutral pH (Ahmed, Lu, Yuan, Bond, & Guo, 2020). Thus, the photo-Fenton process caused by visible irradiation can be a commercially viable method for successfully damaging ARB and reducing ARG abundance by harnessing solar energy. However, just a few experiments employing the photo-Fenton technique have been published (Giannakis, Le, Entenza, & Pulgarin, 2018); consequently, further study is needed.

6.5.2.2.2 UV/H_2O_2

UV/H_2O_2, a common AOP, may produce *OH˙*, which has been used to eliminate a variety of organic micropollutants. The elimination effectiveness of UV/H_2O_2 for ARBs and ARGs has recently piqued the interest of researchers. When *OH˙* attacks oxidized sugar backbone products, the microbial DNA and ARGs composition may be harmed (Sharma et al., 2019). Yoon et al. (2017) employed UV/H_2O_2 to investigate the extracellular and intracellular harmful efficiency of plasmid-encoded ARGs. The influence of COOH on the degradation of iARGs was shown to be modest since the COOH was mostly scavenged by cell organelles (Yoon et al., 2017). ARB and ARGs have also been shown by Sanganyado and Gwenzi (2019) and Sharma et al. (2019) to enhance the elimination of ARB and ARGs (Sanganyado & Gwenzi, 2019; Sharma et al., 2019). Furthermore, two key limitations may limit the method's utility: limited UV light in sunlight and uncontrolled turbidity in wastewater, which restricts UV transmission.

6.5.2.2.3 Ozonation

Ozone (O_3) is a powerful oxidant (2.07 V) that may react with a wide spectrum of chemical contaminants in water. Pollutants can be removed in two ways using the ozonation process: (1) a direct technique (O_3 oxidation), and (2) an indirect way (via a free radical reaction, such as the formation of *OH˙* from

O_3 breakdown) (Chen & Wang, 2019). The effectiveness of ozonation is often strongly dependent on the O_3 dosage and reaction time. According to recent research, ARB inactivation and ARG elimination were only significantly enhanced at a lower ozone concentration (27 mg/L), but ARG removal was substantially higher when the ozone concentration was greater than before at 61 mg/L. Moreover, when the O_3 concentration was set at 2 mg/L, the removal efficacy of two sulfonamide resistance genes (sul1 and sul2) and five tetracycline resistance genes (tetA, tetM, tetO, tetQ, and tetW) in secondary wastewater ranged from 34.5% to 49.2% (Zheng et al., 2017). Meanwhile, O_3 utilization efficiency remains insufficient due to a slow mass transfer rate and low solubility and could be improved further by integrating an effective catalyst to speed up the O_3 breakdown process (Li et al., 2021).

6.5.2.2.4 Photocatalysis

Due to its low cost, great solar utilization, and ease of operation, photocatalysis is among the most advanced technologies for eliminating toxins from aquatic environments (Mishra et al., 2019; Pattanayak et al., 2021). Despite the fact that photocatalysis is routinely employed to remove many pollutants, only a few studies examining ARGs' elimination effectiveness have been published. In their study, Guo et al. (2017) discovered that in the presence of TiO_2, mecA and ampC were lowered by 5.8 log and 4.7 log, respectively, under 120 mJ/cm^2 of UV-254 (TiO_2; a common photocatalytic semiconductor) (Guo et al., 2017). In wastewater, TiO_2-immobilized/H_2O_2/UVA has been shown to improve pathogen killing and contaminant removal (Jiménez-Tototzintle, Ferreira, da Silva Duque, Guimarães Barrocas, & Saggioro, 2018). Photocatalysts made of nanocomposites might be used to inactivate bacterial plasmids, reducing the risk of AR transfer (Saha, Visconti, Desipio, & Thorpe, 2020). However, Karaolia et al. (2018) showed that the photocatalyst TiO_2-rGO does not eliminate all ARGs, with some ARGs, like sul1 and ermB in *P. aeruginosa*, remaining persistent. Graphitic carbon nitride (g-C_3N_4), on the other hand, is a common polymeric photocatalyst having low cytotoxic effects and high photocatalytic activity under visible irradiation that has attracted a lot of attention for eliminating ARB with exceptionally fast ARG removal (Ding et al., 2019). The ARGs can be swiftly removed with Ag/AgBr/g-C_3N_4 as a visible light active photocatalyst; in particular, the elimination efficacy of tetA, tetM, tetQ, and intl1 was 49%, 86%, 69%, and 86%, respectively (Yu, Zhou, Li, & Yan, 2020). Moreover, a photocatalytic membrane using TiO_2 was developed to remove ARB and ARGs from secondary effluent, indicating the total elimination effectiveness of ARGs (e.g., plasmid-assisted floR, sul1, and sul2) was achieved at 98% under UV irradiation(Ren et al., 2018). As a result, photocatalysis appears to have a wide variety of uses in the near future if some of the present difficulties can be overcome.

6.5.3 Biological approach for removal of resistance

6.5.3.1 Wastewater treatment plants (WWTPs)

As previously stated, WWTPs are thought to be the primary hotspot for ARG spread, posing a serious and unseen ecological danger. As a result, increasing the effectiveness of ARG removal in WWTPs is crucial (Kucukunsal & Icgen, 2020). In order to determine the significance of operational factors in AR eradication, researchers looked at the variables that impact the concentration and release of ARGs during wastewater treatment. Following treatment, an interesting phenomenon was identified in which ARGs were decreased by 1.8–2.7 log while the predominance of broad plasmids (IncP-1) grew considerably (Pallares-Vega et al., 2019). Furthermore, Mahfouz et al. (2018) revealed that the genomes of *E. coli* from WWTP inlet and outlet were substantially identical, implying that WWTPs may not be able to afford to eradicate the probable hazard of microbial AR generation (Hiller, Hübner, Fajnorova, Schwartz, & Drewes, 2019; Mahfouz et al., 2018). Some investigations have also discovered that ARB and ARGs are typically present in significant amounts in WWTP effluent even after disinfection (Pattanayak, Mallick, Thakur, & Pal, 2020). Furthermore, while WWTP methods may remove some ARGs, the vast majority of the removed ARGs frequently coexist with active sludge (Xue et al., 2019). This might result in additional environmental concerns throughout the sludge disposal process, as well as a significant rise in operating costs. Taken as a whole, the WWTP approach has to be immediately developed in order to satisfy the demand for extremely efficient ARG removal.

6.5.3.2 Constructed wetlands

Constructed wetlands (CWs) have recently gained popularity as a sustainable and clean option for dealing with pollutants, owing to their inherent advantages of becoming ecologically friendly, less expensive, and easy to manage (Yi, Tran, Yin, He, & Gin, 2017). Fig. 6.3 depicts a typical approach for removing ARB and ARG from wastewater using an artificially CW. Furthermore, one study used CWs to treat wastewater containing antibiotic residue and investigated the status of ARGs, finding that the ARG concentration in the discharge was considerably decreased and the microorganisms were almost completely eradicated due to the vertical up-flow properties of CWs (Song et al., 2018). Furthermore, Chen et al. (2019) demonstrated that when using domestic sewage spiked with antibiotics as the experimental water, various mesocosm-scale CWs can remove approximately 87.8%–99.1% of ARGs, implying that component biosorption and biochemical reactions in the CWs are the main culprits (Chen et al., 2019). Furthermore, several ARGs were successfully reduced in ventilated conditions, whereas the prevalence of other microbial pathogens lacking ARGs increased (Liu et al., 2019). Nonetheless, a range of factors, including the vertically

FIGURE 6.3 The Schematic diagram of ARB or ARG removal in constructed wetlands.

up-flow pattern, aging, variety, feed water dynamics, and quality of water, might have a substantial impact on the ARG removal efficacy of CWs (Zhang et al., 2020), which should be further investigated in the future phase.

6.5.3.3 Anaerobic membrane bioreactors

Membrane filtering is a common method for eliminating a variety of new pollutants, and it might potentially be used to remove ARGs (Pei et al., 2019). Anaerobic membrane bioreactor (AnMBR), which combines anaerobic digestion with membrane separation, is a well-known biological process. Researchers (Wang et al., 2020) have identified this technique as one to minimize the total discharging burden of ARGs and ARBs into the surrounding ecosystem. ARGs in primary clarifier discharge, for example, have been reported to be reduced by 3.3–3.6 log in an AnMBR at 20°C, and the intI1 gene encoding has been estimated and proven as an HGT strategy (Kappell et al., 2018). Zarei-Baygi, Harb, Wang, Stadler, and Smith (2020) discovered that the bacterial population found in AnMBR wastewater varied substantially over time, demonstrating that the presence of certain infectious taxa in the discharge was linked to the abundance of specified resistance genes like sul1, intI1 (Zarei-Baygi et al., 2020). On the other hand, thorough research examining the links between the bacterial population constitution and the AR-pattern in AnMBR seepage, on the other hand, is currently absent and should be investigated further. Furthermore, one clear limitation of AnMBR's application is membrane biofouling. Cheng and Hong (2017) demonstrated that the capacity to remove ARGs reduced significantly as membrane fouling increased, showing that the foulant layer is the most important element influencing ARG removal effectiveness (H. Cheng & Hong, 2017). To address this problem, certain newly redesigned hybrid bioreactors were introduced in stages to boost pollutant removal efficacy while simultaneously slowing membrane clogging development (Guo et al., 2020). On the other hand, the effectiveness and application of these substantially upgraded AnMBRs for ARG repair, on the other hand, are uncertain.

6.6 Some broad steps in the antibiotic-resistance fight

Several national and international policies and campaigns have been developed in recent years to prevent antimicrobial resistance, which is finally being predicted as severe obscurity at the highest political levels, as seen by the 2016 UN statement on antimicrobial resistance. The 2016 report of the Review on Antimicrobial Resistance, commissioned by the UK government, emphasizes the global dimension of the problem and proposes action on several fronts to reduce antibiotic demand and enhance antibiotic production. Strategies encompass the following:

6.6.1 Antimicrobial management

Global control of antibiotic usage, especially in emerging nations, must be enhanced. Stopping the use of over-the-counter antibiotics in these nations, as well as educating prescribers about antimicrobial resistance, may assist in controlling usage. To limit excessive demand, increased worldwide public awareness is also required. Agricultural applications should be restricted to the treatment of diseased animals rather than growth promotion. Antibiotic usage and resistance surveillance must be significantly improved in order to enable efficient antimicrobial management. In the United Kingdom, antibiotic usage surveillance has been established in recent years to provide more feedback to prescribers, as well as to link antibiotic prescribing quality indicators to financial incentives.

6.6.2 Research and advancement

To avoid untreatable infections, it is necessary to identify and produce new antibiotics, particularly those with novel modes of action. Public funding and international collaboration are required to offer economic incentives for pharmaceutical companies to produce antibiotics in a commercially viable manner, rewarding new medication innovation while discouraging needless usage. It is vital to have a better grasp of antibiotics' economic and social value. Biomedical Advanced Research and Development Authority in the United States and Innovative Medicines Initiative in the European Union fund several research activities into new antibiotics. It is also critical to do research and create strategies to prevent infection, thus minimizing the demand for antibiotics. In the subject of diagnostics, there is a lot of room for progress.

6.6.3 Public consciousness

The Antibiotic Action Project of the British Society for Antimicrobial Chemotherapy and the international network React are two public initiatives aimed at raising public knowledge and support for appropriate antimicrobial resistance action, including global collaboration.

6.7 Conclusion

Antibiotics, ARBs, and ARGs may all be found in rivers and other aquatic habitats across the world. Antibiotics contaminating the environment is a worldwide problem that necessitates a variety of strategic measures to control-related dangers. ARB or ARGs in food and drinking water can have catastrophic repercussions for children. These methods must be equally successful in eliminating ARBs and ARGs in order to lessen the environmental and human health implications of microbial resistance. AR in bacteria is a challenging problem that needs to be tackled with next-generation therapeutic methods. Biological treatment like aerobic and anaerobic treatment reactors has been shown to remove high quantities of different ARGs from home wastewater. Throughout the years, CWs with various flow patterns or types of plants have been erected and are considered appealing waste management solutions for eliminating ARGs from basic domestic sewage. The majority of studies on the disinfection of ARG have employed chlorination. However, several investigations have been broadened to include UV irradiation, facilitating comparisons of the two techniques in terms of effectiveness and mechanism. Recently, nanoparticles have been identified as antimicrobial agents capable of removing ARG and acting as new protection against ARB when combined with antibiotics. Overall, a very well-purified system that integrates WWTP may reduce ARB and ARGs in receiving streams significantly. Treatments for microbial resistance prevention and control must be associated with methods for regular inspection of ARB prevalence in environmental matrices, upgrading or establishing restrictions for hospital effluents, appropriate control of feces from animal production units, and programs to regulate antibiotic use in farmed animals and the agriculture sector. Furthermore, strict regulations and systems for controlling medical medication usage and managing human antibiotic waste are essential.

Research challenge and future perspectives

AR among pathogens emphasizes the need for innovative treatment techniques. Despite tremendous efforts to discover therapeutic approaches for ARBs and ARGs, there remain important gaps to be filled:

- Most of the research was carried out on a small scale in the lab or in a pilot setting over a short period of time. Substantial analysis of real environmental samples is essential. Conduct risk assessment research to determine the precise levels of ARB and ARG prevalence in WWTP effluent that does not endanger survival.

- Put more emphasis on companies that produce high-quality ARG emissions (e.g., the husbandry, pharmaceuticals, and other sectors) and the production of novel ARG contamination management measures. For the reduction of ARG emissions, cost-efficient and successful on-site techniques should be supported.
- Evaluating the possibility of introducing changes to enhance current water and wastewater handling amenities in order to boost ARBs and ARGs in plant runoff.
- More study is needed to determine the future of ARBs and ARGs, which transmit a wider range of AR in municipal wastewater.
- Future studies investigating the effects of anxiety on ARG horizontal transmission and ARG fate in aerobic and anaerobic units are needed.
- Take into consideration seasonal variations in ARB and ARG levels and perform trials over a long enough length of time while using CWs as a therapy method. Because ARG variation and patterns are influenced by the factors impacting ARG trends in CW system sediments, they must be investigated.
- In order to develop an acceptable and cost-effective method of eliminating ARGs from WWTP wastewater, researchers are looking into complex treatment systems and integrated disinfection techniques. Because current WWTP disinfection systems can only remove ARGs to a limited extent.
- Given the aforementioned limitations in ARB and ARG therapy options, further research is needed in this field to increase the overall efficiency of existing treatments or develop new ones.

Declarations Acknowledgments

The work greatly acknowledges the Department of Chemical Engineering of the National Institute of Technology Raipur for providing research opportunities.

Author contributions

The study's inception and final preparation were supported by all of the authors. Dhruti Sundar Pattanayak conducted a literature review and drafted the text. Dr. Dharm Pal, Dr. Chandrakant Thakur, and Dr. Awanish Kumar worked on the draught correction and critical revision. The final manuscript was reviewed and approved by all authors.

Conflicts of Interest

There are no conflicting interests declared by the authors.

References

Adefisoye, M. A., & Okoh, A. I. (2016). Identification and antimicrobial resistance prevalence of pathogenic Escherichia coli strains from treated wastewater effluents in Eastern Cape. *South Africa Microbiology Open, 5*(1), 143–151. https://doi.org/10.1002/mbo3.319.

Ahmed, Y., Lu, J., Yuan, Z., Bond, P. L., & Guo, J. (2020). Efficient inactivation of antibiotic resistant bacteria and antibiotic resistance genes by photo-Fenton process under visible LED light and neutral pH. *Water Research, 179*. doi:10.1016/j.watres.2020.115878.

Anthony A, A., Adekunle C, F., & Thor, A. S. (2018). Residual antibiotics, antibiotic resistant superbugs and antibiotic resistance genes in surface water catchments: Public health impact. *Physics and Chemistry of the Earth, 105*, 177–183. https://doi.org/10.1016/j.pce.2018.03.004.

Antunes, P., Machado, J., Sousa, J. C., & Peixe, L. (2005). Dissemination of sulfonamide resistance genes (sul1, sul2, and sul3) in Portuguese Salmonella enterica strains and relation with integrons. *Antimicrobial Agents and Chemotherapy, 49*(2), 836–839. https://doi.org/10.1128/AAC.49.2.836-839.2005.

Arola, K., Van der Bruggen, B., Mänttäri, M., & Kallioinen, M. (2019). Treatment options for nanofiltration and reverse osmosis concentrates from municipal wastewater treatment: A review. *Critical Reviews in Environmental Science and Technology, 49*(22), 2049–2116. https://doi.org/10.1080/10643389.2019.1594519.

Baaloudj, O., Assadi, I., Nasrallah, N., El Jery, A., Khezami, L., & Assadi, A. A. (2021). Simultaneous removal of antibiotics and inactivation of antibiotic-resistant bacteria by photocatalysis: A review. *Journal of Water Process Engineering, 42*, 102089. https://doi.org/10.1016/j.jwpe.2021.102089.

Baker-Austin, C., Wright, M. S., Stepanauskas, R., & McArthur, J. V. (2006). Co-selection of antibiotic and metal resistance. *Trends in Microbiology, 14*(4), 176–182. https://doi.org/10.1016/j.tim.2006.02.006.

Barancheshme, F., & Munir, M. (2018). Strategies to combat antibiotic resistance in the wastewater treatment plants. *Frontiers in Microbiology, 8*. https://doi.org/10.3389/fmicb.2017.02603.

Bengtsson-Palme, J., Kristiansson, E., & Larsson, D. G. J. (2018). Environmental factors influencing the development and spread of antibiotic resistance. *FEMS Microbiology Reviews, 42*(1), 68–80. https://doi.org/10.1093/femsre/fux053.

Berglund, B. (2015). Environmental dissemination of antibiotic resistance genes and correlation to anthropogenic contamination with antibiotics. *Infection Ecology & Epidemiology, 5*(1), 28564. https://doi.org/10.3402/iee.v5.28564.

Carvalho, G., Fouchet, D., Danesh, G., Godeux, A. S., Laaberki, M. H., Pontier, D., et al. (2020). Bacterial transformation buffers environmental fluctuations through the reversible integration of mobile genetic elements. *mBio, 11*(2). https://doi.org/10.1128/mBio.02443-19.

Cassini, A., Högberg, L. D., Plachouras, D., Quattrocchi, A., Hoxha, A., Simonsen, G. S., et al. (2019). Attributable deaths and disability-adjusted life-years caused by infections with antibiotic-resistant bacteria in the EU and the European Economic Area in 2015: A population-level modelling analysis. *The Lancet Infectious Diseases, 19*(1), 56–66. https://doi.org/10.1016/S1473-3099(18)30605-4.

Chattopadhyay, M. K., Chakraborty, R., Grossart, H. P., Reddy, G. S., & Jagannadham, M. V. (2015). Antibiotic resistance of bacteria. *BioMed Research International, 2015*. https://doi.org/10.1155/2015/501658.

Chen, H., & Wang, J. (2019). Catalytic ozonation of sulfamethoxazole over Fe3O4/Co3O4 composites. *Chemosphere, 234*, 14–24. https://doi.org/10.1016/j.chemosphere.2019.06.014.

Chen, J., Deng, W. J., Liu, Y. S., Hu, L. X., He, L. Y., Zhao, J. L., et al. (2019). Fate and removal of antibiotics and antibiotic resistance genes in hybrid constructed wetlands. *Environmental Pollution, 249*, 894–903. https://doi.org/10.1016/j.envpol.2019.03.111.

Cheng, D., Ngo, H. H., Guo, W., Chang, S. W., Nguyen, D. D., Liu, Y., et al. (2020). Removal process of antibiotics during anaerobic treatment of swine wastewater. *Bioresource Technology, 300*, 122707. https://doi.org/10.1016/j.biortech.2019.122707.

Cheng, H., & Hong, P. Y. (2017). Removal of antibiotic-resistant bacteria and antibiotic resistance genes affected by varying degrees of fouling on anaerobic microfiltration membranes. *Environmental Science and Technology, 51*(21), 12200–12209. https://doi.org/10.1021/acs.est.7b03798.

Devarajan, N., Laffite, A., Graham, N. D., Meijer, M., Prabakar, K., Mubedi, J. I., et al. (2015). Accumulation of clinically relevant antibiotic-resistance genes, bacterial load, and metals in freshwater lake sediments in central Europe. *Environmental Science and Technology, 49*(11), 6528–6537. https://doi.org/10.1021/acs.est.5b01031.

Ding, N., Chang, X., Shi, N., Yin, X., Qi, F., & Sun, Y. (2019). Enhanced inactivation of antibiotic-resistant bacteria isolated from secondary effluents by g-C3N4 photocatalysis. *Environmental Science and Pollution Research, 26*(18), 18730–18738. https://doi.org/10.1007/s11356-019-05080-7.

Dodd, M. C. (2012). Potential impacts of disinfection processes on elimination and deactivation of antibiotic resistance genes during water and wastewater treatment. *Journal of Environmental Monitoring, 14*(7), 1754–1771. https://doi.org/10.1039/c2em00006g.

E. implementation WHO report on Surveillance of Antibiotic Consumption, WHO Report on Surveillance of Antibiotic Consumption. E. implementation WHO report on Surveillance of Antibiotic Consumption, WHO Report on Surveillance of Antibiotic Consumption (2018).

Ezeuko, A. S., Ojemaye, M. O., Okoh, O. O., & Okoh, A. I. (2021). Potentials of metallic nanoparticles for the removal of antibiotic resistant bacteria and antibiotic resistance genes from wastewater: A critical review. *Journal of Water Process Engineering, 41*. doi:10.1016/j.jwpe.2021.102041.

Fogg, P. C. M. (2019). Identification and characterization of a direct activator of a gene transfer agent. *Nature Communications, 10*(1). https://doi.org/10.1038/s41467-019-08526-1.

Gat, D., Mazar, Y., Cytryn, E., & Rudich, Y. (2017). Origin-dependent variations in the atmospheric microbiome community in Eastern Mediterranean Dust Storms. *Environmental Science and Technology, 51*(12), 6709–6718. https://doi.org/10.1021/acs.est.7b00362.

Ghaly, T. M., Geoghegan, J. L., Tetu, S. G., & Gillings, M. R. (2020). The Peril and Promise of Integrons: Beyond Antibiotic Resistance. *Trends in Microbiology, 28*(6), 455–464. https://doi.org/10.1016/j.tim.2019.12.002.

Ghernaout, D., & Elboughdiri, N. (2020). Is Not It Time to Stop Using Chlorine for Treating Water? *OALib, 07*(01), 1–11. https://doi.org/10.4236/oalib.1106007.

Giannakis, S., Le, T. T. M., Entenza, J. M., & Pulgarin, C. (2018). Solar photo-Fenton disinfection of 11 antibiotic-resistant bacteria (ARB) and elimination of representative AR genes. Evidence that antibiotic resistance does not imply resistance to oxidative treatment. *Water Research, 143*, 334–345. https://doi.org/10.1016/j.watres.2018.06.062.

Grehs, B. W. N., Lopes, A. R., Moreira, N. F. F., Fernandes, T., Linton, M. A. O., Silva, A. M. T., et al. (2019). Removal of microorganisms and antibiotic resistance genes from treated urban wastewater: A comparison between aluminium sulphate and tannin coagulants. *Water Research, 166*. doi:10.1016/j.watres.2019.115056.

Guo, C., Wang, K., Hou, S., Wan, L., Lv, J., Zhang, Y., et al. (2017). H2O2 and/or TiO2 photocatalysis under UV irradiation for the removal of antibiotic resistant bacteria and their antibiotic resistance genes. *Journal of Hazardous Materials, 323*, 710–718. https://doi.org/10.1016/j.jhazmat.2016.10.041.

Guo, W., Cheng, D., Ngo, H. H., Chang, S. W., Nguyen, D. D., Nguyen, D. P., et al. (2020). *Anaerobic Membrane Bioreactors for Antibiotic Wastewater Treatment* (pp. 219–239). Netherlands: Elsevier BV. https://doi.org/10.1016/b978-0-12-819852-0.00009-9.

Herraiz-Carboné, M., Cotillas, S., Lacasa, E., Sainz de Baranda, C., Riquelme, E., Cañizares, P., et al. (2021). A review on disinfection technologies for controlling the antibiotic resistance spread. *Science of the Total Environment, 797*, 149150. https://doi.org/10.1016/j.scitotenv.2021.149150.

Hiller, C. X., Hübner, U., Fajnorova, S., Schwartz, T., & Drewes, J. E. (2019). Antibiotic microbial resistance (AMR) removal efficiencies by conventional and advanced wastewater treatment processes: A review. *Science of the Total Environment, 685*, 596–608. https://doi.org/10.1016/j.scitotenv.2019.05.315.

Huijbers, P. M. C., Blaak, H., De Jong, M. C. M., Graat, E. A. M., Vandenbroucke-Grauls, C. M. J. E., & De Roda Husman, A. M. (2015). Role of the environment in the transmission of antimicrobial resistance to humans: A review. *Environmental Science and Technology, 49*(20), 11993–12004. https://doi.org/10.1021/acs.est.5b02566.

Hutchings, M., Truman, A., & Wilkinson, B. (2019). Antibiotics: Past, present and future. *Current Opinion in Microbiology, 51*, 72–80. https://doi.org/10.1016/j.mib.2019.10.008.

Jennings, M. C., Minbiole, K. P. C., & Wuest, W. M. (2016). Quaternary ammonium compounds: An antimicrobial mainstay and platform for innovation to address bacterial resistance. *ACS Infectious Diseases, 1*(7), 288–303. https://doi.org/10.1021/acsinfecdis.5b00047.

Jia, S., Wu, J., Ye, L., Zhao, F., Li, T., & Zhang, X. X. (2019). Metagenomic assembly provides a deep insight into the antibiotic resistome alteration induced by drinking water chlorination and its correlations with bacterial host changes. *Journal of Hazardous Materials, 379*. https://doi.org/10.1016/j.jhazmat.2019.120841.

Jiménez-Tototzintle, M., Ferreira, I. J., da Silva Duque, S., Guimarães Barrocas, P. R., & Saggioro, E. M. (2018). Removal of contaminants of emerging concern (CECs) and antibiotic resistant bacteria in urban wastewater using UVA/TiO2/H2O2 photocatalysis. *Chemosphere, 210*, 449–457. https://doi.org/10.1016/j.chemosphere.2018.07.036.

Kappell, A. D., Kimbell, L. K., Seib, M. D., Carey, D. E., Choi, M. J., Kalayil, T., et al. (2018). Removal of antibiotic resistance genes in an anaerobic membrane bioreactor treating primary clarifier effluent at 20°C. *Environmental Science: Water Research and Technology, 4*(11), 1783–1793. https://doi.org/10.1039/c8ew00270c.

Karkman, A., Do, T. T., Walsh, F., & Virta, M. P. J. (2018). Antibiotic-resistance genes in waste water. *Trends in Microbiology, 26*(3), 220–228. https://doi.org/10.1016/j.tim.2017.09.005.

Karkman, A., Johnson, T. A., Lyra, C., Stedtfeld, R. D., Tamminen, M., Tiedje, J. M., et al. (2016). High-throughput quantification of antibiotic resistance genes from an urban wastewater treatment plant. *FEMS Microbiology Ecology, 92*(3). doi:10.1093/femsec/fiw014.

Kohanski, M. A., DePristo, M. A., & Collins, J. J. (2010). Sublethal antibiotic treatment leads to multidrug resistance via radical-induced mutagenesis. *Molecular Cell, 37*(3), 311–320. https://doi.org/10.1016/j.molcel.2010.01.003.

Krzeminski, P., Feys, E., Anglès d'Auriac, M., Wennberg, A. C., Umar, M., Schwermer, C. U., et al. (2020). Combined membrane filtration and 265 nm UV irradiation for effective removal of cell free antibiotic resistance genes from feed water and concentrate. *Journal of Membrane Science, 598*. doi:10.1016/j.memsci.2019.117676.

Kucukunsal, S., & Icgen, B. (2020). Removal of antibiotic resistance genes in various water resources recovery facilities. *Water Environment Research, 92*(6), 911–921. https://doi.org/10.1002/wer.1286.

Kumar, A., & Pal, D. (2018). Antibiotic resistance and wastewater: Correlation, impact and critical human health challenges. *Journal of Environmental Chemical Engineering, 6*(1), 52–58. https://doi.org/10.1016/j.jece.2017.11.059.

Lan, L., Kong, X., Sun, H., Li, C., & Liu, D. (2019). High removal efficiency of antibiotic resistance genes in swine wastewater via nanofiltration and reverse osmosis processes. *Journal of Environmental Management, 231*, 439–445. https://doi.org/10.1016/j.jenvman.2018.10.073.

Langbehn, R. K., Michels, C., & Soares, H. M. (2021). Antibiotics in wastewater: From its occurrence to the biological removal by environmentally conscious technologies. *Environmental Pollution, 275*. https://doi.org/10.1016/j.envpol.2021.116603.

Lee, J., Jeon, J. H., Shin, J., Jang, H. M., Kim, S., Song, M. S., et al. (2017). Quantitative and qualitative changes in antibiotic resistance genes after passing through treatment processes in municipal wastewater treatment plants. *Science of the Total Environment, 605–606*, 906–914. https://doi.org/10.1016/j.scitotenv.2017.06.250.

Leibold, M. A., Holyoak, M., Mouquet, N., Amarasekare, P., Chase, J. M., Hoopes, M. F., et al. (2004). The metacommunity concept: A framework for multi-scale community ecology. *Ecology Letters, 7*(7), 601–613. https://doi.org/10.1111/j.1461-0248.2004.00608.x.

Levin, B. R., Baquero, F., & Johnsen, P. J. (2014). A model-guided analysis and perspective on the evolution and epidemiology of antibiotic resistance and its future. *Current Opinion in Microbiology, 19*(1), 83–89. https://doi.org/10.1016/j.mib.2014.06.004.

Li, N., Sheng, G. P., Lu, Y. Z., Zeng, R. J., & Yu, H. Q. (2017). Removal of antibiotic resistance genes from wastewater treatment plant effluent by coagulation. *Water Research, 111*, 204–212. https://doi.org/10.1016/j.watres.2017.01.010.

Li, S., Zhang, C., Li, F., Hua, T., Zhou, Q., & Ho, S. H. (2021). Technologies towards antibiotic resistance genes (ARGs) removal from aquatic environment: A critical review. *Journal of Hazardous Materials, 411*. doi:10.1016/j.jhazmat.2021.125148.

Liu, X., Guo, X., Liu, Y., Lu, S., Xi, B., Zhang, J., et al. (2019). A review on removing antibiotics and antibiotic resistance genes from wastewater by constructed wetlands: Performance and microbial response. *Environmental Pollution, 254*, 112996. https://doi.org/10.1016/j.envpol.2019.112996.

Lu, J., Zhang, Y., Wu, J., Wang, J., & Cai, Y. (2020). Fate of antibiotic resistance genes in reclaimed water reuse system with integrated membrane process. *Journal of Hazardous Materials, 382*, 121025. https://doi.org/10.1016/j.jhazmat.2019.121025.

Mahfouz, N., Caucci, S., Achatz, E., Semmler, T., Guenther, S., Berendonk, T. U., et al. (2018). High genomic diversity of multi-drug resistant wastewater Escherichia coli. *Scientific Reports, 8*(1). doi:10.1038/s41598-018-27292-6.

Manaia, C. M., Rocha, J., Scaccia, N., Marano, R., Radu, E., Biancullo, F., et al. (2018). Antibiotic resistance in wastewater treatment plants: Tackling the black box. *Environment International, 115*, 312–324. https://doi.org/10.1016/j.envint.2018.03.044.

Mattner, F., Bange, F. C., Meyer, E., Seifert, H., Wichelhaus, T. A., & Chaberny, I. F. (2012). Prävention der ausbreitung von multiresistenten gramnegativen erregern: Vorschläge eines expertenworkshops der deutschen gesellschaft für hygiene und mikrobiologie. *Deutsches Arzteblatt International, 109*(3), 39–45. https://doi.org/10.3238/arztebl.2012.0039.

Mishra, J., Pattanayak, D. S., Das, A. A., Mishra, D. K., Rath, D., & Sahoo, N. K. (2019). Enhanced photocatalytic degradation of cyanide employing Fe-porphyrin sensitizer with hydroxyapatite palladium doped TiO2nano-composite system. *Journal of Molecular Liquids, 287*. doi:10.1016/j.molliq.2019.04.098.

Mittler, R. (2002). Oxidative stress, antioxidants and stress tolerance. *Trends in Plant Science, 7*(9), 405–410. https://doi.org/10.1016/S1360-1385(02)02312-9.

Mohanty, L., Pattanayak, D. S., & Dash, S. K. (2021). An efficient ternary photocatalyst Ag/ZnO/g-C3N4 for degradation of RhB and MG under solar radiation. *Journal of the Indian Chemical Society, 98*(11). doi:10.1016/j.jics.2021.100180.

Mohanty, L., Sundar Pattanayak, D., Singhal, R., Pradhan, D., & Kumar Dash, S. (2022). Enhanced photocatalytic degradation of rhodamine B and malachite green employing BiFeO3/g-C3N4 nanocomposites: An efficient visible-light photocatalyst. *Inorganic Chemistry Communications, 138*, 109286. https://doi.org/10.1016/j.inoche.2022.109286.

Oh, J., Salcedo, D. E., Medriano, C. A., & Kim, S. (2014). Comparison of different disinfection processes in the effective removal of antibiotic-resistant bacteria and genes. *Journal of Environmental Sciences (China), 26*(6), 1238–1242. https://doi.org/10.1016/S1001-0742(13)60594-X.

Pal, C., Bengtsson-Palme, J., Kristiansson, E., & Larsson, D. G. J. (2016). The structure and diversity of human, animal and environmental resistomes. *Microbiome, 4*. doi:10.1186/s40168-016-0199-5.

Pallares-Vega, R., Blaak, H., van der Plaats, R., de Roda Husman, A. M., Hernandez Leal, L., van Loosdrecht, M. C. M., et al. (2019). Determinants of presence and removal of antibiotic resistance genes during WWTP treatment: A cross-sectional study. *Water Research, 161*, 319–328. https://doi.org/10.1016/j.watres.2019.05.100.

Pattanayak, D. S., Mallick, N., Thakur, C., & Pal, D. (2020). Plant mediated green synthesis of silver nanoparticles for antimicrobial application: Present status. *Journal of the Indian Chemical Society, 97*(7), 1108–1114. https://indianchemicalsociety.com/portal/uploads/journal/2020_07_28_Extended_1605501950.pdf.

Pattanayak, D. S., Mishra, J., Nanda, J., Sahoo, P. K., Kumar, R., & Sahoo, N. K. (2021). Photocatalytic degradation of cyanide using polyurethane foam immobilized Fe-TCPP-S-TiO2-rGO nano-composite. *Journal of Environmental Management, 297*. doi:10.1016/j.jenvman.2021.113312.

Pei, M., Zhang, B., He, Y., Su, J., Gin, K., Lev, O., et al. (2019). State of the art of tertiary treatment technologies for controlling antibiotic resistance in wastewater treatment plants. *Environment International, 131*, 105026. https://doi.org/10.1016/j.envint.2019.105026.

Prudhomme, M., Attaiech, L., Sanchez, G., Martin, B., & Claverys, J. P. (2006). Antibiotic stress induces genetic transformability in the human pathogen streptoccus pneumoniae. *Science, 313*(5783), 89–92. https://doi.org/10.1126/science.1127912.

Rawson, T. M., Moore, L. S. P., Zhu, N., Ranganathan, N., Skolimowska, K., Gilchrist, M., et al. (2020). Bacterial and fungal coinfection in individuals with coronavirus: A rapid review to support COVID-19 antimicrobial prescribing. *Clinical Infectious Diseases, 71*(9), 2459–2468. https://doi.org/10.1093/cid/ciaa530.

Ren, S., Boo, C., Guo, N., Wang, S., Elimelech, M., & Wang, Y. (2018). Photocatalytic reactive ultrafiltration membrane for removal of antibiotic resistant bacteria and antibiotic resistance genes from wastewater effluent. *Environmental Science and Technology, 52*(15), 8666–8673. https://doi.org/10.1021/acs.est.8b01888.

Rizzo, L., Manaia, C., Merlin, C., Schwartz, T., Dagot, C., Ploy, M. C., et al. (2013). Urban wastewater treatment plants as hotspots for antibiotic resistant bacteria and genes spread into the environment: A review. *Science of the Total Environment, 447*, 345–360. https://doi.org/10.1016/j.scitotenv.2013.01.032.

Rodriguez-Mozaz, S., Chamorro, S., Marti, E., Huerta, B., Gros, M., Sànchez-Melsió, A., et al. (2015). Occurrence of antibiotics and antibiotic resistance genes in hospital and urban wastewaters and their impact on the receiving river. *Water Research, 69*, 234–242. https://doi.org/10.1016/j.watres.2014.11.021.

Sabbagh, P., Rajabnia, M., Maali, A., & Ferdosi-Shahandashti, E. (2021). Integron and its role in antimicrobial resistance: A literature review on some bacterial pathogens. *Iranian Journal of Basic Medical Sciences, 24*(2), 136–142. https://doi.org/10.22038/ijbms.2020.48905.11208.

Safdari, R., GhaziSaeedi, M., Masoumi-Asl, H., Rezaei-Hachesu, P., Mirnia, K., & Samad-Soltani, T. (2017). A national framework for an antimicrobial resistance surveillance system within Iranian healthcare facilities: Towards a global surveillance system. *Journal of Global Antimicrobial Resistance, 10*, 59–69. https://doi.org/10.1016/j.jgar.2017.03.016.

Sagar, S., Kaistha, S., Das, A. J., & Kumar, R. (2019). Antibiotic Resistant Bacteria: A Challenge to Modern Medicine. Singapore: Springer. https://doi.org/10.1007/978-981-13-9879-7.

Saha, D., Visconti, M. C., Desipio, M. M., & Thorpe, R. (2020). Inactivation of antibiotic resistance gene by ternary nanocomposites of carbon nitride, reduced graphene oxide and iron oxide under visible light. *Chemical Engineering Journal, 382*. doi:10.1016/j.cej.2019.122857.

Sanganyado, E., & Gwenzi, W. (2019). Antibiotic resistance in drinking water systems: Occurrence, removal, and human health risks. *Science of the Total Environment, 669*, 785–797. https://doi.org/10.1016/j.scitotenv.2019.03.162.

Sano, D., Wester, A. L., Schmitt, H., Amarasiri, M., Kirby, A., Medlicott, K., et al. (2020). Updated research agenda for water, sanitation and antimicrobial resistance. *Journal of Water and Health, 18*(6), 858–866. https://doi.org/10.2166/wh.2020.033.

Sharma, V. K., Johnson, N., Cizmas, L., McDonald, T. J., & Kim, H. (2016). A review of the influence of treatment strategies on antibiotic resistant bacteria and antibiotic resistance genes. *Chemosphere, 150*, 702–714. https://doi.org/10.1016/j.chemosphere.2015.12.084.

Sharma, V. K., Yu, X., McDonald, T. J., Jinadatha, C., Dionysiou, D. D., & Feng, M. (2019). Elimination of antibiotic resistance genes and control of horizontal transfer risk by UV-based treatment of drinking water: A mini review. *Frontiers of Environmental Science and Engineering, 13*(3). https://doi.org/10.1007/s11783-019-1122-7.

Sillanpää, M., Ncibi, M. C., Matilainen, A., & Vepsäläinen, M. (2018). Removal of natural organic matter in drinking water treatment by coagulation: A comprehensive review. *Chemosphere, 190*, 54–71. https://doi.org/10.1016/j.chemosphere.2017.09.113.

Singh, R., Singh, A. P., Kumar, S., Giri, B. S., & Kim, K. H. (2019). Antibiotic resistance in major rivers in the world: A systematic review on occurrence, emergence, and management strategies. *Journal of Cleaner Production, 234*, 1484–1505. https://doi.org/10.1016/j.jclepro.2019.06.243.

Slipko, K., Reif, D., Wögerbauer, M., Hufnagl, P., Krampe, J., & Kreuzinger, N. (2019). Removal of extracellular free DNA and antibiotic resistance genes from water and wastewater by membranes ranging from microfiltration to reverse osmosis. *Water Research, 164*. doi:10.1016/j.watres.2019.114916.

Somensi, C. A., Souza, A. L. F., Simionatto, E. L., Gaspareto, P., Millet, M., & Radetski, C. M. (2015). Genetic material present in hospital wastewaters: Evaluation of the efficiency of DNA denaturation by ozonolysis and ozonolysis/sonolysis treatments. *Journal of Environmental Management, 162*, 74–80. https://doi.org/10.1016/j.jenvman.2015.07.039.

Song, H. L., Zhang, S., Guo, J., Yang, Y. L., Zhang, L. M., Li, H., et al. (2018). Vertical up-flow constructed wetlands exhibited efficient antibiotic removal but induced antibiotic resistance genes in effluent. *Chemosphere, 203*, 434–441. https://doi.org/10.1016/j.chemosphere.2018.04.006.

Stedt, J., Bonnedahl, J., Hernandez, J., Waldenström, J., McMahon, B. J., Tolf, C., et al. (2015). Carriage of CTX-M type extended spectrum β-lactamases (ESBLs) in gulls across Europe. *Acta Veterinaria Scandinavica, 57*(1). https://doi.org/10.1186/s13028-015-0166-3.

Sun, L., Shi, P., He, N., Zhang, Q., & Duan, X. (2019). Antibiotic resistance genes removal and membrane fouling in secondary effluents by combined processes of PAC/BPAC-UF. *Journal of Water and Health, 17*(6), 910–920. https://doi.org/10.2166/wh.2019.160.

Tadesse, B. T., Ashley, E. A., Ongarello, S., Havumaki, J., Wijegoonewardena, M., González, I. J., et al. (2017). Antimicrobial resistance in Africa: A systematic review. *BMC Infectious Diseases, 17*(1). https://doi.org/10.1186/s12879-017-2713-1.

Takeuchi, N., Kaneko, K., & Koonin, E. V. (2014). Horizontal gene transfer can rescue prokaryotes from Muller's ratchet: Benefit of DNA from dead cells and population subdivision. *G3: Genes, Genomes, Genetics, 4*(2), 325–339. https://doi.org/10.1534/g3.113.009845.

Tandukar, M., Oh, S., Tezel, U., Konstantinidis, K. T., & Pavlostathis, S. G. (2013). Long-term exposure to benzalkonium chloride disinfectants results in change of microbial community structure and increased antimicrobial resistance. *Environmental Science and Technology, 47*(17), 9730–9738. https://doi.org/10.1021/es401507k.

Von Wintersdorff, C. J. H., Penders, J., Van Niekerk, J. M., Mills, N. D., Majumder, S., Van Alphen, L. B., et al. (2016). Dissemination of antimicrobial resistance in microbial ecosystems through horizontal gene transfer. *Frontiers in Microbiology, 7*. doi:10.3389/fmicb.2016.00173.

Wang, G., Deng, D., Hu, C., Lou, L., Luo, L., He, J., et al. (2020). More effective removal of antibiotic resistance genes from excess sludge by microwave integrated fenton treatment. *International Biodeterioration & Biodegradation, 149*, 104920. https://doi.org/10.1016/j.ibiod.2020.104920.

Wang, M., Xu, J., Chai, Y., Guo, Y., Liu, X., & Yue, M. (2020). Differentiation of Environmental Conditions Promotes Variation of Two Quercus wutaishanica. *Community Assembly Patterns. Forests, 11*(1), 43. https://doi.org/10.3390/f11010043.

Wang, Y., Lu, J., Mao, L., Li, J., Yuan, Z., Bond, P. L., et al. (2019). Antiepileptic drug carbamazepine promotes horizontal transfer of plasmid-borne multi-antibiotic resistance genes within and across bacterial genera. *ISME Journal, 13*(2), 509–522. https://doi.org/10.1038/s41396-018-0275-x.

Wang, Z., Han, M., Li, E., Liu, X., Wei, H., Yang, C., et al. (2020). Distribution of antibiotic resistance genes in an agriculturally disturbed lake in China: Their links with microbial communities, antibiotics, and water quality. *Journal of Hazardous Materials, 393*, 122426. https://doi.org/10.1016/j.jhazmat.2020.122426.

Windels, E. M., Michiels, J. E., Fauvart, M., Wenseleers, T., Van den Bergh, B., & Michiels, J. (2019). Bacterial persistence promotes the evolution of antibiotic resistance by increasing survival and mutation rates. *ISME Journal, 13*(5), 1239–1251. https://doi.org/10.1038/s41396-019-0344-9.

Wu, D., Huang, Z., Yang, K., Graham, D., & Xie, B. (2015). Relationships between antibiotics and antibiotic resistance gene levels in municipal solid waste leachates in Shanghai. *China. Environmental Science and Technology, 49*(7), 4122–4128. https://doi.org/10.1021/es506081z.

Xue, G., Jiang, M., Chen, H., Sun, M., Liu, Y., Li, X., et al. (2019). Critical review of ARGs reduction behavior in various sludge and sewage treatment processes in wastewater treatment plants. *Critical Reviews in Environmental Science and Technology, 49*(18), 1623–1674. https://doi.org/10.1080/10643389.2019.1579629.

Yi, X., Tran, N. H., Yin, T., He, Y., & Gin, K. Y. H. (2017). Removal of selected PPCPs, EDCs, and antibiotic resistance genes in landfill leachate by a full-scale constructed wetlands system. *Water Research, 121*, 46–60. https://doi.org/10.1016/j.watres.2017.05.008.

Yoon, Y., Chung, H. J., Wen Di, D. Y., Dodd, M. C., Hur, H. G., & Lee, Y. (2017). Inactivation efficiency of plasmid-encoded antibiotic resistance genes during water treatment with chlorine, UV, and UV/H2O2. *Water Research, 123*, 783–793. https://doi.org/10.1016/j.watres.2017.06.056.

Yu, P., Zhou, X., Li, Z., & Yan, Y. (2020). Inactivation and change of tetracycline-resistant Escherichia coli in secondary effluent by visible light-driven photocatalytic process using Ag/AgBr/g-C3N4. *Science of the Total Environment, 705*, 135639. https://doi.org/10.1016/j.scitotenv.2019.135639.

Yu, W., Zhan, S., Shen, Z., Zhou, Q., & Yang, D. (2017). Efficient removal mechanism for antibiotic resistance genes from aquatic environments by graphene oxide nanosheet. *Chemical Engineering Journal, 313*, 836–846. https://doi.org/10.1016/j.cej.2016.10.107.

Zarei-Baygi, A., Harb, M., Wang, P., Stadler, L. B., & Smith, A. L. (2020). Microbial community and antibiotic resistance profiles of biomass and effluent are distinctly affected by antibiotic addition to an anaerobic membrane bioreactor. *Environmental Science: Water Research and Technology, 6*(3), 724–736. https://doi.org/10.1039/c9ew00913b.

Zhang, C.-M., Xu, L.-M., Mou, X., Xu, H., Liu, J., Miao, Y.-H., et al. (2019). Characterization and evolution of antibiotic resistance of Salmonella in municipal wastewater treatment plants. *Journal of Environmental Management, 251*, 109547. https://doi.org/10.1016/j.jenvman.2019.109547.

Zhang, G., Li, W., Chen, S., Zhou, W., & Chen, J. (2020). Problems of conventional disinfection and new sterilization methods for antibiotic resistance control. *Chemosphere, 254*, 126831. https://doi.org/10.1016/j.chemosphere.2020.126831.

Zhang, S., Lu, Y.-X., Zhang, J.-J., Liu, S., Song, H.-L., & Yang, X.-L. (2020). Constructed wetland revealed efficient sulfamethoxazole removal but enhanced the spread of antibiotic resistance genes. *Molecules (Basel, Switzerland), 25*(4), 834. https://doi.org/10.3390/molecules25040834.

Zhang, Y., Zhuang, Y., Geng, J., Ren, H., Xu, K., & Ding, L. (2016). Reduction of antibiotic resistance genes in municipal wastewater effluent by advanced oxidation processes. *Science of the Total Environment, 550*, 184–191. https://doi.org/10.1016/j.scitotenv.2016.01.078.

Zhang, Y., Zhuang, Y., Geng, J., Ren, H., Zhang, Y., Ding, L., et al. (2015). Inactivation of antibiotic resistance genes in municipal wastewater effluent by chlorination and sequential UV/chlorination disinfection. *Science of the Total Environment, 512–513*, 125–132. https://doi.org/10.1016/j.scitotenv.2015.01.028.

Zheng, J., Su, C., Zhou, J., Xu, L., Qian, Y., & Chen, H. (2017). Effects and mechanisms of ultraviolet, chlorination, and ozone disinfection on antibiotic resistance genes in secondary effluents of municipal wastewater treatment plants. *Chemical Engineering Journal, 317*, 309–316. https://doi.org/10.1016/j.cej.2017.02.076.

Zhuang, Y., Ren, H., Geng, J., Zhang, Y., Zhang, Y., Ding, L., et al. (2015). Inactivation of antibiotic resistance genes in municipal wastewater by chlorination, ultraviolet, and ozonation disinfection. *Environmental Science and Pollution Research, 22*(9), 7037–7044. https://doi.org/10.1007/s11356-014-3919-z.

Chapter 7

Do's and don'ts of wastewater treatment, their reuse, and future directions

M. Narayana Rao[a], A.D. Prasad[b] and K.V.S.G. Murali Krishna[c]
[a]*National Institute of Technical Teachers Training and Research, Chennai, India,* [b]*Civil Engineering Department, National Institute of Technology Raipur, Raipur, India,* [c]*Civil Engineering Department, Jawaharlal Nehru Technological University, Kakinada, India*

7.1 Introduction

The self-purification processes in nature are normally slow. The artificial treatment processes are, therefore, designed to intensify these processes and accomplish the desired degree of treatment in a comparatively shorter time and in smaller space.

Wastewater treatment plants accomplish a major part of the cleaning work and when the effluents are disposed into water sources, land, etc., nature completes the work and gets rid of water pollution.

With increased demand for water, reuse of treated wastewater became a (pls. check word) necessity and the need of the hour (Santos Pereira et al., 2009). Depending upon the purpose for which the treated water is reused and socio-economical, technological, and environmental conditions prevailing in the region, various treatment processes have been developed over the years by different countries.

This chapter explains the generation, characteristics, and treatment of wastewater from various sources, technologies for recovery of nutrients, energy and reusable water from wastewater, and their limitations.

The wastewater utilization states in various sector, permissible limits of pollutants, risks and challenges in recycling, and the precaution in the reuse of wastewater are discussed at length in this chapter.

7.2 Wastewater generation

7.2.1 Wastewater generation status in different countries

Many countries in the world are using recycled water for different purposes. The general wastewater chain is represented as in Fig. 7.1. According to the

Antimicrobial Resistance in Wastewater and Human Health.
DOI: https://doi.org/10.1016/B978-0-323-96124-0.00004-0

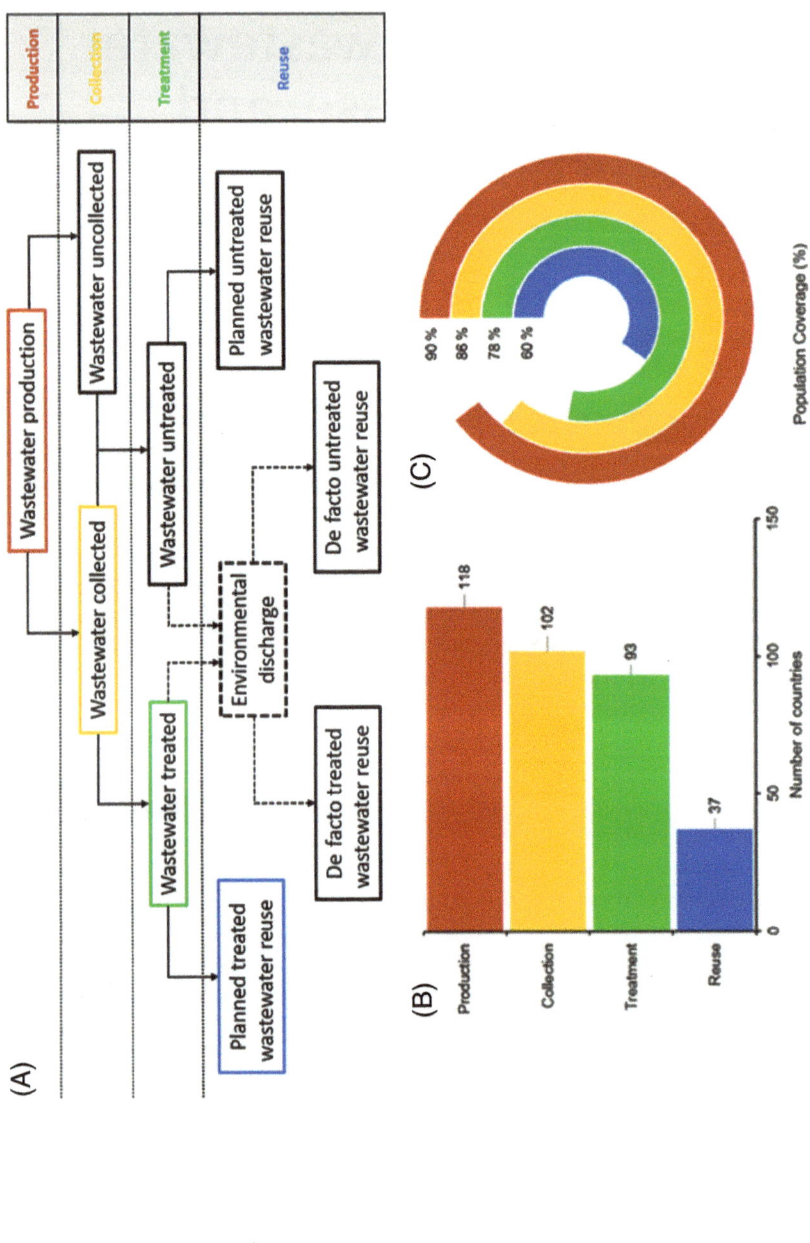

FIGURE 7.1 The wastewater chain (A), including wastewater data availability with number of countries for which wastewater data are available (B) and the percentage of population coverage (i.e., the proportion of the global population for which wastewater data are available) (C). (Source: (Jones et al., 2021).

2017 WHO and USEP census, the states reusing treated wastewater for drinking are Australia, Belgium, California, Jordan, Kuwait, Namibia, Singapore, South Africa, Texas, and the UK. Projects for reuse are under consideration in Brazil and India (Jones et al., 2021).

Recent research studies state that globally 359.4 × 109 m^3/year (358.0 × 109 – 361.4 × 109 m^3/year) of wastewater is produced annually, with a global average of 49.0 m^3/year (48.8–49.2 m^3/year) per capita. Global annual wastewater collection and treatment is estimated at 225.6 × 109 m^3/year (224.4 × 109–226.9 × 109 m^3/year) and 188.1 × 109 m^3/year (186.6 × 109–189.3 × 109 m^3/year), respectively. These values indicate that approximately 63% and 52% of globally produced wastewater is collected and treated, respectively, with approximately 84% of collected wastewater undergoing a treatment process. Region-wise wastewater data sources are presented in Table 7.1.

7.2.2 Wastewater generation status in India

Just as the application of fertilizers to the farmlands helps the growth of crops, nutrients entering water bodies like lakes and rivers feed the growth of bacteria, algae, and other organisms. Excess nutrient dose results in excessive growth of algae. Decomposition and die-off of algal blooms can reduce dissolved oxygen in the water and kill fish and other aquatic life.

7.2.3 Wastewater reuse in different sectors

7.2.3.1 Wastewater reuse for agricultural irrigation

Agriculture is the most common application of reclaimed water with a 30% share; since a shortage of water for irrigation can impact food security, nutrition, and other socio-economic aspects. This practice reduces the need for synthetic chemical fertilizers when used optimally. However, there are some risks associated with this practice (Table 7.2). The risks involved are dependent on the following factors:

a) Composition of wastewater used for reuse.
b) Treatment, shortage, and distribution methods adopted.
c) Type of irrigation practice and techniques of irrigation used.
d) Soil and groundwater conditions.
e) Climatic conditions.

Potential exposure pathways of cleaning agents, antibiotics, biocides, etc., have persistent bioaccumulative and toxic (PBT) properties on the environment.

These factors should be taken into consideration for the effective reuse of wastewater for agricultural irrigation.

Advanced technologies can help to lower the energy footprint. However, public perception in water reuse and biosolid application to land is a daunting

TABLE 7.1 Wastewater data sources and population coverage by regional and economic aspects, including the number of unique countries (in square brackets). Method for standardization of wastewater data to 2015 and the method for compiling wastewater data from multiple sources into a single quantification per country.

				Regional aspects		Economic aspects	
	Data sources[a]	**Standardisation to 2015**	**Data complling method**	**Region**	**Population coverage[b]**	**Classification**	**Population coverage[c]**
Production	GWI [94]	Divide by GDP (USD) in reporting year, multiply by sources.	Average of all available sources.	North America	100% [2]	High	99.4% [48]
	FAO [98]			Latin America and Caribbean	93.9% [19]		
	UNSD [23]	GDP (USD) in 2015.		Western Europe	99.8% [19]	Uppr middle	98.0 % [34]
	Eurostat [20]			Middle East and North Africa	98.8% [19]		
				Sub-Saharan Africa	49.6& [17]	Lower middle	98.0% [31]
				South Asia	96.4% [4]		
				Eastern Europe and Central Asia	89.4% [23]	Low	13.3% [5]
				East Asia and Pacific	95.3% [15]		

(continued on next page)

Collection	GWI [95]	Divide by GDP per capita (USD per capita) in reporting year, multiply with GDP per capita (USD per capita) in 2015.	GWI data prioritised. FAO data if unavailable.	North America	100% [2]	High	99.4% [47]
	FAO [55]			Latin America and Caribbean	96.7% [20]		
	UNSD [23]	GDP (USD) in 2015.		Western Europe	99.8% [18]	Uppr middle	97.7 % [29]
	Eurostat [20]			Middle East and North Africa	88.3% [17]		
				Sub-Saharan Africa	61.1% [13]	Lower middle	81.0% [21]
				South Asia	96.4% [4]		
				Eastern Europe and Central Asia	69.9% [16]	Low	34.9% [5]
				East Asia and Pacific	83.6% [12]		
Treatment	GWI [76]	Divide by GDP per capita (USD per capita) in reporting year, multiply with GDP per capita (USD per capita) in 2015.	GWI data prioritised. FAO or UNSD where unavailable (most recent reporting year prioritised).	North America	100% [2]	High	98.4% [46]

(continued on next page)

TABLE 7.1 Wastewater data sources and population coverage by regional and economic aspects, including the number of unique countries (in square brackets). Method for standardization of wastewater data to 2015 and the method for compiling wastewater data from multiple sources into a single quantification per country—cont'd

				Regional aspects		Economic aspects	
	Data sources[a]	Standardisation to 2015	Data complling method	Region	Population coverage[b]	Classification	Population coverage[c]
	FAO [78]			Latin America and Caribbean	90.0% [17]		
	UNSD [23]	GDP (USD) in 2015.		Western Europe	99.8% [19]	Uppr middle	91.2 % [27]
	Eurostat [20]			Middle East and North Africa	65.9% [13]		
				Sub-Saharan Africa	25.7% [8]	Lower middle	69.4% [15]
				South Asia	95.2% [3]		
				Eastern Europe and Central Asia	73.4% [21]	Low	27.1% [5]
				East Asia and Pacific	80.2% [10]		
Reuse	GWI [20]	Wastewater production normalised to reporting year of wastewater reuse based on GDP (USD), percentage reuse calculated, applied to 2015 production data.	GWI data prioritised. FAO data if unavailable.	North America	90% [1]	High	68.7% [19]

(*continued on next page*)

	FAO [32]			Latin America and Caribbean	67.2% [5]		
	UNSD [23]	GDP (USD) in 2015.		Western Europe	42.5% [3]	Uppr middle	77.7 % [10]
	Eurostat [20]			Middle East and North Africa	83.0% [13]		
				Sub-Saharan Africa	21.5% [6]	Lower middle	48.7% [4]
				South Asia	74.9% [1]		
				Eastern Europe and Central Asia	0.6% [2]	Low	24.8% [4]

[a] *Abbreviations for the data sources are as follows: Global Water Intelligence (GWI), Food and Agriculture Organization of the United Nations (FAO), United Nations Statistics Department (UNSD), European Union statistics office (Eurostat).*
[b] *Data availability per region expressed as a percentage of the total population. Geographic region followed by the total number of countries per region in square brackets: East Asia and Pacific [38], eastern Europe and Central Asia [30], Latin America and Caribbean [41], Middle East and North Africa [21], North America [3], South Asia [8], sub-Saharan Africa [48], and western Europe [26].*
[c] *Data availability per economic classification expressed as a percentage of the total population. Total number of countries per economic classification are: high [76], upper-middle [56], lower -middle [52] and low [31] income.*
(Source: E. R. Jones et al., Earth Syst. Sci. Data, 13, 237–254, 2021).

TABLE 7.2 National status of wastewater generation & treatment.

States/UTs	Sewage generation (in MLD)	Installed capacity (MLD)	Proposed capacity (MLD)	Total treatment capacity (in MLD) including planned/ proposed	Operational treatment capacity (in MLD)
Andaman & Nicobar Islands	23	0	0	0	0
Andhra Pradesh	2882	833	20	853	443
Arunachal Pradesh	62	0	0	0	0
Assam	809	0	0	0	0
Bihar	2276	10	621	631	0
Chandigarh	188	293	0	293	271
Chhattisgarh	1203	73	0	73	73
Dadra & Nagar Haveli	67	24	0	24	24
Goa	176	66	38	104	44
Gujarat	5013	3378	0	3378	3358
Haryana	1816	1880	0	1880	1880
Himachal Pradesh	116	136	19	155	99
Jammu & Kashmir	665	218	4	222	93
Jharkhand	1510	22	617	639	22
Karnataka	4458	2712	0	2712	1922
Kerala	4256	120	0	120	114
Lakshadweep	13	0	0	0	0
Madhya Pradesh	3646	1839	85	1924	684
Maharashtra	9107	6890	2929	9819	6366

(continued on next page)

TABLE 7.2 National status of wastewater generation & treatment—cont'd

States/UTs	Sewage generation (in MLD)	Installed capacity (MLD)	Proposed capacity (MLD)	Total treatment capacity (in MLD) including planned/proposed	Operational treatment capacity (in MLD)
Manipur	168	0	0	0	0
Meghalaya	112	0	0	0	0
Mizoram	103	10	0	10	0
Nagaland	135	0	0	0	0
NCT of Delhi	3330	2896	0	2896	2715
Orissa	1282	378	0	378	55
Pondicherry	161	56	3	59	56
Punjab	1889	1781	0	1781	1601
Rajasthan	3185	1086	109	1195	783
Sikkim	52	20	10	30	18
Tamil Nadu	6421	1492	0	1492	1492
Telangana	2660	901	0	901	842
Tripura	237	8	0	8	8
Uttar Pradesh	8263	3374	0	3374	3224
Uttarakhand	627	448	67	515	345
West Bengal	5457	897	305	1202	337
Total	**72368**	**31841**	**4827**	**36668**	**26869**

Notes: *(i) Swage generation is estimated based on water supply @ 85lpcd and rate of sewage generation as 80%. (ii) Sewage generation for NCT of Delhi is estimated based on their 80% of water supply of 925 MGD. State-wise sewage generation and treatment capacity of urban centers—India (As of 17.06.2021). Source: (National Inventory of Sewage Treatment Plants 2021)CPCB, India* problem. It is difficult to convince the public regarding emerging contaminants like pharmaceuticals and antibiotic-resistant bacteria.

7.2.3.2 Groundwater recharge

Groundwater recharge is practiced mainly in countries where seawater intrusion into freshwater aquifers is a major threat. Appropriate hydrogeological conditions are needed to practice this. Groundwater subsidence control methods should be ensured.

7.2.4 Permissible limits of pollutants

Water quality standards are required for the safe and sustainable practice of the reuse of water. While using treated wastewater for irrigation, the possible accumulation of harmful elements for plant growth and potential damage to soil should be kept in mind (Guidelines for Water Reuse, 2012).

7.2.4.1 Salinity

High salinity in soil can affect the crop in absorbing nutrients and can also specific ion toxicity. Total dissolved solids (TDS) and electrical conductivity are normally noticed below 500 mL/L. Similarly, chlorine residuals less than 1 mg/L pose no problem to the crops. Salinity increases soil compaction and heavy metal leaching.

7.2.5 Energy production through wastewater

Sustainable development is possible if production systems become autogenerated is converted into raw material. This is called the cradle-to-cradle concept (circular economy). Wastewater treatment technologies now are developing plants where organics, energy, heavy metals, etc., are recovered.

Examples include the usage of metal-reducing and oxidizing organisms for metal recovery and the use of microalgae, photosynthetic bacteria, and terrestrial plants for the recovery of organics.

The conversion of organic-rich wastewater into biofuels in anaerobic systems has been practiced for a long time. Biogas, biohydrogen, biodiesel, biopolymers, single-cell protein (yeast, edible microorganisms with high protein content) heavy metal absorption, microbial fuel cells, and nutrient recovery from wastewater streams are subjects of active research.

7.2.6 Precautions in the reuse of reclaimed water

If wastewater cannot be treated adequately, some important steps are to be taken to mitigate the risks involved in the reuse of wastewater. Some steps include:

a) Crop restrictions and standards for use in irrigation.
b) Use of stabilization farms on-farm.

c) Appropriate practices that limit risks in irrigation, harvesting, public health, and human exposure.

d) Best practices in irrigation techniques and suitable standards, guidelines, and regulations that are suitable to place and are economically and administratively enforceable.

7.3 Wastewater treatment

The precautions we need to take to ensure effective and efficient treatment of wastewater for reuse at various levels of treatment (namely, primary, secondary, and tertiary levels) are pointed out below as do's and don'ts for treatment.

7.3.1 Do's and don'ts in primary treatment

Do's and Don'ts in primary treatment

Do's	Don'ts
Proper homogenization should be carried out.	Do not flush unwanted materials down the drain. Flushing flammable and toxic products is hazardous, whereas other things, such as rags, paper towels, newspaper, cans, cigarettes, coffee grounds, eggshells, baby wipes, sanitary napkins, condoms, large amounts of hair, and cooking grease are an annoyance in maintenance and entail regular pumping of the chamber.

7.3.2 Do's and don'ts in secondary treatment

Do's and don'ts in secondary treatment

Do's	Don'ts
Do have your septic tank pumped out regularly.	Do not use any septic tank supplements. They have not been favorable and could damage your disposal system. Active bacteria are naturally occurring in sewage. Also with supplements, regular pumping of solids is still necessary.
Do make sure your tank is compliant with the new septic tank regulation that came into force in January 2020.	Do not enter the septic tank without adequate ventilation. Another person must be present above the ground as well as other requirements by law are fulfilled in the confined spaces. Sewer gases could be lethal.
Needs to maintain the appropriate service plan by a qualified assistance provider for the lifespan of the system. It is critical that aerobic systems are receiving routine maintenance.	Does not plant all over the waste disposal field except grass. Particularly don't cover tank or field with asphalt or concrete or other impermeable substances.

(continued on next page)

Do's and don'ts in secondary treatment	
Do's	**Don'ts**
Need to keep system available for check-ups and pumping; however, shielded from the unauthorized entrance, ensure your service provider has solution.	Do not place into a stand-alone pipe to carry wash waters to a side trench or woods. These "greywaters" additionally comprise disease-carrying organisms.
Need to call a service specialist every time that you're having problems with your system.	Foremost do not expect signs of malfunction. System check required regularly.
Possess detailed records on the subject of aerobic system, including locality map, model name, capacity, date of installation, service agreement, previous service schedules, and the maintenance performed.	Do not permit anybody to drive over or park on any part of the system.
Will have to save water in order to prevent overloading the system. Do not forget to fix any leaking.	Do not enable access to irrigation disposal zone, which includes the domestic animals.
Should redirect the ceiling and the surface water with an aerobic tank and irrigation disposal zone.	Do not permit unapproved repairs to aerobic system without acquiring approval from the concern personnel.
Should understand how is running, and as it looks, sounds, and smells.	Do not use: Bleachings, sanitizers, chemical cleaners, insecticides.

7.3.3 Do's and don'ts in tertiary treatment

Do's and don'ts in tertiary treatment	
Do's	Don'ts
You must continue to preserve the quality of water.	Don't allow any activity that impacts the quality of treated water.
The storage tank should be maintained properly.	Don't keep open tank for storage of treated water.
UV treatment may be given for the removal of microorganism.	

7.4 Wastewater reuse

7.4.1 Wastewater utilization status in different countries

Many countries in the world are using recycled water for different purposes. According to 2017 WHO and USEP census, the states reusing treated wastewater for drinking are Australia, Belgium, California, Jordan, Kuwait, Namibia, Singapore, South Africa, Texas, and the UK. Projects for reuse are under consideration in Brazil and India.

Delaware has used reclaimed water for irrigation since the 1970s without adverse effects. Florida reuses about 50% of its treated wastewater effectively.

Smart irrigation systems using sensors measure soil and adapt the amount of nutrients contained in the recycled water used. Agricultural reuse is widely practiced in Egypt, Yemen, and Jordan while Tunisia, Morocco, and United Arab Emirates use recycled wastewater for green space irrigation in urban areas.

The public utility board Singapore set up and tested membrane technology for reclamation of wastewater for public purposes in 1998. A full-scale plant to produce 10,000 cubic meters of water per day was commissioned in 2000. High-grade water well within the norms prescribed by the Who and USEPA and is named NEWater. Microfiltration, reverse osmosis, and ultraviolet disinfection were adopted to convert treated used water to portable water.

Several factors have contributed to the success of this project. Technology demonstration in the local environment to create public acceptance, endorsement by experts through tests and audits, and political will are the main factors.

The world's largest system for indirect portable reuse is in Orange County water district, California. Treated wastewater is purified using microfiltration, reverse osmosis, and ultraviolet light with hydrogen peroxide in this system. This groundwater replenishment system is operational since 2008 and is treating 100 MGD since 2015. Their Green Acres project produce 7.5 MGD of recycled water serving 100 different sites for providing landscape irrigation.

In Israel, over 85% of wastewater is purified and renewed for agricultural irrigation and other purposes. In Japan, reclaimed wastewater is used for nonportable purposes. The Governments' policy of "sewage vision 2100" aims to create round water cycles using reclaimed water effectively and efficiently. Australia used treated water for augmenting groundwater supplies as well as for agricultural irrigation. Belgium uses reclaimed water for augmenting the aquifers. Germany, Holland, Italy, Poland, and Hungary mostly use recycled water for irrigation. In metropolitan cities like Bangalore, Chennai, Delhi, and Mumbai, treated gray water is used for nonportable purposes like toilet flushing in major condominiums. Secondary treated wastewater is used as cooling water makeup in major industries like fertilizer plants and power plants in Chennai. Delhi and Mumbai international airport also use recycled water for cooling purposes. Kolkata treated wastewater is used in fish farms. Fish hatcheries in Vietnam and Bangladesh also use recycled wastewater.

Australia, Africa, Kuwait, Israel, Mexico, Egypt, and Abu Dhabi utilize treated wastewater for watering urban forests, public gardens, and shrubs. Indirect recharge of impoundments is done in Berlin in Germany, NEWater project in Singapore, Windhoek in Namibia, and Meguro river area in Japan. Coach cleaning and building construction activities use reclaimed water in Seoul.

7.4.2 Nutrients in fresh water and wastewater

Just as application of fertilizers to the farmlands helps growth of crops, nutrients entering water bodies like lakes and rivers feed growth of bacteria, algae, and other organisms. Excess nutrient dose results in excessive growth of algae.

Decomposition and die off of algal blooms can reduce dissolved oxygen in oxygen in water and kill fish and other aquatic life.

Nitrogen (N) and phosphorous (P) are the primary nutrients that cause pollution when present in excess amounts. Human activities like urbanization, residential development, sewage discharges, and agricultural accelerate the process of degradation of water quality. Wastewater contains metals that are also toxic to aquatic life and need to be removed.

Removal methods are energy-intensive and produce sludge which is toxic. Researchers are developing technologies using membrane systems to recover precious metals from water.

Microalgae-based processes for nutrient recovery are developed and attempts to reduce the hydraulic retention time in these systems are in progress. The growth of microalgae and production of biomass depend on the availability of solar radiation and photosynthetic efficiency. This technology will therefore be more suitable to tropical countries.

With eutrophication becoming a wider phenomenon all over the world, nutrient management and nutrient recovery assumed a great importance.

Agricultural land management practices like manure storage and changes in tillage reduce nutrient loading. Diversion of external nutritional loadings to lakes can restore eutrophic lakes. Lower algal biomass can be maintained by washing away with large quantities of nutrient-poor water.

7.4.3 Wastewater reuse in different sectors

7.4.3.1 Wastewater reuse for agricultural irrigation

Agriculture is the most common application of reclaimed water with a 30% share; since shortage of water for irrigation can impact food security, nutrition, and other socio-economic aspects. This practice reduces the need for synthetic chemical fertilizers when used optimally. However, there are some risks associated with this practice. The risks involved are dependent on the following factors:

a) Composition of wastewater used for reuse.
b) Treatment, shortage, and distribution methods adopted.
c) Type of irrigation practice and techniques of irrigation used.
d) Soil and groundwater conditions.
e) Climatic conditions.

Potential exposure pathways of cleaning agents, antibiotics, biocides, etc., have PBT properties on the environment.

These factors should be taken into consideration for effective reuse of wastewater for agricultural irrigation (Tripathi et al., 2019).

Advanced technologies can help to lower energy footprint. However public perception in water reuse and biosolid application to land is a daunting problem. It is difficult to convince the public regarding emerging contaminants like pharmaceuticals and antibiotic-resistant bacteria.

7.4.3.2 Groundwater recharge

Groundwater recharge is practiced mainly in countries where seawater intrusion into freshwater aquifers is a major threat. Appropriate hydrogeological conditions are needed to practice this. Groundwater subsidence control methods should be ensured.

7.4.3.3 Landscape irrigation

Landscape irrigation is widely used in urban areas and typically involves the spray irrigation of residential lawns, freeway medians, golf course, school grounds, and parks. Irrigation of areas not accessed by the public-like highway medians has limited potential for creating risks while other areas may pose problems of microbial pollution (Jeong et al., 2016). Hence there is a need to maintain proper disinfection in landscape irrigation.

7.4.3.4 Toilet flushing, fine protection, vehicle washing

Reuse of wastewater in areas where the public would come in contact with it should be taken to avoid cross-connection.

7.4.3.5 Industrial use

Wastewater can be reused in industrial applications such as cooling, boiler feed, stock scrubbing, process water with appropriate treatment to meet the standards prescribed for the purpose (Chaabane et al., 2017).

7.4.3.6 Recreational impoundment

In ornamental and recreational impoundments of water, human beings may experience full-body contact with water. The treatment of wastewater must be most stringent involving dual systems, nutrient removal to prevent algal growth, and other methods to protect the quality and the sensitivity of the species.

7.4.4 Risks and challenges in recycling wastewater

Increasing quantity of reclaimed water use poses several challenges and risks. There is a need for higher degree of competence in handling such projects (Singh et al., 2019). The costs involved are higher than using conventional water supplies. The operating systems must be highly reliable to prevent possible health risks (Puyol et al., 2017). Higher degree of treatment is required to take care of the emerging contaminants and their levels should be monitored regularly. Public perception, social acceptance, and political will are the other factors affecting successful recycling projects.

7.4.4.1 Research trends in wastewater recycling

Identifying the cost-effective treatment to reuse the water for a particular purpose popularly known as "fit-for-purpose water reuse" is a subject of many researchers now. Studies to use this recycled water as an alternative resource along with stormwater, rainwater, and surface water resources and integrate wastewater in an urban water cycle are also in progress. Management of biosolid, and sludge to generate energy is also being researched widely.

In industrial wastewater treatment systems, zero liquid discharge options biological treatment processes, and modeling techniques are widely used. Automation, instrumentation control, and use of sensors are getting due research attention. Researchers are on high-end treatment processes such as oxidation method and membrane filters to make them more affordable to the consumers. Alternate energy production methods from wastewater and sludge processing such as biogas are also being researched by scientists.

We should ensure the following procedure to take up a wastewater reuse project:

a) Social acceptance of the project.
b) Technological feasibility to make it implementable.
c) Environmental safety and protection of health and wellbeing of the people.
d) Reliability of operation and maintenance to produce reclaimed water that meets the environmental regulations prescribed for target reuse.
e) Cost effectiveness.

7.4.5 Permissible limits of pollutants

Water quality standards are required for safe and sustainable practice of reuse of water. While using treated wastewater for irrigation, possible accumulation of harmful elements for plant growth and potential damage to soil should be kept in mind.

7.4.5.1 Salinity

High salinity in soil can affect the crop in absorbing nutrients and can also specific ion toxicity. TDS and electrical conductivity are normally noticed below 500 mL/L. Similarly, chlorine residuals less than 1 mg/L poses no problem to the crops. Salinity increases soil compaction and heavy metal leaching.

7.4.5.2 Nutrients

Nitrogen and phosphorous are the main nutrients for the growth of crops. However, they can give a negative effect when applied in excess doses. They cause crops like paddy rice to overgrow which leads to lodging. Excessive nutrients can also cause eutrophication in coastal areas and lakes and contamination of groundwater. The secondary nutrients include calcium, magnesium, and sulfur. Micronutrients like boron, copper, iron, chloride, manganese, molybdenum, and zinc are elements essential to plant growth and are known as trace elements.

7.4.5.3 Organic matter

Biochemical oxygen demand (BOD) is normally used as an index for concentration of organic matter in water. When the BOD used is very high, dissolved oxygen in water is consumed for the decomposition of organic matter. In this anaerobic state, oxides in soil like SO42-, Fe3+, Mw5+ consume oxygen to lower the oxidation reduction potential (Read et al., 2001). Eventually, the produced sulfide, Fe, and Mn together with organic acids can disrupt the capacity of paddy rice to soak up nutrients. Organic matter affects the cation exchange capacity of the soil and soil structure stabilization.

7.4.5.4 Hydrogen ion concentration

The pH value represents the hydrogen ion concentration. Generally, the pH of irrigation water is in the range from 6.5 to 8.8. Low pH will be able to accelerate the corrosion process of the essential amenities while high pH can lower the efficiency of the trickle irrigation system. The pH outside the normal range can cause imbalance of nutrients. Several research studies have reported variations in soil pH due to wastewater reuse. It is mainly connected with (1) type of soil cover, (2) soil texture, and (3) period of irrigation.

7.4.5.5 Trace elements

For good crop growth trace elements are necessary. But they can be harmful when present in excess. Copper (Cu) may cause leaf chlorosis and suppression of root growth, aluminum (Al) in acid soil may reduce productivity, zinc (Zn) and arsenic (As) have side effects of stem chlorosis, lead (Pb) and cadmium (Cd) when dissolved in water or soil can get accumulated in the crop and can harm the consumers. The quality guidelines, therefore, have set allowable limits of micronutrients in irrigation water.

7.4.5.6 Water quality

Quality water standards applicable to agricultural reuse are more conservative in California than the standards prescribed by EPA and every state government establish their own water quality standard compared to WHO guidelines, the US standards are more stringent.

WHO recommended the guidelines on wastewater irrigation that takes into account the human health risk through epidemiological studies and microbial risk assessment (QMRA), a procedure for assessing risk of exposure to microbes.

Cultivate root crops that are typically eaten raw and uncooked, permissible limit of *E. coli.* (cfu/100 mL) (Arithmetic mean) of $\leq 10^3$ is prescribed while the standard is $\leq 10^5$ cfu/100 mL for fruit trees and olives.

The risk assessment can be done using (1) quantitative microbial risk assessment (QMRA) studies, (2) microbial laboratory tests, and (3) epidemiological studies. The combined study using all three methods would yield better results in evaluating the risk associated with agricultural reuse of treated wastewater.

TABLE 7.3 Threshold levels of trace elements for crop production.

Element	Recommended max concentration (mg/L)	Remarks
Aluminum (Al)	5.0	In alkaline soils it precipitates the ion and eliminate any toxicity
Arsenic (Ar)	0.10	
Cadmium (Cd)	0.01	Gets accumulated in plants
Cobalt (Co)	0.05	
Copper (Cu)	0.20	Toxic to several plants
Manganese (Mn)	0.20	
Selenium (Se)	0.02	Toxic to plants even in 0.025 mg/L concentration
Zinc (Zn)	2.0	Toxic to numerous plants at ptt.6.0, toxicity is low

Guidelines for reuse of wastewater for irrigation are as follows:

a) Salinity (electrical conductivity in Deci siemens per meter at 25°C) - 0.7 to 0.3 or TDS = 450 – 2000 mg/L.
b) Infiltration: For a given sodium absorption ratio (SAR value) (Table 7.3) increasing total salt concentration may raise soil permeability. Thus, soil permeability hazards triggered by sodium in irrigation water could not have been predicted independently.
c) Specific ion toxicity municipal wastewater effluents contain heavy metals and toxic elements. Threshold levels of trace elements for crop production are given in Table 7.4.

7.4.5.7. Industrial reuse

Reclaimed water is used in a wide variety of industries that range from electronics to food processing. The power generation industry reuses it extensively. Cooling systems use it to absorb process heat and transmit heat by evaporation. Dissolved solids and minerals which remain in water need to be removed to prevent their accumulation in the equipment. Water quality requirements for boiler make-up water include control of hardness, silica. alumina, calcium, and magnesium. For steam generation, TDS levels need to be less than 0.2 parts per million (ppm).

The use of reclaimed water in semiconductor industry is of recent origin. Intel uses ultrapure water to clean silicon wafers during fabrication. From there, it is used in cooling towers and scrubbers.

TABLE 7.4 SAR limits.

SAR	Normal range permissible
0–3 and EC	0.7–0.2 ds/m
3–6 and EC	1.2–0.3 ds/m
6–12 and EC	1.9–0.5 ds/m
12–20 and EC	9.9–1.3 ds/m
20–40 and EC	5.0–2.9 ds/m
Sodium (Na) SAR = 3.9	
Chlorine (Cl) = 4–10 me/L	
Boron (B) = 0.7–3.0 mL/L	

EC is the electrical conductivity in deci siemens per meter at 25°C.

7.4.5.8 Food manufacturing

Food and beverage industry needs large quantities of water for cleaning bottles, cans, containers, and sometimes for transporting through water flumes. Companies like Coca-Cola have developed process water recovery systems using MBR ultrafiltration, RO, ozonization, and UV disinfection.

7.4.6 Energy production through wastewater

Sustainable development is possible if production systems become autogenerated is converted into raw material. This is called cradle-to-cradle concept (circular economy). Wastewater treatment technologies now are developing plants where organics, energy, heavy metals, etc., are recovered.

Examples include use of metal-reducing and oxidizing organism for metal recovery and use of microalgae, photosynthetic bacteria, and terrestrial plants for recovery of organics.

The conversion of organic-rich wastewater into biofuels in anaerobic systems has been practiced for a long time. Biogas, biohydrogen, biodiesel, biopolymers, single cell protein (SCP) (yeast, edible microorganisms with high protein content) heavy metal absorption, microbial fuel cells, and nutrient recovery from wastewater streams are subjects of active research.

7.4.7 Precautions in reuse of reclaimed water

If wastewater cannot be treated adequately, some important steps to be taken to mitigate the risks involved in reuse of wastewater. Some of the steps include:

a) Crop restrictions and standards for use in irrigation.
b) Use of stabilization farms on-farm.

c) Appropriate practices that limit risks in irrigation, harvesting, public health, and human exposure.
d) Best practices in irrigation techniques and suitable standards, guidelines, and regulations that are suitable to place and are economically and administratively enforceable.

7.5 Future directions

In our attempts to recover materials and energy from wastewater streams, three groups of targets can be pursued in future:

a) Biofuels, plastics.
b) High-value chemicals like antibiotics, pesticides, or herbicides.
c) Metal complexes can serve as fertilizers or a source of valuable metals.

Risks of human health through exposure to microbes should be assessed thoroughly through epidemiological studies before adopting a reverse technology for wastewater. Revival of SCP production promises economic feasibility in the future.

The current production systems need to be changed to enable a more environmental technological culture. The circular economy concepts make the production system autoregenerative and the wastes generated in the biological cycles get transformed into the raw matter.

References

2012 Guidelines for Water Reuse, AR-1530, EPA/600/R-12/618 | September 2012. (2012).

Chaabane, S., Riahi, K., Hamrouni, H., & Thayer, B. B. (2017). Suitability assessment of grey water quality treated with an upflow-downflow siliceous sand/marble waste filtration system for agricultural and industrial purposes. *Environmental Science and Pollution Research, 24*(11), 9870–9885. https://doi.org/10.1007/s11356-016-7471-x.

Jeong, H., Kim, H., & Jang, T. (2016). Irrigation water quality standards for indirect wastewater reuse in agriculture: A contribution toward sustainable wastewater reuse in South Korea. *Water, 8*(4), 169–187. https://doi.org/10.3390/w8040169.

Jones, E. R., Van Vliet, M. T. H., Qadir, M., & Bierkens, M. F. P. (2021). Country-level and gridded estimates of wastewater production, collection, treatment and reuse. *Earth System Science Data, 13*(2), 237–254. https://doi.org/10.5194/essd-13-237-2021.

National Inventory of Sewage Treatment Plants. (2021). Central Pollution Control Board. https://cpcb.nic.in/openpdffile.php?id=UmVwb3J0RmlsZXMvMTIyOF8xNjE1MTk2MzIyX211ZGlhcGhvdG85NTY0LnBkZg.

Puyol, D., Batstone, D. J., Hülsen, T., Astals, S., Peces, M., & Krömer, J. O. (2017). Resource recovery from wastewater by biological technologies: Opportunities, challenges, and prospects. *Frontiers in Microbiology, 7*:2106, 1–23. doi:10.3389/fmicb.2016.02106.

Read, J. F., John, J., MacPherson, J., Schaubel, C., & Theriault, A. (2001). The kinetics and mechanism of the oxidation of inorganic oxysulfur compounds by potassium ferrate. *Inorganica Chimica Acta, 315*(1), 96–106. https://doi.org/10.1016/s0020-1693(01)00331-0.

Santos Pereira, L., Cordery, I., & Iacovides, I. (2009). *Coping with Water Scarcity: Addressing the Challenges* (pp. 1–382). Springer, Springer Dordrecht. https://doi.org/10.1007/978-1-4020-9579-5.

Singh, R. P., Kolok, A. S., & Bartelt-Hunt, S. L. (2019). *Water Conservation, Recycling and Reuse: Issues and Challenges* (pp. 1–276). Singapore: Springer. https://doi.org/10.1007/978-981-13-3179-4.

Tripathi, M. P., Bisen, Y., & Tiwari, P. (2019). Reuse of wastewater in agriculture. *Water Conservation, Recycling and Reuse: Issues and Challenges* (pp. 231–258). Singapore: Springer. https://doi.org/10.1007/978-981-13-3179-4_13.

Chapter 8

Impact of waste treatment through genetic modification and reuse of treated water on human health

Hemant Kumar[a] and Aradhana Sharma[b]

[a] *Department of Biotechnology, Govt. V.Y.T. P.G. Auto. College, Durg, Chhattisgarh,* [b] *Kalyan P.G. College, Bhilai Nagar, Chhattisgarh, India*

8.1 Introduction

Water plays an essential role in life (Karimi-Maleh et al., 2020a; Orooji, Ghanbari, Amiri, & Salavati-Niasari, 2020). Water succeeded as a social and economic asset in various parts of the globe (Hassandoost, Pouran, Khataee, Orooji, & Joo, 2019; Karimi-Maleh et al., 2020). Globally 71% of water is present, only 2.5% of freshwater is available (Bhat, 2014). The Earth's environment and life depend upon fresh water and its resources like lakes, groundwater, ponds, streams, and rivers (James, 2002; Xu et al., 2020a, 2020b). The two unusable resources of fresh water, glaciers, and groundwater get reduced due to over explosion of population and industrialization. Climate change, agricultural wastes, clinical wastes, and their use in the generation of energy are the other factors that cause the depletion and scarcity of fresh water (Murshed & Kaluarachchi, 2018; Teodosiu, Gilca, Barjoveanu, & Fiore, 2018). Freshwater scarcity became one of the crucial issues related to the environment. Worldwide, humans influenced or utilized water known as wastewater. Water contamination and shortage are one of the major issues around the world. Recovery of water from used water or wastewater or finding a substitute for water resources for life is necessary to overcome water scarcity-related problems.

Globally, 92% of the water is currently applicable for agricultural activities (Clemmens, Allen, & Burt, 2008; Hoekstra & Mekonnen, 2012; Tanji & Kielen, 2002). Freshwater consumption for irrigation purposes is near 70% of total agricultural water (World Resources Institute (WRI), 2020). According to statistics, water crises are a serious concern throughout the world. In total, 40%

Antimicrobial Resistance in Wastewater and Human Health.
DOI: https://doi.org/10.1016/B978-0-323-96124-0.00001-5

of the population globally reported water crises for irrigation processes because they are all situated in heavy water-stressed basins (Shen et al., 2014). Industrial wastewater contains contaminations such as pathogens, dyes, and heavy metals, which are highly toxic and shows a higher risk for life (Chaoua, Boussaa, El Gharmali, & Boumezzough, 2019; World Health Organization, 2006). The wastewater directly gets mixed with freshwater bodies and is contaminated, causing several problems (Chen, Lu, Pan, & Jiao, 2013). Heavy metals contain chemical elements like Cu, Mn, Zn, Cr, Fe, Ni, Zn, etc., which shows the toxic effect on flora and fauna (Mahfooz et al., 2020). The half-life of heavy material is too long as well as nonbiodegradable. Heavy metals directly enter animal and human bodies through the consumption of leafy vegetables because plant roots absorb these metals from water directly. They also enter through contaminated soil inhalation and drinking polluted water (Mahmood et al., 2014). The assessment of health risk is necessary of this kind of wastewater before using it for regular works (Mehmood et al., 2019; Njuguna et al., 2019; Xiao, Wang, Li, Wang, & Zhang, 2017). The adequate recycling of wastewater is necessary for the environment and public health. There is various kind of wastewater, and related treatment to wastewater depends upon the type and characteristics of wastewater.

Wastewater treatment plants highly used techniques in the past few decades to treat wastes. But unfortunately, these technique does not reduce 100% waste. Various kinds of chemical and microbial contamination should be get involved during the process. To overcome these problems, genetically engineered or modified organisms (plants and microorganisms) are one of the possible approaches to treat wastes. Genetically modified organisms (GMOs) reduce toxic matters, chemicals, and heavy metals and further utilize them for their biochemical pathways and change their physicochemical properties. Environmental pollutions at a high concentration are depleting through using GMOs were reported. Treated wastewater shows an adverse effect on the life of the earth. The amount of freshwater and used water may differ from country to country for their routine work in different fields or sectors (Table 8.1).

8.2 Generation of waste and impact on human health

Waste is an inevitable outcome of nearly all human activity (UN ESCAP, 2000) Solid wastes are the abandoned solid substances generated from numerous sources like residential, industrial, biomedical, etc., in a specified area. Population explosion, people moving from rural to urban areas for improved education and lifestyle intensifies the generation of solid waste and water (Table 8.2), which results in increased pollution (CPCB, 2004; Gangadhar, Ravi, & Ramakrishna Naidu, 2017; Sharma & Shah, 2005; Shekdar, Krishnaswamy, Tikekar, & Bhide, 1992).

With meteoric growth or expansion in agriculture, industry, healthcare facilities most of the Asian and Pacific region is absorbing a consequential amount of

TABLE 8.1 Utilization of freshwater and treated wastewater status in different countries.

Country	Sectors utilizing freshwater	Reusing of treated wastewater	References
USA	Agricultural irrigation—37% Freshwater thermoelectric plants—40% Aquaculture and livestock—3% Domestic—14% Industries—6%	Agricultural irrigation—37% Landscape irrigation—17% Groundwater recharge—12% Geothermal energy—2% Golf course irrigation—7% Recreational impoundment—4% Industrial and commercial—8% Wildlife habitat and wetlands—4% Seawater intrusion barrier—7% Other—2%	Kenny et al. (2009); SWRCB (2011)
India	Energy—2% Domestic—4% Agriculture—87% Industrial—7%	Industrial use—12% Thermal power plant—4% Groundwater recharge and artificial Lakes—6% Agricultural irrigation—78% Industrial use—12%	Jindal and Kamat (2011)
Europe	Energy generation and Industry—40% Agriculture—44% Public water supply—16%	Landscape irrigation—20% Groundwater Recharge—2.2% Recreational—6.8% Indirect potable uses—2.3% Agriculture irrigation—32% Nonpotable urban uses—8.3% Industrial—19.3% Environmental enhancement—8% Other—1.5%	EEA CSI (2018), GWI/PUB Water Reuse Inventory (2009)
South Africa	Domestic—27% Industrial—3% Agriculture—60% Power—4% Mining—3% Other—3%	Agriculture—43% Irrigation of sports field and landscape and -9% Industry—48%	Adewumi, Ilemobade, and Van Zyl (2010)

(*continued on next page*)

TABLE 8.1 Utilization of freshwater and treated wastewater status in different countries—cont'd

Country	Sectors utilizing freshwater	Reusing of treated wastewater	References
Japan	Agricultural—67% Industrial—14% Domestic—19%	Agricultural irrigation—5.8% Landscape irrigation—21.6% Industrial—1.2% Snow melting—20.2% Directly supply industrial use—9.9% Recreational use—2.1% River flow augmentation—35% Toilet flushing—4.1%	Takeuchi and Tanaka (2020), Tembata and Takeuchi (2018)
Greece	Animal husbandry—1.3% Public use or potable—13% Irrigation—83% Industry—2.2% Other—1.2%	Firefighting and forested land irrigation—17.7% Agricultural irrigation—58.38% Landscape irrigation—23.92%	Frontistis et al. (2011) Tsagarakis, Tsoumanis, Chartzoulakis, and Angelakis (2001)

TABLE 8.2 Generation of waste based on different aspects.

According to its origin	Domestic
	Commercial
	Industrial
	Medical
	E waste
According to its contents	Organic material
	Glass
	Metal
	Plastic
	Chemicals
According to hazard potential	Toxic
	Flammable
	Radioactive
	Infectious

toxic chemicals and thus give rise to illimitable hazardous waste (UN ESCAP, 2000).

In this day and age, the municipal solid waste discarding operation followed by many municipal corporations in different cities and towns is unscientific and unsystematic. Dumping sites are usually rural areas or low-lying areas. The majority of the sites where disposal of the wastes takes place are uncontrolled slag heaps. Those slag heaps are an amalgamation of domestic, industrial, commercial, hospital waste, etc., thrown away at a single dumping site which initiates more air pollution, soil water contamination and becomes a hotbed for various frightful disease-causing pathogens. Various types of solid waste generated from different sources (Table 8.3). To maintain good public health, it is essential to have a well-driven solid waste management system. Various surveys organized by UN-Habitat reveal that in such areas where the collection of waste is infrequent, the prevalence of diarrhea almost doubled and acute respiratory infections are six times larger than in the areas where a routine collection of waste is done (UN-Habitat, 2009; Hoornweg & Bhada-Tata,2012). Further incognizance of people, there is a common opinion which shows the standard service of collection and disposal of wastes are gradually declining (Gangadhar et al., 2017). In the present day, however, a major concern has emerged as for the potential spread of pathogens, additionally giving rise to environmental contamination owing to the inappropriate or unprofessional handling and control of clinical and biomedical remains or leavings (Ogawa, Ohtsubo, Tsuda, & Tsuji, 1993; Schomaker, 1997; WHO, 1996). The amount of solid waste generation and their processing value were reported in different states and UT-wise in India (Table 8.4).

It is fact-based that in numerous cities and towns nearly half of the waste rests unattended. Municipal solid waste management requires numerous processes (Fig. 8.1) (Gangadhar et al., 2017).

Solid waste management is associates with potential risks to human health for the most part with inappropriate operation and not following a systematic process of waste management. For the most part, the workers who are working at the ground level and who pick rags are most exposed and at high health risks and it is required to protect them from such exposure to hazardous and toxic wastes. The speed breeding of Infectious vectors like mosquitoes, flies, rats, etc., is of the main concern to the health of general people. Population living in such areas where they lack waste management system or have poor management, workers working for such waste management system, children, people living in close proximity to waste dumps, and animals all are unshielded and at high risk to the unfortunate impacts related to unmanaged wastes. Unrestrained raw hazardous wastes originating from innumerable industries are muddle up with municipal wastes hence creating higher possible risks to humankind. The interconnection between municipal waste and industrial effluents carrying heavy metals ejecting to the sewage system or unsecured dumps of municipal solid wastes somehow break into the food chain thereby retaining a vicious cycle

TABLE 8.3 Types of solid waste and generators.

Types of solid wastes	Source	Typical waste generators
Food wastes, paper, cardboard, plastics, textiles, leather, yard wastes, wood, glass, metals, ashes, special wastes (e.g., bulky items, consumer electronics, white goods, batteries, oil, tires), and household hazardous wastes (e.g., paints, aerosols, gas tanks, waste containing mercury, motor oil, cleaning agents), e-wastes (e.g., computers, phones, TVs)	Residential	Single and multifamily dwellings
Housekeeping wastes, packaging, food wastes, construction and demolition materials, hazardous wastes, ashes, special wastes	Industrial	Light and heavy manufacturing, fabrication, construction sites, power and chemical plants (excluding specific process wastes if the municipality does not oversee their collection)
Paper, cardboard, plastics, wood, food wastes, glass, metals, special wastes, hazardous wastes, e-wastes	Commercial	Stores, hotels, restaurants, markets, office buildings
Same as commercial	Institutional	Schools, hospitals (nonmedical waste), prisons, government buildings, airports
Wood, steel, concrete, dirt, bricks, tiles	Construction and demolition	New construction sites, road repair, renovation sites, demolition of buildings
Street sweepings; landscape and tree trimmings; general wastes from parks, beaches, and other recreational areas, sludge	Municipal services	Street cleaning, landscaping, parks, beaches, other recreational areas, water, and wastewater treatment plants
Industrial process wastes, scrap materials, off-specification products, slag, tailings	Process	Heavy and light manufacturing, refineries, chemical plants, power plants, mineral extraction and processing

(*continued on next page*)

TABLE 8.3 Types of solid waste and generators—cont'd

Types of solid wastes	Source	Typical waste generators
Infectious wastes (bandages, gloves, cultures, swabs, blood and body fluids), hazardous wastes (sharps, instruments, chemicals), radioactive waste from cancer therapies, pharmaceutical waste	Medical waste	Hospitals, nursing homes, clinics
Spoiled food wastes, agricultural wastes (e.g., rice husks, cotton stalks, coconut shells, coffee waste), hazardous wastes (e.g., pesticides)	Agricultural waste	Crops, orchards, vineyards, dairies, feedlots, farms

Adapted from Hoornweg and Thomas (1999).

(Gangadhar et al., 2017). About 60% of all communicable diseases in human beings are zoonotic, as are 75% of all appearing communicable diseases. Generally, every four months, one new communicable disease comes to light in humans (Frontiers, 2016).

Normally it is assumed that polymers in plastic are inert or inactive and of slight concern to human health, although various kind of supplements and remnant monomers perhaps reserved from these plastic polymers is accountable for the health risks (Table 8.5) (Araujo, Sayer, Giudici, & Poço, 2004). Largely additives that are present in plastic are possible endocrine disrupters and carcinogens. Skin contact, ingestion, and inhalation are the major passage of exposure of these additives to humankind (Fig 8.2). Because of the plastic containing harmful additives, reports of skin irritation and dermatitis came to light (Brydson, 1999). Microplastics remain the utmost contaminants that can bioaccumulate in the food chain following ingestion by a diversity of marine and freshwater lives causing high risk to human health (Galloway, 2015). Consumption of such water bodies and animals who are subjected to plastic additives and microplastics can be dangerous (Fig 8.2). Biomonitoring of human tissues unveils that elements of plastics remain on the human population throughout the evaluation of environmental contaminants (Galloway, 2015).

8.3 Waste treatment

There are two approaches widely used for the treatment of waste according to their physical properties—solid waste treatment and water waste treatment.

TABLE 8.4 State/UT-wise status of solid waste generated and processed up to November 2018.

Sl. no.	State/UT	Total waste generation (MTPA)	Total waste processing (%)
1	Maharashtra	8,238,050	44%
2	Uttar Pradesh	6,132,000	57%
3	Tamil Nadu	5,601,655	55%
4	NCT of Delhi	3,832,500	55%
5	Gujarat	3,702,925	57%
6	Karnataka	3,650,000	32%
7	West Bengal	2,810,500	5%
8	Telangana	2,690,415	73%
9	Rajasthan	2,372,500	55%
10	Madhya Pradesh	2,344,760	65%
11	Andhra Pradesh	2,330,160	29%
12	Haryana	1,647,610	17%
13	Punjab	1,496,500	33%
14	Odisha	992,800	12%
15	Jharkhand	849,335	42%
16	Bihar	828,915	43%
17	Chhattisgarh	601,885	84%
18	Uttarakhand	513,190	38%
19	Jammu & Kashmir	501,510	8%
20	Assam	413,910	35%
21	Kerala	227,760	60%
22	Chandigarh UT	172,280	85%
23	Tripura	153,300	45%
24	Puducherry UT	127,750	10%
25	Himachal Pradesh	124,830	40%
26	Nagaland	124,830	52%
27	Meghalaya	97,820	58%
28	Goa	94,900	65%
29	Mizoram	73,365	4%
30	Arunachal Pradesh	66,065	20%
31	Manipur	64,240	50%
32	Andaman & Nicobar Islands	36,500	52%

(*continued on next page*)

TABLE 8.4 State/UT-wise status of solid waste generated and processed up to November 2018—cont'd

Sl. no.	State/UT	Total waste generation (MTPA)	Total waste processing (%)
33	Sikkim	32,485	66%
34	Dadra & Nagar Haveli	12,775	0%
35	Daman & Diu	11,680	65%
	Total/ Average	52,971,700	43%

Adapted from Babul Supriyo (2019).

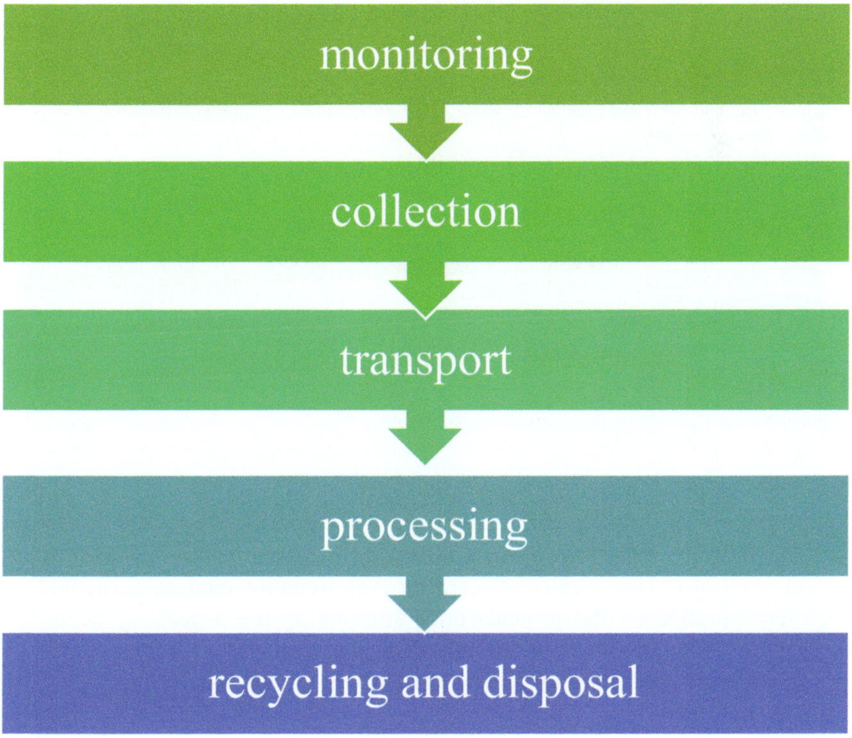

FIGURE 8.1 Numerous processes involved in Municipal solid waste management.

8.3.1 Solid waste treatment

Solid waste treatment involves a distinct range of activities involving deduction, recycling, segregation (separation), conversion, treatment, and discarding at varying degrees of levels of civilization. The industry uncovers its lineages in waste removal utilizing reasonable procedures such as restricted terrestrial

TABLE 8.5 Different additives used in plastic production, their effects, and the plastic types.

Toxic additives	Plastic types	Uses	Public health effect(S)
Bisphenol A	Polyvinyl chloride (PVC), Polycarbonate (PC)	Plasticizers, can liner	Mimics oestrogen, Ovarian disorder
Phthalates	Polystyrene (PS), Polyvinyl chloride (PVC).	Plasticizers, artificial fragrances	Interference with testosterone, sperm motility
Persistent organic pollutants (POPs)	All plastics	Pesticides, flame retardants, etc.	Possible neurological and reproductive damage
Dioxins	All plastics	Formed during low-temperature combustion of PVC	Carcinogen, interferes with Testosterone
Polycyclic aromatic hydrocarbon (PAHs)	All plastics	Use in making pesticides	Developmental and reproductive toxicity
Polychlorinated biphenyls (PCBs)	All plastics	Dielectrics in electrical equipment	Interferes with thyroid hormone
Styrene monomer	Polystyrene	Breakdown product	Carcinogen can form DNA adducts
Nonylphenol	PVC	Antistatic, antifog, surfactant (in detergents)	Mimics oestrogen

Adapted from Alabi, Ologbonjaye, Awosolu, and Alalade (2019).

dumping (landfill), tossing out into both fresh and oceanic water, and disorderly burning, none of which propose health-safe plus hazard-free waste administration solutions. Historically, discarding was demonstrated as a method of treatment practice on the grounds after eliminating the aspects of deposited waste frequently modified as an effect of degradation, a phenomenon that largely increases the polluting chance of various wastes. The primary purpose of waste treatment is to stabilize rather by accelerated degradation so that the last remnants generated are non-noxious and inadequate of distinct modification, that is, they are entirely mineralized, or able to discover inclined passage into the numerous natural biogeochemical (elemental) cycles that govern elements cycling in the habitat, without resulting in contortion in any way term relates to another (Hamer, 2003).

A few of the chief sets of solid wastes are:

- Community solid waste (residential, commerce, and business waste);
- Development industry and destruction waste;

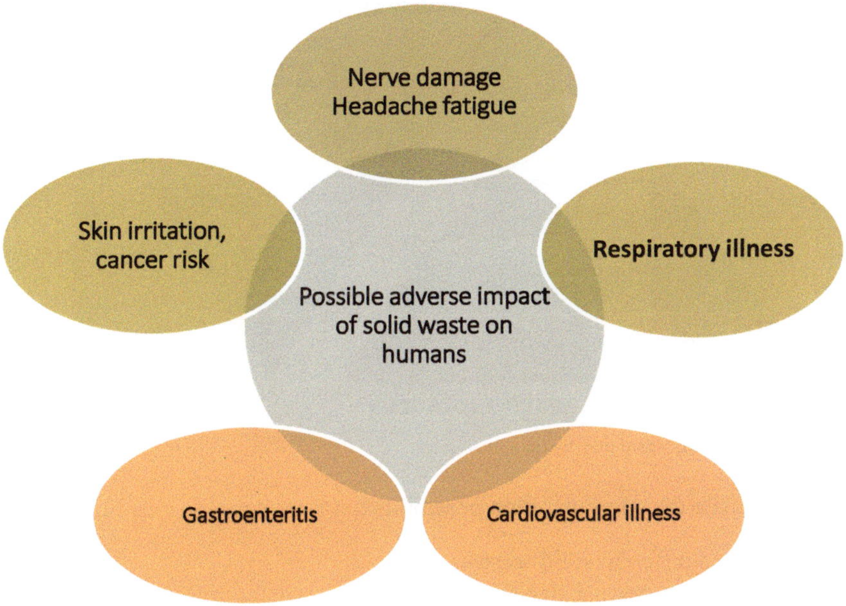

FIGURE 8.2 Impact of solid waste on human health.

- Combustible product and energy-generation waste;
- Food, beverages, and agricultural waste;
- Catering business waste;
- Forestry and forest produced waste;
- Convenience sector and garden waste;
- Slurries from intense animal husbandry (animal manures);
- Slaughterhouse solid waste and diseased carcasses;
- Waste sewerage sludge (treated or untreated);
- Biomedical waste (Hamer, 2003).

Integrated solid waste management (ISWM) indicates the requirement to address solid waste comprehensively with a particular assortment and maintained application of relevant technology, operating conditions, and establishing a "social license" within the population and assigned waste management authorizations (generally local government). ISWM is based upon both an extraordinary measure of professionalism in place of solid waste supervisors; and on the recognition of the significant role that the population, representatives, and regional (and more global) ecosystems have inefficient SWM. ISWM should be guided by definite purposes and should be established on the hierarchy of waste management: reduce, reuse, recycle—oftentimes appending a fourth "R" for recovery. Those deviation selections of wastes are then accompanied by landfill and incineration, or different disposal alternatives (Hoornweg & Bhada-Tata, 2012). Solid waste management involves various steps (Fig. 8.3).

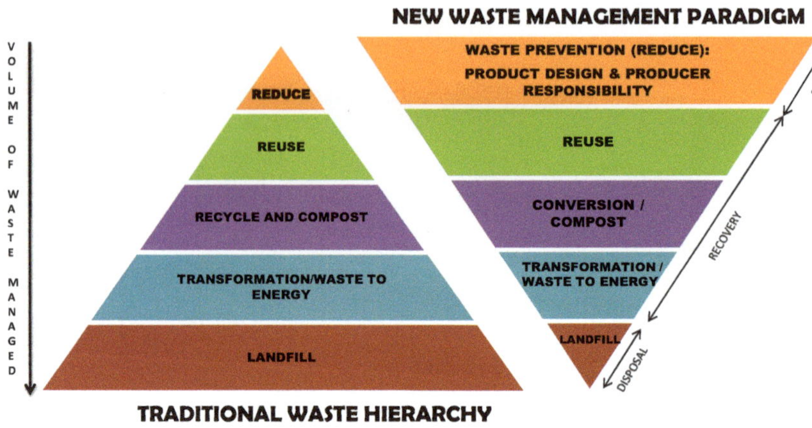

FIGURE 8.3 Solid waste management hierarchy.

To estimate the administration possibilities it is essential to examine:

(i) Origin of different solid wastes.
(ii) Reduction in raw substances utilization.
(iii) Reduction in the amount of solid wastes.
(iv) Reuse of waste matters.
(v) Restoring.

(i) **Origin of solid wastes:** The generation of wastes takes place from mining, municipal solid waste, agricultural waste, biomedical waste, industrial waste, construction, and demolition waste, E-waste.

(ii) **Reducing the number of raw materials: Reduction in waste** can be accomplished by more conventional methods as condemning the old-fashioned machinery whose performance is low developing the output by choosing more suitable technology.

(iii) **Reduction in quantities of solid wastes: Reduction in quantities of solid waste** can be achieved by reusing or recycling some of the wastes produced so that the amount of waste produced is considerably decreased per tone of the product originated.

(iv) **Segregation and recovery:** Separate the solid waste to collect relevant materials like metals. Even secondary wastes as "Bagasse" in a sugar millhouse can be made use of as a combustible to produce steam in industrial process or for making pulp in papermaking. Coconut fibers can be utilized in the coir industry.

(v) **Salvaging:** Organic wastes obtained from different sources can be used for the extraction of good organic manure. Food content waste adequately can be utilized as animal feed.

Engineered systems for solid waste management.

The exercises associated with the administration of solid wastes from the time of generation to ultimate disposal are:

(i) Waste accumulation into dust bins.
(ii) On-site handling, storage.
(iii) Transportation.
(iv) Processing.
(v) Disposal.

(i) Waste accumulation into dust bins/containerization.

Collection of solid trash from the time of production (residential, commercial, industrial, institutional) till the end of treatment or disposal is waste collection. The solid waste matter should be collected into trash bins/recycling bins given by the Municipal authorities. These trash bins may be entirely covered or partially open so that, they do not attract insects and stray dogs. It should be kept clean between the definite intervals, that is, the capacity of the trash bin must be larger than the waste produced. Mode of the solid waste collection can be house to house, self-delivered, community bins, or by contracted service. Accumulated solid waste can be distinguished or mixed, influenced by local regulations. Waste generators should separate their waste at source, for example, into "wet" (nonrecyclable) and "dry" (recyclables) (Hoornweg & Bhada-Tata, 2012; Srinivas, 2008).

(ii) On-site handling and storage.

On-site handling involves differentiating, compressing, and incinerating the contents of the trash bin to lessen the mass of refused materials to be carried for ultimate disposal (Srinivas, 2008). Storage of solid wastes at the origin to a great extent is required in most of the areas. Sometimes both decomposable and nondecomposable waste is disposed of at a common bin at a public disposal point. The movable storage bins lack durability but are flexible for transportation, while the fixed trash bins cannot be transported once constructed but are durable (Malviya, Chaudhary, & Buddhi, 2002; Nema, 2004; Sharholy, Ahmad, Mahmood, & Trivedi, 2008).

(iii) Transportation.

The large municipal trash bins or recyclable bins are unloaded into bigger carrier vehicles and these vehicles draw the solid waste to the area of disposal. Transport is the most expensive of all the processes and should be thoughtfully engineered. The carrier vehicle should be closed type to restrict the outspread of the foul smell of waste (Hoornweg & Bhada-Tata, 2012; Srinivas, 2008).

(iv) Processing.

Processing is the separation and classification of solid wastes to:

(a) Recover substances that can be used as pieces of paper, glass, wood, metal scrap, plastics, etc.
(b) Segregate waste that can be easily burned.

(c) Separate recyclables from nonrecyclables that cannot be disintegrated even after a significant period.
(d) Salvage organic material that can be utilized in the formation of hog feed and poultry feed.

Processing further includes mechanical compressing that can reduce the mass considerably (Srinivas, 2008).

(v) Disposal.

8.3.1.1 Land disposal

Land disposal is nonengineered and uncontrolled dumping of solid waste on the earth's surface. Areas far from social habitation which are low lying are usually utilized for disposal. Inert wastes obtained from demolition and construction sites are ideally suited for landfilling. It is the most affordable method used by municipalities that are resorting to this. But lightweight solid matters scatter across a wide area, creating hideous conditions which produce undesirable odor out of decaying waste and became a hotbed for various pathogens. Hence this method is undesirable (Hoornweg & Bhada-Tata, 2012; Srinivas, 2008).

8.3.1.2 Landfills

Landfills are the general dumping sites where residue or waste left after other processes is dumped and which needs to be engineered and wisely utilized to save the environment and human health. Conventional land filling is oftentimes lacking, particularly in developing countries. Landfill gases originated from sanitary land filling of organic waste, and methane (about 50% of landfill gas) can be recovered to lessen greenhouse gas emissions. In the aerobic method, the air is made to flow through several layers of the landfill by giving interconnected wells within landfill layers. No such aeration is given in the anaerobic method and methane produced is delicately accumulated else it will lead to catching fire or giving rise to greenhouse gases accountable for global warming (Hoornweg & Bhada-Tata, 2012; Srinivas, 2008).

8.3.1.3 Composting

Municipal solid waste consists of more than 70% organic materials. Agricultural waste, waste from paper Industries, and food processing wastes comprise almost 100% organic matter, which after decomposition can yield a large source of manure (Srinivas, 2008).

8.3.1.4 Incineration

Incineration is the combustion of solid waste under controlled conditions in an appliance specifically designed for the purpose. By incineration, the amount of the waste is decreased by around 90% and weight by 75 to 80%. Open burning increases the risk of air pollution and need not be **recommended because**

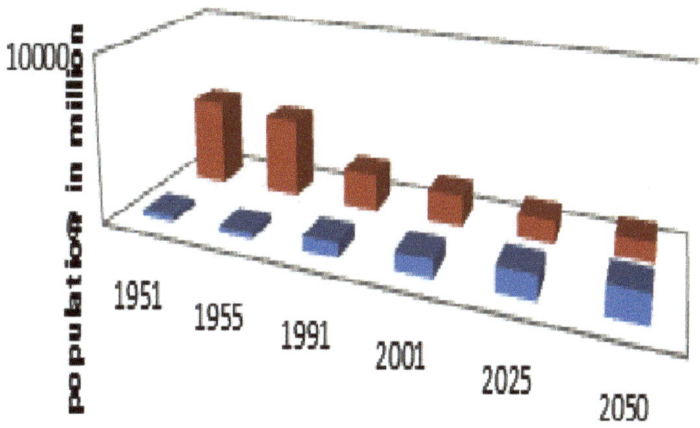

	1951	1955	1991	2001	2025	2050
Population in millions	361	395	846	1027	1394	1640
Per capita water availability(m3/year)	5177	4732	2209	1820	1341	1140

FIGURE 8.4 Per capita water availability in India (Adapted from Mehta, 2015).

of possible environmental effects whereas incineration does not involve open burning which reduces the chance of environmental effects (Morrison & Munro, 1997).

Retrieving energy value immersed in waste before ultimate disposal is acknowledged superior to straight land filling considering pollution administration requirements and expenses are acceptably addressed. Generally, incineration lacking energy recovery is not preferred due to cost-effectiveness and pollution (UN ESCAP, 2000; Hoornweg & Bhada-Tata, 2012).

8.3.2 Water waste treatment

"JAL HI JEEVAN HAI" we all have gone through using this expression but sadly very rare have realized and executed it. The outcome of its incomprehension is now crystal clear. Now we are standing an acute scarcity of freshwater and unexpectedly even drinking water in most of the countries. The chief causes are the rapid population growth (Fig. 8.4), urbanization, and current lifestyle with extreme extraction of groundwater, etc. (Chalkhure et al., 2020).

TABLE 8.6 Where water is found and the percentage.

Oceans	97.20%
Ice caps/glaciers	2.00%
Groundwater[a]	0.62%
Freshwater lakes	0.01%
Inland seas/salt lakes	0.01%
Atmosphere	0.00%
Rivers	0.00%
Total	99.84%

[a]*Adapted From-Bureau of Reclamation California-Great Basin.*

India's primary water use includes irrigation, environmental water, and industrial uses. Nevertheless, the unavailability of well-grounded data obstructs the evaluation of domiciliary wastewater in India. The improvement in the living standards with improving lifestyle also increases domestic water use (Joseph, Ryu, Malano, George, & Sudheer, 2021; Kumar, Singh, & Sharma, 2005).

Around 71% of the earth's surface is submerged in water, 97% of which is oceans (too salty for consumption, irrigation, and other uses in industries except for cooling). About 3% of the water available on earth is fresh, out of which 2.5% is unavailable: polar ice caps, locked in glaciers, atmosphere, highly polluted in soil, or lies too underneath the earth's surface which cannot be extracted at a budget-friendly cost. Only 0.5% of the earth's fresh water is available for direct use (except direct consumption) (Table 8.6).

Bureau of Reclamation California-Great Basin, several kinds of activities like natural, and man-made (domestic, agricultural, industrial, etc.) give rise to water pollution. Water pollution is categorized by several noticeable changes in the normal qualities and uses of freshwater which include foul odors, reduction in aquatic life, inadequate taste, and uncontrolled growth of aquatic weeds.

WATER POLLUTION

Groundwater pollution

Groundwater is commonly regarded as safe and can be used for drinking, irrigation, and industrial use. Groundwater is less likely to get contaminated. Yet, in later years, it is accepted that water contamination with arsenic, fluoride, and nitrate in the groundwater creates a serious warning to social health.

Surface water Pollution

The rivers, reservoirs, and lakes are extremely sensitive to contamination because of natural and human-made activities. The routes from which pollutants can enter water sources include sewage outlets, industrial outfalls, and nuclear stations exits. Surface water contamination is a principal warning for the continuance of life.

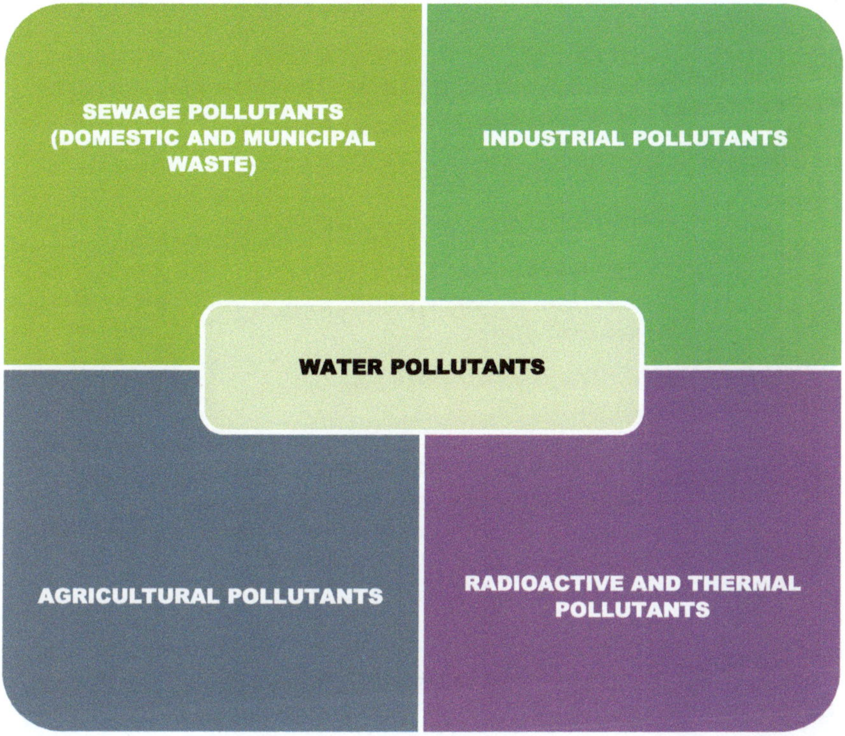

FIGURE 8.5 Sources of water pollutants.

Hence it is essential that routine monitoring of different sources of water pollution should be prepared and protective measures must be adopted to reduce the pollution (Fig. 8.5).

Water contaminants are mainly categorized as organic contaminants, inorganic contaminants, pathogens, suspended solids, agriculture pollutants, radioactive, macroscopic pollutants, and other pollutants. The waste from sewage and industrial effluents discharged into water bodies are the main contributors of organic and inorganic contaminants (Wasewar, Singh, & Kansal, 2020; (151)).

8.3.3 Classification of wastewater treatment processes

The wastewater treatment methods may be classified into four methods (Fig. 8.6).

8.3.3.1 Preliminary/physical treatment

Preliminary/physical treatment involves the elimination of floating substances (papers, wood, plastic, polythene, leaves, rags, animals, etc.), settleable solids

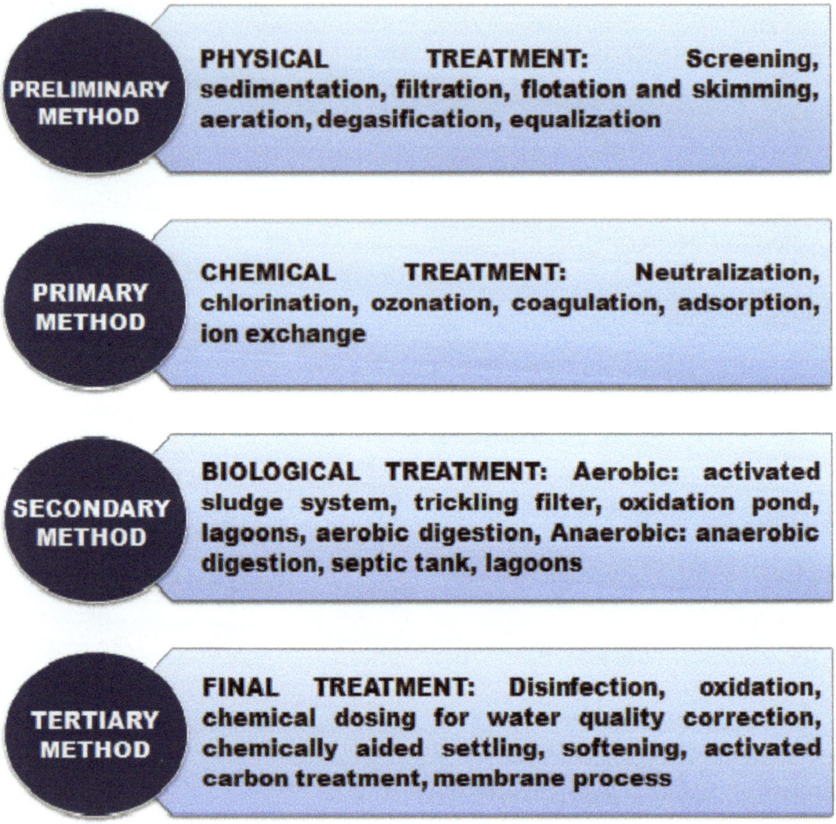

FIGURE 8.6 Wastewater treatment processes.

(grit, sand, etc.), and oils, grease, and fats. The elimination of suspended solid materials by primary methods before ensuing secondary treatment will significantly decrease the biological oxygen demand of the effluent. Sometimes less polluted wastewater is discharged without further treatment. The waste volume can be overcome by mechanical dewatering by a filter press or belt press then, the squeezed waste is incinerated or thrown away in a landfill site (Satyanarayana, 2010; Stanbury et al., 2003; Poerio, Piacentini, & Mazzei, 2019).

Physical processes established for primary wastewater treatment involve the following steps (Table 8.7):

1. Screens for removing large floating and suspended matter.
2. Comminutors to cut short the particle size.
3. Grit removal channels to limit destruction to the plant in succeeding processes.
4. Sedimentation tanks by which fine/lightweight suspended particles can be removed (Stanbury et al., 2003).

8.3.3.2 Primary/chemical treatment

In this method of wastewater treatment, chemicals are used, based on chemical reactions treatment like chlorination, neutralization, precipitation, disinfection, coagulation, etc., (Table 8.8) takes place. Nowadays the biological system is considered as secondary treatment and chemical treatment as the final step to reduce the use of chemicals.

8.3.3.3 Secondary/biological treatment

Biological treatments (Table 8.9) are used to fix the finely distributed and dissolved organic matter in wastewater into flocculent organic and inorganic solids. In the biological method, microorganisms, usually bacteria, modify the dissolved and colloidal carbonaceous organic matter into a combination of gases that can be eliminated in sedimentation tanks (Raouf, Maysour, & Farag, 2019).

8.3.3.4 Tertiary/final treatment

Tertiary treatment works in addition to primary and secondary treatment. It may include filtration, coagulation, precipitation, and flocculation, ion exchange, membrane processes, adsorption, nitrification, or denitrification, and other processes (Table 8.10). These methods may be combined with the secondary methods or added onto the secondary effluent. Advance treatment technologies are frequently mentioned as Best Available Technology Economically Achievable and are utilized to match stream measures or to meet with total maximum daily loads requirements. Advance treatment is applied to eliminate toxins, nonconventional pollutants, persistent organics, nutrients, etc. (Englande, Krenkel, & Shamas, 2015).

8.4 Genetically modified organisms

GMO, totally developed in laboratories through using genetic engineering or recombinant DNA technology. In GMOs, genetic material gets altered of an organism. GMO plants, animals, and microorganisms, used in various sectors for various purposes. Currently, in the treatment of wastewater, genetically modified plants and microorganisms are used.

8.4.1 Genetically modified plants

The management of waste through plants is known as phytoremediation. It is a diverse collection of technologies based on plants – natural or genetically modified, to clean environmental contaminants (Flathman & Lanza, 1998). It is cost-effective, simple, clean, and nonhazardous toward the environment (Wei et al., 2004). The by-products generated by phytoremediation (by plants) are used for other purposes (Truong, 1999).

TABLE 8.7 Different types of methods involved in preliminary wastewater treatment.

Pretreatment/ primary treatment	Screening	Screening is done for the removal of large floating or solid settleable matter which should be eradicated periodically to limit flow hindrance. The treatment of extracted materials is done accordingly to their nature (buried or incinerated)	Englande et al. (2015)
	Sedimentation	In the sedimentation tank, the wastewater is introduced in a large tank and allowed to settle by gravity. The supernatant flows over channels and moves to secondary treatment Primary treatment further acts as a blockage for grease and oil to limit operational obstacles during succeeding operations	Kyzas and Matis (2018)
	Comminution	Comminutor acts like a huge "garbage grinder" which can be used to blend the solids as a substitute for screening. To protect the machines, the comminutor usually follows grit extraction	Englande et al. (2015)
	Grit removal	Grit removal is the removal of small coarse particles like gravel, sand, or other minute material. Grit is eliminated to limit the damage of equipment and to manage tank capacity. Grit may be eliminated in a chamber wherever the outlay of air is quite adequate to maintain the lighter matter in suspension and provide the heavier matter to settle. Grit may further be eliminated by examining the velocity of water flow into a chamber so that gritty elements will sort out and organics rest suspended. The grit eliminated is usually washed and disposed of	Englande et al. (2015)

(*continued on next page*)

TABLE 8.7 Different types of methods involved in preliminary wastewater treatment—cont'd

	Flotation and skimming	Flotation is the separation method, which is based on the introduction of air bubbles as a carrier medium. Suspended matter, signifying hydrophobic or conditioned to be so, is then attached to the air bubbles and progress toward the water surface, that is, contradictory to the path of gravity and as the air pressure is applied in the opposite direction it pushes the contaminants to coagulate and compress the greasy and oily materials, these materials are then pushed to the stilling chamber from where the sludge blanket or floating contaminants can be removed mechanically or manually (skimming). Usually, the flotation and skimming methods work parallelly	Englande et al. (2015), Kyzas and Matis (2018)

Waste generated from various sectors (industrial, medical, agricultural, domestic, etc.) includes mixed pollutions (herbicides, pesticides, heavy metals, etc.). Phytoremediation is a method to overcome pollutants from wastes. More than 500 species of plants were reported as accumulators of heavy metals (Sarma, 2011) Some of the plant species which possess an accumulation of heavy metals are presented below (Table 8.11).

Genetically modified plants are beneficial for the environment they may either; reduce the chemicals used in the fields or agricultural lands and more efficiently act as contaminated area remediation. Transgenic plants involve two separate units, removal of pollution and prevention of pollution for the protection of the environment (Macek et al., 2008).

Some genetically modified plants treat contaminants of explosive, mercury, phenolics, selenium, chlorinated solvents, etc. (Mackova, Dowling, & Macek, 2006; McCutcheon & Schnoor, 2003; Meagher, 2000; Macek, Pavlikova, & Mackova, 2004; Macek et al., 2005). These modified plants contain transgenes, which accelerate inorganic compounds accumulation or great metabolized organic compounds (Table 8.13) (Macek et al., 2002).

TABLE 8.8 Different types of methods involved in primary wastewater treatment.

Primary/ chemical treatment	Chlorination	In the chlorination process, chlorine is added to drinking water to destroy bacteria, parasites, and viruses. Various methods can be applied to deliver safe levels of chlorine in drinking water. Levels of Chlorine up to 4 mg/L or 4 ppm are considered safe in drinking. At this level, harmful health effects are unlikely to occur	https://www.cdc.gov/healthywater/drinking/public/water_disinfection.html. (1) https://www.epa.gov/dwreginfo/drinking-water-regulations (2)
	Ozonation	The ozonation method has been broadly implemented for wastewater treatment, like disinfection, for the breakdown of harmful organic pollutants. Nevertheless, the implementation efficacy of ozone (O_3) is quite low and mineralization of organic pollutants with the ozone oxidation is inefficient, and few harmful disinfection byproducts (DBPs) possibly be developed through the ozonation process. Ozone in marine aquaculture has been used limitedly as it can potentially form bromate when the oxidation of naturally occurring bromide by ozone takes place. Since bromate is a known human carcinogen, it is also a topic of concern by its chronic impact on aquatic health. The catalytic ozonation process may be used to overcome the problems to a remarkable extent, which has gained growing recognition in modern times. When catalytic ozonation takes place, catalysts promote O_3 decay and produce active free radicals, which may improve the degeneration and mineralization of organic pollutants	Wang and Chen (2020), Tango and Gagnon (2003)

(*continued on next page*)

TABLE 8.8 Different types of methods involved in primary wastewater treatment—cont'd

	Neutralization	Neutralization becomes necessary usually for industrial wastewater treatment, to neutralize wide fluctuations in pH and concentrations of contaminants in the influent to the treatment plant. The equalization process assists in maintaining the uniform flow of influent perhaps the most significant system procedure in the treatment chain for industrial wastewater. Neutralization normally serves equalization so the acidic and alkaline waters can be partially neutralized. The main function of neutralization is to achieve pH of 6.5–8.5 which is generally needed before biological treatment	Englande et al. (2015)
	Coagulation	Coagulation is the incorporation of chemicals that rapidly gets mixed with water and usually involves the use of polymers or the oxides of calcium, aluminum, or iron. Once coagulation is achieved, then the flocculation process allows neutralization, aggregation, and adsorption of the floc particles which is then followed by sedimentation	Englande et al. (2015)
	Adsorption	Adsorption is one of the most prominent technologies used for wastewater treatment because it is an economical, effective, and eco-friendly treatment. For physical adsorption, the forces between adsorbed particles and the solid surface are Van der-Waals forces of attraction and Desorption occurs when they act weak. It is sufficient to understand water reuse responsibility and huge runoff figures in some industries. Adsorption is a mass transfer method with which the metal ions are transferred from solution to sorbent surface and become bounded by physical and chemical interactions. The functional groups play an essential function in deciding the selectivity, capacity, effectiveness, and reusability of these adsorbents	Raouf et al. (2019)

TABLE 8.9 **Different types of methods involved in secondary wastewater treatment.**

Secondary/ biological treatment	Aerobic process: activated sludge system	Activated sludge treatment delivers the operation of adsorbing, assimilating, and flocculating waste materials from wastewater. In this process, the fine suspended/colloidal soluble organic materials are induced into close association with a biologically activated sludge sustained in suspension in the chamber through injecting air which not only helps to maintain turbulence and highest contact but also supplies oxygen required for the metabolism of microorganisms	Englande et al. (2015)
	Trickling filtration	A trickling filter is like an activated sludge process in principle. Though, rather than water running within the suspended sludge carrying the microorganisms, the waste substance runs over a suitable platform to which the microorganisms hold on. Meanwhile, wastewater passes over the surface of the fixed film, growth of bacteria and different microorganisms take place and then these microorganisms form a slimy, gelatinous layer/film which carries the material kept in suspension, both colloidal as well in solution, to the microorganisms, that remove food substrate required for the growth and transport back into the liquid for decomposition to the end products, including carbon dioxide, nitrates, and sulfates. The microorganisms acting in the trickling filter process require a continual food supply, suitable support, sufficient oxygen, and proper nutrients	Englande et al. (2015)

(*continued on next page*)

TABLE 8.9 Different types of methods involved in secondary wastewater treatment—cont'd

	Anaerobic process: waste stabilization/ anaerobic ponds	Waste stabilization ponds are artificial, shallow sinks that comprise single or various sets of anaerobic, facultative ponds utilized for wastewater treatment. The wastewater treatment processes as components are separated by sedimentation or transformed by biological and chemical methods. Meanwhile, for facultative ponds, organic material is divided further into carbon dioxide, nitrogen, and phosphorous utilizing oxygen created through algae in the pond. Waste Stabilization Ponds are an important process that has better outcomes. The most advanced examination achieved that techniques like Waste Stabilization Ponds improve the property of wastewater with the least cost, simplistic, and more beneficial for pathogen removal. An extra advantage of these types of techniques is that they "don't need to pay for regular maintenance" as the waste stabilization ponds don't require to be aerated hence the treatment is sometimes called natural wastewater treatment. Anaerobic ponds are used for the heavily loaded treatment that no aerobic zone exists. Usually, they are used for the pretreatment of strong agricultural or industrial wastes	Fahad, Mohamed, Radhi, and Al-Sahari (2019), Englande et al. (2015)
	Septic tank	Biological treatment of wastewater would not be concluded without discussing septic tanks, so far a large section of the population is yet assisted by septic tanks. In this process, the solids accumulate, and if not periodically pumped out will release objectionable material. Usually, discharge from septic tanks flows into an intended drain area and eventually into the surface water or groundwater. A practically designed sewerage system within suitable soil is necessary as the septic tank itself behaves alone as a settler and the discharge persists with high soluble organic and microbial contaminants. Decent hydro-geologic requirements are accordingly significant in the effective sewerage field treatment	Englande et al. (2015)

TABLE 8.10 Different types of methods involved in tertiary wastewater treatment.

Tertiary/ final treatment	Ion exchange	Ion-exchange requires the dislocation shift of one ion by different. The replacement happens within the ions of insoluble/ion exchange material and the different ions in solution (i.e., wastewater for high-level of treatment). It is been practiced for several years for the elimination of hardness of water and specific ions like certain metals. The higher valancy, higher molecular weight elements substitute the minor ones The method exercises benefit of the strength of several typical and artificial substances to exchange an ion held in the body for another held in the water crossing over it. For example, if liquid carrying calcium ions is transferred within an ion-exchange mechanism that preferred calcium ions rather of the sodium ions which were previously attached to it, the calcium ions will hold on to exchange medium while the sodium ions would pass-out with the effluent	Englande et al. (2015)
	Disinfection	Disinfection is the process that aims to eradicate or inhibit the outgrowth of microbes. It aims to inactivate the microbial growth by physical, chemical, or biological processes and this inactivation is accomplished by modifying or damaging vital structures or roles within the microbe. The most commonly ways used for disinfection include: 1. Physical factors like heat and light. 2. Mechanical means like filtration, screening, sedimentation. 3. Radiation treatment is mainly gamma rays. 4. Chemical agents.	Raouf et al. (2019)

(*continued on next page*)

TABLE 8.10 Different types of methods involved in tertiary wastewater treatment—cont'd

	Oxidation	Chemical oxidation is usually applied for the treatment of contaminants which may be toxic, inhibitory to microbial growth, nonbiodegradable, and odor causing. Oxidizing agents commonly used are ozone (O_3); permanganate (MnO_4); hydrogen peroxide (H_2O_2); various chlorine compounds; or even oxygen. These reactions generally need catalysts to be value efficient. UV and Ozone plus several catalysts might be mixed for particular waste in advanced oxidation methods. In the 1980s for potable water treatment Advanced oxidation processes (AOPs) were proposed, which are defined as oxidation involving the formation of hydroxyl radicals ($OH^{\cdot}$) in adequate amount to influence water disinfection. Later, AOP idea enlarged to the oxidative methods with sulfate radicals (SO_4^-)	Englande et al. (2015), Deng and Zhao (2015)
	Activated carbon treatment	In activated carbon adsorption soluble substances are collected within a solution upon a precise surface. Wastewater treatment with activated carbon adsorption on solid interface normally mimics natural biological procedure and is trained at eliminating a segment of the residual dissolved organic matter Activated carbon maintains the quality of pulling various contaminants to stick to its surface via adsorption. It is highly porous with much high surface-area-to-unit-weight ratio and thus holds great adsorption potential for the hydrophobic soluble organics in wastewater Powdered activated carbon (PAC) has been widely used in the aeration container to adsorb VOCs and harmful compounds	Raouf et al. (2019), Englande et al. (2015)

(*continued on next page*)

	Filtration	Membrane filtration: The common membrane filtration methods used to remove contaminants from wastewater are: Ultrafiltration, reverse osmosis, nanofiltration, and electro-dialysis Ultrafiltration (UF), is a membrane system operating at low transmembrane pressures for the elimination of dissolved and colloidal materials. Ultrafiltration (UF) represents a reasonable choice for wastewater treatment as it allows to attain high-quality drinking water in an economic behavior thanks to high separation efficiency, less energy consumption, and compact plant size Reverse osmosis (RO) utilizes a semipermeable membrane, enabling the fluid that is being filtered to pass through it while expelling the contaminants. One of the popularly used techniques which removes a wide variety of dissolved species from water is reverse osmosis. It values higher than 20% of the world's desalination ability Nano-filtration (NF) is the common process between reverse osmosis and ultrafiltration. Nanofiltration is an effective technology for the removal of heavy metal ions like copper, chromium, nickel, and arsenic from wastewater. The achievement of various nanofiltration and reverse osmosis membranes in eliminating noxious lead ions from wastewater has been examined. The impact of operational variables, that is, feed solution pH, feed solution concentration, and the pressure applied, on the ability of the nanofiltration membranes to eliminate ions was estimated. Electro-dialysis is based on the use of electric potential as a driving force and an ion switch membrane is utilized between cathode and anode. The membrane is described as a particular discriminating barrier within two sides and it can be made from natural and artificial material including ceramic, organic and inorganic polymer, and metal material. The membranes are essential of two main types: cation exchange and anion-exchange membranes. Electro-dialysis has been usually used for drinking water production treatment of industrial effluents, and process water from the sea, recovery of useful materials from effluents	Englande et al. (2015), Molinari et al. (2009), Molinari et al. (2019), Raouf et al. (2019), Moslehyani, Ismail, Matsuura, Rahman, and Goh (2019)

TABLE 8.11 List of some plant species and their role in metals/other remediation.

Plant species	Common name	Metals/other	References
Crotalaria dactylon	Rattlebox or rattlepod	Nickel and chromium	Saraswat and Rai (2009)
Crotalaria juncea	Indian hemp, brown hemp, sunn hemp, or Madras hemp	Nickel and chromium	Saraswat and Rai (2009)
Arabidopsis thaliana	Mouse-ear cress, or thale cress	Zinc and cadmium	Saraswat and Rai (2009)
Lemma gibba	Fat duckweed, swollen duckweed, or gibbous duckweed	Arsenic	Mkandawire and Dudel (2005)
Hydrocotyle vulgaris	Marsh pennywort	Nitrogen, phosphate, and ammonium nitrate	Duan, Zhao, Xue, and Yang (2016)
Rorippa globosa	Globe yellow-cress	Cadmium	Sun, Jin, and Zhou (2010)
Sedum alfredii		Iron, zinc, or lead	Sun, Ye, Wang, and Wong (2005)
Brassica napus	Annual rape, canola, colza, oilseed rape, rapeseed, rape, etc.	Cadmium	Selvam and Wong (2009), Selvam & Wong (2008)
Tamarix smyrnensis	Smyrna-tamariske, Smyrna Tamarisk	Cadmium	Manousaki, Kadukova, Papadantonakis, and Kalogerakis (2008)
Sesbania drummondi	Poison Bean	Lead	Sharma, Gardea-Torresdey, Parsons, and Sahi (2004)
Pteris vittata	Chinese Brake Fern, Chinese Ladder Fern, or Brake Fern	Nickel, copper, arsenic, and zinc	Ma et al. (2001), Dong et al. (2005)

(*continued on next page*)

TABLE 8.11 List of some plant species and their role in metals/other remediation—cont'd

Plant species	Common name	Metals/other	References
Thlaspi caerulescens	Alpine pennygrass or Alpine Penny-cress	Nickel, zinc, cadmium, and lead	Assunção, Schat, and Aarts (2003), Banasova, Horak, Nadubinska, Ciamporova, and Lichtscheidl (2008)
Hemidesmus indicus	Indian Sarsaparilla, Anantamul, etc.	Lead	Sekhar, Kamala, Chary, Balaram, and Garcia (2005)
Arabis gemmifera		Zinc and cadmium	Kubota and Takenka (2003)
Helianthus annuus	Sunflower	Lead, zinc, cadmium, copper, and chromium	Boonyapookana, Parkpian, Techapinyawat, DeLaune, and Jugsujinda (2005), Jadia and Fulekar (2008)
Juncus effusus L.	Soft rush or common rush	Reduction of nitrogen, COD, BOD, TSS, fecal coliforms, and phosphate	Coleman et al. (2001)
Brassica juncea	Indian mustard, Chinese mustard, or Leaf mustard, Brown mustard	Selenium, nickel, and copper	Ellis et al. (2004), Ebbs and Kochian (1997)
Alyssum wulfenianum	Madwort	Nickel	Reeves and Brooks (1983)
Pelargoniuna spp.	Storksbills, geraniums, or pelargoniums	Cadmium	Dan, KrishnaRaj, and Saxena (2002)
Arabidopsis hallerii		Cadmium	Bert et al. (2003)
Astraguls bisulcatus	Silver-leafed milkvetch or two-grooved milkvetch	Selenium and sulfur	Sors et al. (2005)

(*continued on next page*)

TABLE 8.11 List of some plant species and their role in metals/other remediation—cont'd

Plant species	Common name	Metals/other	References
Eichhornia crassipes	Common water-hyacinth	Nitrogen, phosphorus	Brumer (2000), Jacquez and Walner (1985)
Azolla pinnata	Mosquitofern, water velvet, or feathered mosquitofern	Chromium and Copper	Jain, Vasudevan, and Jha (1990)
Piptatherum miliaceum	Smilograss	Lead	Garcıa, Faz, and Cunha (2004)
Chengiopanax sciadophylloide	Koshiabura	Manganese	Mizuno et al. (2008)
Oenanthe javanica	Chinese celery, Indian pennywort, Japanese parsley,	Influences dissolved oxygen, pH, and temperature wastewater purification and nutrient uptake	Zhou and Wang (2010), Zhu, Li, and Ketola (2011)
Psychotria douarrei	Wild coffee	Nickel	Davis, Pritchard, Boyd, and Prior (2001)
Potentilla griffithii	Cinquefoils	Zinc and cadmium	Hu et al. (2009)
Spartina plants	Cordgrass	Mercury	Tian, Zhu, Yang, and He (2004)
Pistia startiotes	Water lettuce, shellflower, water cabbage, or Nile cabbage	Cadmium, zinc, mercury, lead, copper, chromium, nickel, and silver	Odjegba and Fasidi (2004)
Amanita muscaria	Fly amanita, fly agaric, or muscimol mushroom	Mercury	Falandysz et al. (2003)

8.4.2 Genetically modified microorganism

The diverse group of microorganisms like bacteria, fungi, and algae was used in the biological treatment of wastewater (Table 8.12). Liquid waste in high amounts is generated due to the production of high scale fermentation products via the fermentation industry. The wastewater generated through the fermentation industry involves the treatment, removal of nitrogen and organic contents from the wastewater before discharging into water bodies (Nielsen, Johnsen, Bensasson, & Daffonchio, 2007). The revolutionary manner applied for the elimination of heavy metals from adsorption through bacterial exo-polysaccharides (Gupta & Diwan, 2017). Genetically engineered microorganisms used for the treating wastes (Table 8.13). An analysis of risk toward the environment when the releasing of genetically modified microorganisms should be evaluated depending upon the process. Such analysis or monitoring is necessary for taking information about genetically engineered microorganisms released in the environment, it helps in searching out wild-type microorganisms before their release into the environment (Thakur, Kennedy, & Karanth, 1991).

8.5 Reuse of treated wastewater

Globally water resources management involves the reuse of wastewater. Reuse of wastewater (Fig. 8.7) considers public health, conservation of water, control of water pollution, and management of water utility (Henry & Jerry, 1982) Table 8.14. Reuse of wastewater serves as a good management technique for water depletion-related problems (Table 8.14).

Based on treatments wastewater is divided/comes into three forms; fully treated water, partially treated water, and untreated water. Fully treated wastewater is water released after complete tertiary treatment. Treated water can be used for various purposes like; irrigation, industrial, and it may directly release into water bodies. According to various reports, it is not suitable for drinking purposes. It may cause some unwanted diseases or illnesses. But in many countries fully treated water get used for drinking purposes as well.

The skin disorders, parasitic and diarrheal infections, and other systemic infections are related to untreated or partially treated water reused (Amoah, Abubakari, Stenström, Abaidoo, & Seidu, 2016; Anh et al., 2009; Contreras et al., 2017; Ferrer, Nguyen-Viet, & Zinsstag, 2012; Lam, Nguyen-Viet, Tuyet-Hanh, Nguyen-Mai, & Harper, 2015; Okoh, Sibanda, & Gusha, 2010; Trang et al., 2007a; Trang, Molbak, Cam, & Dalsgaard, 2007b) health risk managing is an effective strategy but, it is challenging (Verbyla et al., 2016). It majorly depends upon water quality; behavioral and societal factors as well (Mihelcic et al., 2017; Verbyla, Cairns, Gonzalez, Whiteford, & Mihelcic, 2015). Microbial hazards exposure is not only a health risk, it also includes pharmaceuticals and chemicals elements or waste. The four main groups of the population get exposed to partially or untreated wastewater are as follows:

TABLE 8.12 Microorganism and their role in remediation of wastewater.

Species	Remediation effects	References
Bacteria		
Methylobacterium organophilum	Copper and lead	Kim, Kim, Kim, and Oh (1996)
Pseudomonas luteola	Wastewater decoloration	Chang, Chou, and Chen (2001)
Bacillus subitlis	TOC reduction	Hesnawi, Dahmani, Al-Swayah, Mohamed, and Mohammed (2014)
Mycobacterium sp.	Branched hydrocarbon, benzene, cycloparaffins	Solano-Serena et al. (2000)
Burkholderia pickettii	Quinoline	Jianlong, Xiangchun, Libo, Yi, and Hegemann (2002)
Sphingomonas sp. strain BN6	Azo dye degradation	Russ, Rau, and Stolz (2000)
Shewanella sp. XB	Acid orange 7dye	Wang et al. (2012)
Bacillus laterosponus	Remove halogenated hydrocarbons	Girma (2015)
Corynebacterium sp.	Phenoxyacetate and halogenated hydrocarbons	Girma (2015)
Paenibacillus azoreducens	Decoloration	Meehan, Bjourson, and McMullan (2001)
Herminiimonas arsenicoxydans	Arsenic	Marchal, Briandet, Koechler, Kammerer, and Bertin (2010)
Pseudomonas aeruginosa	Chromium, nickel, and cadmium	Ndeddy Aka & Babalola (2016)
Fungi		
Fusarium flocciferum	Nickel (II) and cadmium (II)	Delgado, Anselmo, and Novais (1998)
Bjerkandera adusta MUT 2295	Detoxification and decolorisation of wastewater	Anastasi et al. (2010), Spina, Anastasi, Prigione, Tigini, & Varese (2012)
Rhizopus arrhizus	Heavy metal biosorption	Sağ (2001)
Candida sp.	Polychlorinated biphenyls	
Trametes versicolor	Removal of humic acid and decolourization of wastewater	Zahmatkesh, Spanjers, and van Lier (2018)

(continued on next page)

TABLE 8.12 Microorganism and their role in remediation of wastewater—cont'd

Species	Remediation effects	References
Penicillium chrysogenum	Cadmium(II)	Volesky (1994)
Aspergillus sp.	Phenols	dos Passos, Michelon, Burkert, Kalil, and Burkert (2010)
Phanerochaete chrysosporium	Aromatic compound degradation	Spina, Anastasi, Prigione, Tigini, & Varese (2012)
Algae		
Botryococcus braunii	Phosphorus, nitrogen, zinc, etc.	Oilgae (2014)
Scenedesmus sp.	Zinc	Nmaya, Agam, Matias-Peralta, Yabagi, and Kimpa (2017)
Sargassum muticum	Methylene blue dye removal	Oilgae (2014)
Dunaliella salina	Zinc, cobalt, copper, cadmium, and removal of hypersalinity.	Oilgae (2014); Santhanam (2009)
Scenedesmus abundans	Copper, cadmium, and cyanide detoxification	Oilgae (2014)
Chlorella sp.	Cyanide detoxification, lead (II), nitrogen, and phosphorus	Oilgae (2014)

a. Farmers and their families.
b. Handlers, operational or technical staff, and crop merchants.
c. Farm products (fruits, vegetables, milk, or meat) utilizing consumers.
d. Wastewater irrigated residentially shows the highest risks including the elderly, children, and individuals (immune-compromised) (Embrey, 2003).

The untreated wastewater is hazardous to the environment. It is mixed with water bodies and shows an adverse effect on the water ecosystem completely, while treated water did not show any adverse interference with the water ecosystem. Partially treated water somehow disturbs the ecosystem of water. Toxic elements or compounds present in partially and untreated wastewater blocks the growth of aquatic weeds by altering direct sunlight toward them and also changes in dissolved oxygen concentration.

8.6 Future prospects

With the growing population, we are also developing various industries and are surrounding ourselves with highly toxic chemicals and environment. These toxic

TABLE 8.13 List of some genetically modified organisms and their role in waste treatment.

Transgenic	Plant or microorganism	Remediation	Expression of transgenic protein or gene	References
Achromobacter sp AO22	Bacteria	Mercury	mer gene	Ng, Davis, Polombo, and Bhave (2009)
Arabidopsis thaliana	Plant	Selenium accumulation	Selenocysteine lyase	Pilon et al. (2003)
Arabidopsis thaliana	Plant	Hg(II) volatilization	Mercuric reductase	Rugh et al. (1996)
Arabidopsis thaliana	Plant	Cd(II) and Sb(II)	Yeast cadmium factor (YFC1)	Song et al. (2003)
Arabidopsis thaliana	Plant	Arsenic accumulation	Arsenate reductase and c-glutamylcysteine synthetase	Dhankher et al. (2002)
Arabidopsis thaliana	Plant	Methylmercury detoxification	Organomercurial lyase	Bizily, Rugh, Summers, and Meagher (1999)
Arabidopsis thaliana	Plant	TNT degradation	Nitroreductase	Rosser et al. (2001)
Astragalus bisulcatus	Plant	Selenium	selenocysteine methyltransferase	Ellis et al. (2004)
Brassica juncea	Plant	Cadmium accumulation	γ-glutamylcysteine synthetase	Zhu et al. (1999)
Brassica juncea	Plant	Accumulation of selenium and other metals	ATP sulfurylase	Wangeline et al. (2004)
Brassica juncea	Plant	Selenium volatilization	Cystathionine-it c-synthase	Van Huysen et al. (2003)

(continued on next page)

Brassica juncea	Plant	Accumulation of cadmium and other metals	γ-glutamylcysteine synthetase, glutathione synthetase	Bennett et al. (2003)
Deinococcus radiodurans	Bacteria	Mercury and cadmium	merA	Brim et al. (2000)
E. coli strain	Bacteria	Arsenic accumulation	Metalloregulatory protein ArsR	Kostal, Yang, Wu, Mulchandani, and Chen (2004)
E .coli JM109	Bacteria	Mercury	Hg2+ transporter	Zhao et al. (2005)
E .coli strain	Bacteria	Mercury	Organomurcurial lyase	Murtaza, Dutt, and Ali (2002)
E .coli strain	Bacteria	Cadmium and Cd(II)	SpPCS	Kang et al. (2007)
Liriodendron tulipifera	Plant	Methylmercury detoxification	Organomercurial lyase	Bizily, Rugh, and Meagher (2000), Bizily, Kim, Kandasamy, and Meagher (2003)
Methylococcus capsulatus	Bacteria	Cr(VI)	CrR	Al Hasin et al. (2010)
Nicotiana tabacum	Plant	Cadmium partitioning	Metallothionein	de Borne, Elmayan, de Roton, de Hys, and Tepfer (1998)

(*continued on next page*)

TABLE 8.13 List of some genetically modified organisms and their role in waste treatment—cont'd

Transgenic	Plant or microorganism	Remediation	Expression of transgenic protein or gene	References
Nicotiana tabacum	Plant	Al(III) tolerance	Citrate synthetase	De la Fuente, Ramírez-Rodríguez, Cabrera-Ponce, and Herrera-Estrella (1997)
Nicotiana tabacum	Plant	Human cytochrome P450 2IE1	Trichloroethylene degradation	Doty et al. (2000)
Oryza sativa	Plant	Mercury volatilization	Mercuric reductase	Heaton, Rugh, Kim, Wang, and Meagher (2003)
P. fluorescens 4F39	Bacteria	Nickel	Phytochelatin synthase (PCS)	López, Lázaro, Morales, and Marqués, 2002; Sriprang et al. (2003)
P. putida strain	Bacteria	Chromium	Chromate reductase (ChrR)	Ackerley, Gonzalez, Keyhan, Blake, and Matin (2004)
Pseudomonas K-62	Bacteria	Mercury	Organomercurial lyase	Kiyono and Pan-Hou (2006)
Ralstonia eutropha CH34,	Bacteria	Mercury and cadmium	merA	Valls, Atrian, de Lorenzo, and Fernández (2000)

TABLE 8.14 Wastewater reuse and their hazardous effects.

Exposure route	Wastewater type	Health hazard	References
Occupational exposure, underground water contamination, resident aerosols exposure	Untreated water	Diarrhea	Contreras et al. (2017)
Direct ingestion of greywater during maintenance	Treated greywater	Gastroenteritis (rotavirus-based)	Dalahmeh, Lalander, Pell, Vinnerås, and Jönsson (2016)
Parasitic infection (Ascaris lumbricoides and Trichuris trichiura	Partially treated and untreated wastewater	Occupational exposure and consumption of vegetable	Pham-Duc et al. (2013)
Gastrointestinal infection (E. coli and rotavirus)	Partially treated wastewater	Consumption of salad crops	Pavione, Bastos, and Bevilacqua (2013)
Diarrhea (Giardia lamblia and Entamoeba histolytica)	Untreated wastewater	Direct exposure	Ferrer et al. (2012)
Gastroenteritis	contaminated Surface water	Swimming, fishing, consuming canal water-irrigated vegetables and ingesting/inhaling water or aerosols while working in canal water-irrigated fields	Tserendorj et al. (2011); Daley et al. (2018)
Parasitic infections (hookworm and G. lamblia)	Partially treated wastewater	Exposure via occupational consumption	Gumbo, Malaka, Odiyo, and Nare (2010)

(*continued on next page*)

Skin infection	Partially treated wastewater	Occupational exposure	Anh et al. (2009)
Skin infection/irritation	Partially treated wastewater	Exposure to infected source	Chary, Kamala, and Raj (2008)
Intestinal parasitic infection	Partially treated and untreated wastewater	Occupational exposure	Ensink, Blumenthal, and Brooker (2008)
Escherichia coli infection (risk)	Untreated wastewater	Occupational exposure	Yajima and Kurokura et al. (2008)
Helminthic infection	Untreated wastewater	Occupational exposure	Trang et al. (2006)
Skin infection	Partially treated wastewater	Occupational exposure	Anh et al. (2007)
Diarrhea	Partially treated wastewater	Children of occupationally exposed farmers	Hien, Scheutz, Cam, Mølbak, and Dalsgaard (2007)
Giardiasis	Untreated wastewater	Occupational exposure	Ensink, van der Hoek, Mukhtar, Tahir, and Amerasinghe (2005)
Helminthic infection	Partially treated wastewater	Occupational exposure	Olsen, Murrell, Dalsgaard, Johansen, and Van De (2006)
Infection of Ascaris, Trichuris	Untreated wastewater	Children resident in wastewater irrigated farmhouse	Amahmid and Bouhoum (2005)
Intestinal parasitic infection	Unknown	Occupational exposure	Verle et al. (2003)

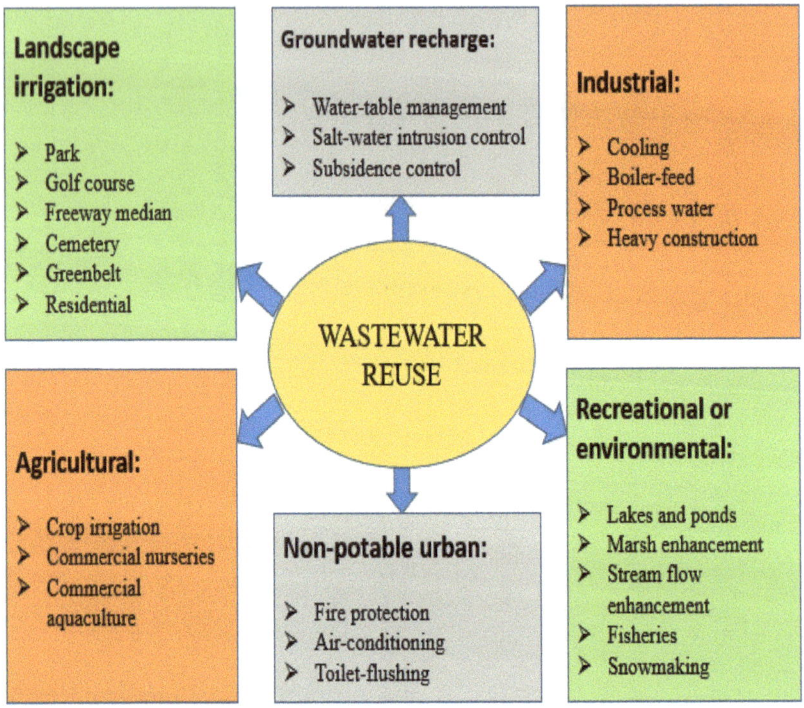

FIGURE 8.7 Applications for reused wastewater.

chemicals are highly dispersed in environment (air, water, and soil) which is not only harmful for humans but also to the animals as well as to aquatic life and plants. In humans such toxic chemicals can cause respiratory disorders, skin allergies or irritations and some are even potential carcinogens that can enter human body via plants or water. With the increasing global warming because of harmful games and chemicals present in our environment glaciers are melting at accelerating rate and we are also loosing forest because of fire and some human stupidity hence resulting less oxygen and water. At this time when everyone is concerned about the future we are irresponsibly wasting natural resources like fresh water air and soil. Country like India where more than half of the population is dependent on agriculture, still we lack the waste management systems to save our land air and water. In the coming age it is much needed that biological modes of waste management should be applied to manage all the waste without or causing less harm to the environment. Hence GMO plays important role in the management of environment. Genetically modified plants because of resistance can grow in high temperatures as well as drought conditions with insect resistance and GMOs specially bacteria plays a huge role in wastewater treatment. To save our environment, it is high time for national and international strict environment laws for waste management, ban plastic, mandatory rain water harvesting at drought areas, proper land allotments away from population for waste disposal. To save our environment and our future, we have to work ourselves rather than blaming others. Every individual has to work on waste management which starts at home by refusing to use plastic and other nonenvironmental friendly products, reducing waste, reusing or restoring products for use, recycling all the recyclables, decomposing household organic products which create compost, planting more trees and air purifying plants, and installing rain water harvesting system at homes as well as in school colleges.

8.7 Conclusion

In conclusion, we can say that wastewater is highly risky for flora and fauna as well as the environment. The direct use of wastewater causes several diseases in humans as well as animals. The treatment of wastewater is a highly used method to reduce the toxicity of the water. By using GMOs (plants and microorganisms), we can increase the concentration of eliminating heavy metals, organic and inorganic wastes from water. There are lots of hazardous effects of using partially treated and untreated wastewater. To overcome the problem related to wastewater-direct release into water bodies and lands should be banned, along with a high amount of accuracy is needed in treatment plants.

References

https://www.epa.gov/dwreginfo/drinking-water-regulations

https://www.cdc.gov/healthywater/drinking/public/water_disinfection.html

Ackerley, D. F., Gonzalez, C. F., Keyhan, M., Blake, R., & Matin, A. (2004). Mechanism of chromate reduction by the Escherichia coli protein, NfsA, and the role of different chromate reductases in minimizing oxidative stress during chromate reduction. *Environmental Microbiology, 6*(8), 851–860.

Adewumi, J. R., Ilemobade, A. A., & Van Zyl, J. E. (2010). Treated wastewater reuse in South Africa: Overview, potential and challenges. *Resources, Conservation and Recycling, 55*(2), 221–231.

Alabi, O. A., Ologbonjaye, K. I., Awosolu, O., & Alalade, O. E. (2019). Public and environmental health effects of plastic wastes disposal: A review. *Journal of Toxicology and Risk Assessment, 5*(021), 1–13.

Al Hasin, A., Gurman, S. J., Murphy, L. M., Perry, A., Smith, T. J., & Gardiner, P. H. (2010). Remediation of chromium (VI) by a methane-oxidizing bacterium. *Environmental Science & Technology, 44*(1), 400–405.

Amoah, I. D., Abubakari, A., Stenström, T. A., Abaidoo, R. C., & Seidu, R. (2016). Contribution of wastewater irrigation to soil transmitted helminths infection among vegetable farmers in Kumasi, Ghana. *PLoS Neglected Tropical Diseases, 10*(12), e0005161.

Anastasi, A., Spina, F., Prigione, V., Tigini, V., Giansanti, P., & Varese, G. C. (2010). Scale-up of a bioprocess for textile wastewater treatment using Bjerkandera adusta. *Bioresource Technology, 101*(9), 3067–3075.

Anh, V. T., van der Hoek, W., Ersbøll, A. K., Thuong, N. V., Tuan, N. D., Cam, P. D., et al. (2007). Dermatitis among farmers engaged in peri-urban aquatic food production in Hanoi, Vietnam. *Tropical Medicine & International Health, 12*, 59–65. doi:10.1111/j.1365-3156.2007.01942.x.

Anh, VT, van der Hoek, W, Ersbøll, AK, Vicheth, C, Cam, PD, & Dalsgaard, A. (2009). Peri-urban aquatic plant culture and skin disease in Phnom Penh, Cambodia. *Journal of Water and Health, 7*, 302–311. doi:10.2166/wh.2009.128.

Araujo, P. H. H., Sayer, C., Giudici, R., & Poço, J. G. R. (2004). Techniques for reducing residual monomer content in polymers: A review, First published in 07 April 2004. https://doi.org/10.1002/pen.11043.

Assunção, A. G., Schat, H., & Aarts, M. G. (2003). Thlaspi caerulescens, an attractive model species to study heavy metal hyperaccumulation in plants. *New Phytologist, 159*(2), 351–360.

Babul Supriyo, Shri (2019). Generation of Waste.pdf - India Environment Portal. http://www.indiaenvironmentportal.org.in

Banasova, V., Horak, O., Nadubinska, M., Ciamporova, M., & Lichtscheidl, I. (2008). Heavy metal content in Thlaspi caerulescens J. et C. Presl growing on metalliferous and non-metalliferous soils in Central Slovakia. *International Journal of Environment and Pollution, 33*(2-3), 133–145.

Bennett, L. E., Burkhead, J. L., Hale, K. L., Terry, N., Pilon, M., & Pilon-Smits, E. A. (2003). Analysis of transgenic Indian mustard plants for phytoremediation of metal-contaminated mine tailings. *Journal of Environmental Quality, 32*(2), 432–440.

Bert, V., Meerts, P., Saumitou-Laprade, P., Salis, P., Gruber, W., & Verbruggen, N. (2003). Genetic basis of Cd tolerance and hyperaccumulation in Arabidopsis halleri. *Plant and Soil, 249*(1), 9–18.

Bhat, T. A. (2014). An analysis of demand and supply of water in India. *Journal of Environment and Earth Science, 4*(11), 67–72.

Bizily, S. P., Kim, T., Kandasamy, M. K., & Meagher, R. B. (2003). Subcellular targeting of methylmercury lyase enhances its specific activity for organic mercury detoxification in plants. *Plant Physiology, 131*(2), 463–471.

Bizily, S. P., Rugh, C. L., & Meagher, R. B. (2000). Phytodetoxification of hazardous organomercurials by genetically engineered plants. *Nature Biotechnology, 18*(2), 213–217.

Bizily, S. P., Rugh, C. L., Summers, A. O., & Meagher, R. B. (1999). Phytoremediation of methylmercury pollution: MerB expression in Arabidopsis thaliana confers resistance to organomercurials. *Proceedings of the National Academy of Sciences, 96*(12), 6808–6813.

Boonyapookana, B., Parkpian, P., Techapinyawat, S., DeLaune, R. D., & Jugsujinda, A. (2005). Phytoaccumulation of lead by sunflower (Helianthus annuus), tobacco (Nicotiana tabacum), and vetiver (Vetiveria zizanioides). *Journal of Environmental Science and Health, 40*(1), 117–137.

Brim, H., McFarlan, S. C., Fredrickson, J. K., Minton, K. W., Zhai, M., Wackett, L. P., & Daly, M. J. (2000). Engineering Deinococcus radiodurans for metal remediation in radioactive mixed waste environments. *Nature Biotechnology, 18*(1), 85–90.

Brumer, L. (2000). *Use of aquatic macrophytes to improve the quality of effluents after chlorination.* Haifa: Technion Israel Institute of Technology Doctoral dissertation, Ph.D. Dissertation.

Brydson, J. A. (1999). *Plastics materials*. Elsevier.

Bureau of Reclamation California-Great Basin, Water facts - worldwide water supply. https://www.usbr.gov/mp/arwec/water-facts-ww-water-sup.html

Central Pollution Control Board, (CPCB), (2004). "Management of municipal solid waste. Ministry of Environment and Forests, New Delhi.

Chalkhure, A. N., Marve, S. R., Wankar, M. S., Bhendale, A. N., Daulatkar, D. P., & Chopade, P. S. (2020). Design of harvestine filter unit.

Chang, J. S., Chou, C., & Chen, S. Y. (2001). Decolorization of azo dyes with immobilized Pseudomonas luteola. *Process Biochemistry, 36*(8-9), 757–763.

Chaoua, S., Boussaa, S., El Gharmali, A., & Boumezzough, A. (2019). Impact of irrigation with wastewater on accumulation of heavy metals in soil and crops in the region of Marrakech in Morocco. *Journal of the Saudi Society of Agricultural Sciences, 18*(4), 429–436.

Chary, S. N., Kamala, C. T., & Raj, D. S. S. (2008). Assessing risk of heavy metals from consuming food grown on sewage irrigated soils and food chain transfer. *Ecotoxicology and Environmental Safety, 69*, 513–524. doi:10.1016/j.ecoenv.2007.04.013.

Chen, W., Lu, S., Pan, N., & Jiao, W. (2013). Impacts of long-term reclaimed water irrigation on soil salinity accumulation in urban green land in Beijing. *Water Resources Research, 49*(11), 7401–7410.

Clemmens, A. J., Allen, R. G., & Burt, C. M. (2008). Technical concepts related to conservation of irrigation and rainwater in agricultural systems. *Water Resources Research, 44*(7), 1–16. doi:10.1029/2007WR006095.

Coleman, J., Hench, K., Garbutt, K., Sexstone, A., Bissonnette, G., & Skousen, J. (2001). Treatment of domestic wastewater by three plant species in constructed wetlands. *Water, Air, and Soil Pollution, 128*(3), 283–295.

Contreras, J. D., Meza, R., Siebe, C., Rodríguez-Dozal, S., López-Vidal, Y. A., CastilloRojas, G., et al. (2017). Health risks from exposure to untreated wastewater used for irrigation in the Mezquital Valley, Mexico: A 25-year update. *Water Research, 23*, 834–850. doi:10.1016/j.watres.2017.06.058.

Dalahmeh, S. S., Lalander, C., Pell, M., Vinnerås, B., & Jönsson, H. (2016). Quality of greywater treated in biochar filter and risk assessment of gastroenteritis due to household exposure during maintenance and irrigation. *Journal of Applied Microbiology, 121*, 1427–1443. doi:10.1111/jam.13273.

Daley, K., Hansen, L. T., Jamieson, R. C., Hayward, J. L., Piorkowski, G. S., Krkosek, W., et al. (2018). Chemical and microbial characteristics of municipal drinking water supply systems in the Canadian Arctic. *Environmental Science and Pollution Research, 25*(33), 32926–32937.

Dan, T. V., KrishnaRaj, S., & Saxena, P. K. (2002). Cadmium and nickel uptake and accumulation in scented geranium (Pelargonium sp.Frensham'). *Water, Air, and Soil Pollution, 137*(1), 355–364.

Davis, M. A., Pritchard, S. G., Boyd, R. S., & Prior, S. A. (2001). Developmental and induced responses of nickel-based and organic defences of the nickel-hyperaccumulating shrub, Psychotria douarrei. *New Phytologist, 150*(1), 49–58.

de Borne, F. D., Elmayan, T., de Roton, C., de Hys, L., & Tepfer, M. (1998). Cadmium partitioning in transgenic tobacco plants expressing a mammalian metallothionein gene. *Molecular Breeding, 4*(2), 83–90.

De la Fuente, J. M., Ramıŕez-Rodrıǵuez, V., Cabrera-Ponce, J. L., & Herrera-Estrella, L. (1997). Aluminum tolerance in transgenic plants by alteration of citrate synthesis. *Science, 276*(5318), 1566–1568.

Delgado, A., Anselmo, A. M., & Novais, J. M. (1998). Heavy metal biosorption by dried powdered mycelium of Fusarium flocciferum. *Water Environment Research, 70*(3), 370–375.

Deng, Y., & Zhao, R. (2015). Advanced oxidation processes (AOPs) in wastewater treatment. *Current Pollution Reports, 1*(3), 167–176.

Dhankher, O. P., Li, Y., Rosen, B. P., Shi, J., Salt, D., Senecoff, J. F., & Meagher, R. B. (2002). Engineering tolerance and hyperaccumulation of arsenic in plants by combining arsenate reductase and γ-glutamylcysteine synthetase expression. *Nature Biotechnology, 20*(11), 1140–1145.

Dong, R., Formentin, E., Losseso, C., Carimi, F., Benedetti, P., Terzi, M., & Schiavo, F. L. (2005). Molecular cloning and characterization of a phytochelatin synthase gene, PvPCS1, from Pteris vittata L. *Journal of Industrial Microbiology and Biotechnology, 32*(11,12), 527–533.

dos Passos, C. T., Michelon, M., Burkert, J. F. D. M., Kalil, S. J., & Burkert, C. A. V. (2010). Biodegradation of phenol by free and encapsulated cells of a new Aspergillus sp. isolated from a contaminated site in southern Brazil. *African Journal of Biotechnology, 9*(40), 6716–6720.

Doty, S. L., Shang, T. Q., Wilson, A. M., Tangen, J., Westergreen, A. D., Newman, L. A., et al. (2000). Enhanced metabolism of halogenated hydrocarbons in transgenic plants containing mammalian cytochrome P450 2E1. *Proceedings of the National Academy of Sciences, 97*(12), 6287–6291.

Duan, J. J., Zhao, J. N., Xue, L. H., & Yang, L. Z. (2016). Nutrient removal of a floating plant system receiving low-pollution wastewater: Effects of plant species and influent concentration. *IOP Conference Series: Earth and Environmental Science*: 41. IOP Publishing.

Ebbs, S. D., & Kochian, L. V. (1997), *Toxicity of zinc and copper to Brassica species: Implications for phytoremediation* (26, pp. 776–781). American Society of Agronomy, Crop Science Society of America, and Soil Science Society of America.

EEA CSI (2018). https://www.eea.europa.eu/themes/water/waterresources/water-use-by-sectors Accessed October 3, 2019.

Ellis, D. R., Sors, T. G., Brunk, D. G., Albrecht, C., Orser, C., Lahner, B., et al. (2004). Production of Se-methylselenocysteine in transgenic plants expressing selenocysteine methyltransferase. *BMC Plant Biology, 4*(1), 1–11.

Embrey, M. (2003). Subpopulations susceptible to waterborne diseases are surprisingly diverse. *Journal American Water Works Association, 95*, 34. doi:10.1002/j.1551-8833.2003.tb10307.x.

Englande, A. J., Jr, Krenkel, P., & Shamas, J. (2015). Wastewater treatment &water reclamation. *Reference Module in Earth Systems and Environmental Sciences*.

Ensink, J. H., Blumenthal, U. J., & Brooker, S. (2008). Wastewater quality and the risk of intestinal nematode infection in sewage farming families in Hyderabad, India. *The American journal of tropical medicine and hygiene, 79*(4), 561.

Ensink, J. H., van der Hoek, W., Mukhtar, M., Tahir, Z., & Amerasinghe, F. P. (2005). High risk of hookworm infection among wastewater farmers in Pakistan. *Transactions of the Royal Society of Tropical Medicine and Hygiene, 99*(11), 809–818.

ESCAP, U. (2000). State of the environment in Asia and the Pacific 2000.

Fahad, A., Mohamed, R. S., Radhi, B., & Al-Sahari, M. (2019). Wastewater and its treatment techniques: An ample review. *Indian Journal of Science and Technology, 12*(25), 1–13.

Falandysz, J., Lipka, K., Kawano, M., Brzostowski, A., Dadej, M., Jedrusiak, A., & Puzyn, T. (2003). Mercury content and its bioconcentration factors in wild mushrooms at Łukta and Morag, northeastern Poland. *Journal of Agricultural and Food Chemistry, 51*(9), 2832–2836.

Ferrer, A., Nguyen-Viet, H., & Zinsstag, J. (2012). Quantification of diarrhea risk related to wastewater contact in Thailand. *Ecohealth, 9*, 49–59. doi:10.1007/s10393-012-0746-x.

Flathman, P. E., & Lanza, G. R. (1998). Phytoremediation: Current views on an emerging green technology. *Journal of Soil Contamination, 7*(4), 415–432.

Frontiers, U. (2016). Report emerging issues of environmental concern. *Emerging Zoonotic Diseases and Links to Ecosystem Health–UNEP Frontiers*.

Frontistis, Z., Xekoukoulotakis, N. P., Hapeshi, E., Venieri, D., Fatta-Kassinos, D., & Mantzavinos, D. (2011). Fast degradation of estrogen hormones in environmental matrices by photo-Fenton oxidation under simulated solar radiation. *Chemical Engineering Journal, 178*, 175–182.

Galloway, T. S. (2015). Micro-and nano-plastics and human health. *Marine Anthropogenic Litter* (pp. 343–366). Cham: Springer.

Gangadhar, B., Ravi, V., & Ramakrishna Naidu, G. (2017). Impact of solid waste on human health and environment in India – an overview.

Garcıa, G., Faz, A., & Cunha, M. (2004). Performance of Piptatherum miliaceum (Smilo grass) in edaphic Pb and Zn phytoremediation over a short growth period. *International Biodeterioration & Biodegradation, 54*(2-3), 245–250.

Girma, G. (2015). Microbial bioremediation of some heavy metals in soils: An updated review. *Egyptian Academic Journal of Biological Sciences, G. Microbiology, 7*(1), 29–45.

Gumbo, J. R., Malaka, E. M., Odiyo, J. O., & Nare, L. (2010). The health implications of wastewater reuse in vegetable irrigation: A case study from Malamulele, South Africa. *International Journal of Environmental Health Research, 20*(3), 201–211.

Gupta, P., & Diwan, B. (2017). Bacterial exopolysaccharide mediated heavy metal removal: A review on biosynthesis, mechanism and remediation strategies. *Biotechnology Reports, 13*, 58–71.

Habitat, U. N. (2009). Fact sheet. *Global Report on Human Settlement.*

Hamer, G. (2003). Solid waste treatment and disposal: Effects on public health and environmental safety. *Biotechnology Advances, 22*(1,2), 71–79.

Hassandoost, R., Pouran, S. R., Khataee, A., Orooji, Y., & Joo, S. W. (2019). Hierarchically structured ternary heterojunctions based on Ce^{3+}/Ce^{4+} modified Fe_3O_4 nanoparticles anchored onto graphene oxide sheets as magnetic visible-light-active photocatalysts for decontamination of oxytetracycline. *Journal of Hazardous Materials, 376*, 200–211.

Heaton, A. C., Rugh, C. L., Kim, T., Wang, N. J., & Meagher, R. B. (2003). Toward detoxifying mercury-polluted aquatic sediments with rice genetically engineered for mercury resistance. *Environmental Toxicology and Chemistry: An International Journal, 22*(12), 2940–2947.

Hesnawi, R., Dahmani, K., Al-Swayah, A., Mohamed, S., & Mohammed, S. A. (2014). Biodegradation of municipal wastewater with local and commercial bacteria. *Procedia Engineering, 70*, 810–814.

Hien, B. T. T., Scheutz, F., Cam, P. D., Mølbak, K., & Dalsgaard, A. (2007). Diarrhoeagenic Escherichia coli and other causes of childhood diarrhoea: A case–control study in children living in a wastewater-use area in Hanoi, Vietnam. *Journal of Medical Microbiology, 56*(8), 1086–1096.

Hoekstra, A. Y., & Mekonnen, M. M. (2012). The water footprint of humanity. *Proceedings of the National Academy of Sciences, 109*(9), 3232–3237.

Hoornweg, D., & Bhada-Tata, P. (2012). *What a waste: A global review of solid waste management. Washington, DC*: World Bank Urban development series *Knowledge Papers (15).*

Hoornweg, D., & Thomas, L. (1999). *What a waste: Solid waste management in Asia.* The World Bank.

Hu, P. J., Qiu, R. L., Senthilkumar, P., Jiang, D., Chen, Z. W., Tang, Y. T., et al. (2009). Tolerance, accumulation and distribution of zinc and cadmium in hyperaccumulator Potentilla griffithii. *Environmental and Experimental Botany, 66*(2), 317–325.

Intelligence, G. G. W. (2009). PUB study: Perspectives of water reuse.

Jacquez, R. B., & Walner, H. Z. (1985). Combining Nutrient Removal with Protein Synthesis Using a Water Hyacinth-Freshwater Prawn Polyculture Wastewater Treatment System. Project No. 1345677. *Water Resources Research.*

Jadia, C. D., & Fulekar, M. H. (2008). Phytoremediation: The application of vermicompost to remove zinc, cadmium, copper, nickel and lead by sunflower plant. *Environmental Engineering & Management Journal (EEMJ), 7*(5), 547–558.

Jain, S. K., Vasudevan, P., & Jha, N. K. (1990). Azolla pinnata R. Br. and Lemna minor L. for removal of lead and zinc from polluted water. *Water Research, 24*(2), 177–183.

James, I. D. (2002). Modelling pollution dispersion, the ecosystem and water quality in coastal waters: A review. *Environmental Modelling & Software, 17*(4), 363–385.

Jianlong, W., Xiangchun, Q., Libo, W., Yi, Q., & Hegemann, W. (2002). Bioaugmentation as a tool to enhance the removal of refractory compound in coke plant wastewater. *Process Biochemistry, 38*(5), 777–781.

Jindal, A., & Kamat, S. (2011). Water recycling and reuse for domestic and industrial sectors. *Chemical Engineering World, 232*(208), 52–62.

Joseph, N., Ryu, D., Malano, H. M., George, B., & Sudheer, K. P. (2021). Estimation of state-wide and monthly domestic water use in India from 1975 to 2015. *Urban Water Journal, 18*(6), 421–432.

Kang, S. H., Singh, S., Kim, J. Y., Lee, W., Mulchandani, A., & Chen, W. (2007). Bacteria metabolically engineered for enhanced phytochelatin production and cadmium accumulation. *Applied and Environmental Microbiology, 73*(19), 6317–6320.

Karimi-Maleh, H., Karimi, F., Malekmohammadi, S., Zakariae, N., Esmaeili, R., Rostamnia, S., et al. (2020). An amplified voltammetric sensor based on platinum nanoparticle/polyoxometalate/two-dimensional hexagonal boron nitride nanosheets composite and ionic liquid for determination of N-hydroxysuccinimide in water samples. *Journal of Molecular Liquids, 310*, 113185.

Karimi-Maleh, H., Orooji, Y., Ayati, A., Qanbari, S., Tanhaei, B., Karimi, F., et al. (2020a). Recent advances in removal techniques of Cr (VI) toxic ion from aqueous solution: A comprehensive review. *Journal of Molecular Liquids, 4*(4), 115062.

Kenny, J. F., Barber, N. L., Hutson, S. S., Linsey, K. S., Lovelace, J. K., & Maupin, M. A. (2009). *Estimated use of water in the United States in 2005*. US Geological Survey (No. 1344).

Kim, S. Y., Kim, J. H., Kim, C. J., & Oh, D. K. (1996). Metal adsorption of the polysaccharide produced from Methylobacterium organophilum. *Biotechnology Letters, 18*(10), 1161–1164.

Kiyono, M., & Pan-Hou, H. (2006). Genetic engineering of bacteria for environmental remediation of mercury. *Journal of Health Science, 52*(3), 199–204.

Kostal, J., Yang, R., Wu, C. H., Mulchandani, A., & Chen, W. (2004). Enhanced arsenic accumulation in engineered bacterial cells expressing ArsR. *Applied and Environmental Microbiology, 70*(8), 4582–4587.

Kubota, H., & Takenaka, C. (2003). Field Note: Arabis gemmifera is a hyperaccumulator of Cd and Zn. *International Journal of Phytoremediation, 5*(3), 197–201.

Kumar, R., Singh, R. D., & Sharma, K. D. (2005). Water resources of India. *Current Science, 6*(14), 794–811.

Kyzas, G. Z., & Matis, K. A. (2018). Flotation in water and wastewater treatment. *Processes, 6*(8), 116.

Lam, S., Nguyen-Viet, H., Tuyet-Hanh, T. T., Nguyen-Mai, H., & Harper, S. (2015). Evidence for public health risks of wastewater and excreta management practices in Southeast Asia: A scoping review. *International Journal of Environmental Research and Public Health, 12*(10), 12863–12885.

López, A., Lázaro, N., Morales, S., & Marqués, A. M. (2002). Nickel biosorption by free and immobilized cells of Pseudomonas fluorescens 4F39: A comparative study. *Water, Air, and Soil Pollution, 135*(1), 157–172.

Ma, L. Q., Komar, K. M., Tu, C., Zhang, W., Cai, Y., & Kennelley, E. D. (2001). A fern that hyperaccumulates arsenic. *Nature, 409*(6820), 579 –579.

Macek, T., Kotrba, P., Svatos, A., Novakova, M., Demnerova, K., & Mackova, M. (2008). Novel roles for genetically modified plants in environmental protection. *Trends in biotechnology, 26*(3), 146–152.

Macek, T., Macková, M., Pavlíková, D., Száková, J., Truksa, M., Singh Cundy, A., & Scouten, W. H. (2002). Accumulation of cadmium by transgenic tobacco. *Acta Biotechnologica, 22*(1,2), 101–106.

Macek, T., Pavlikova, D., & Mackova, M. (2004). Phytoremediation of metals and inorganic pollutants. Applied bioremediation and phytoremediation (pp. 135–157). Berlin, Heidelberg: Springer.

Macek, T., Surá, M., Pavliková, D., Francová, K., Scouten, W. H., Szekeres, M., & Macková, M. (2005). Can tobacco have a potentially beneficial effect to our health? *Zeitschrift fur Naturforschung. C, Journal of biosciences, 60*(3,4), 292–299.

Mackova, M., Dowling, D., & Macek, T. (2006). *Phytoremediation and rhizoremediation*: 9. Springer Science & Business Media.

Mahfooz, Y., Yasar, A., Guijian, L., Islam, Q. U., Akhtar, A. B. T., Rasheed, R., et al. (2020). Critical risk analysis of metals toxicity in wastewater irrigated soil and crops: A study of a semi-arid developing region. *Scientific Reports, 10*(1), 1–10.

Mahmood, A., & Malik, R. N. (2014). Human health risk assessment of heavy metals via consumption of contaminated vegetables collected from different irrigation sources in Lahore, Pakistan. *Arabian Journal of Chemistry, 7*(1), 91–99.

Malviya, R., Chaudhary, R., & Buddhi, D. (2002). Study on solid waste assessment and management-Indore city. *Indian Journal of Environmental Protection, 22*(8), 841–846.

Manousaki, E., Kadukova, J., Papadantonakis, N., & Kalogerakis, N. (2008). Phytoextraction and phytoexcretion of Cd by the leaves of Tamarix smyrnensis growing on contaminated non-saline and saline soils. *Environmental Research, 106*(3), 326–332.

Marchal, M., Briandet, R., Koechler, S., Kammerer, B., & Bertin, P. N. (2010). Effect of arsenite on swimming motility delays surface colonization in Herminiimonas arsenicoxydans. *Microbiology, 156*(8), 2336–2342.

McCutcheon, S. C., & Schnoor, J. L. (2003). Overview of phytotransformation and control of wastes. *Phytoremediation: Transformation and Control of Contaminants, 7*, 3–58.

Meagher, R. B. (2000). Phytoremediation of toxic elemental and organic pollutants. *Current Opinion in Plant Biology, 3*(2), 153–162.

Meehan, C., Bjourson, A. J., & McMullan, G. (2001). Paenibacillus azoreducens sp. nov., a synthetic azo dye decolorizing bacterium from industrial wastewater. *International Journal of Systematic and Evolutionary Microbiology, 51*(5), 1681–1685.

Mehmood, A., Mirza, M. A., Choudhary, M. A., Kim, K. H., Raza, W., Raza, N., et al. (2019). Spatial distribution of heavy metals in crops in a wastewater irrigated zone and health risk assessment. *Environmental Research, 168*, 382–388.

Mehta, K. P. (2015). Design of reverse osmosis system for reuse of waste water from common effluent treatment plant. *International Research Journal of Engineering and Technology (IRJET), 2*(4), 983–991.

Mihelcic, Jr, Naughton, C. C., Verbyla, M. E., Zhang, Q., Schweitzer, R. W., Oakley, S. M., et al. (2017). The grandest challenge of all: The role of environmental engineering to achieve sustainability in the world's developing regions. *Environmental Engineering Science, 34*, 16–41. doi:10.1089/ees.2015.0334.

Mizuno, T., Hirano, K., Kato, S., & Obata, H. (2008). Cloning of ZIP family metal transporter genes from the manganese hyperaccumulator plant Chengiopanax sciadophylloides, and its metal transport and resistance abilities in yeast. *Soil Science and Plant Nutrition, 54*(1), 86–94.

Mkandawire, M., & Dudel, E. G. (2005). Accumulation of arsenic in Lemna gibba L. (duckweed) in tailing waters of two abandoned uranium mining sites in Saxony, Germany. *Science of the Total Environment, 336*(1,3), 81–89.

Molinari, R., Argurio, P., & Poerio, T. (2009). Membrane processes based on complexation reactions of pollutants as sustainable wastewater treatments. *Sustainability, 1*(4), 978–993.

Molinari, R., Lavorato, C., Argurio, P., Szymański, K., Darowna, D., & Mozia, S. (2019). Overview of photocatalytic membrane reactors in organic synthesis, energy storage and environmental applications. *Catalysts, 9*(3), 239.

Morrison, R. J., & Munro, A. (1997). Waste management in small island developing states in the South Pacific: Report of a regional workshop. *Canberra, ACT, Australia, 1*, 232–246 doi:https://doi.org/10.1080/14486563.1999.10648474.

Moslehyani, A., Ismail, A. F., Matsuura, T., Rahman, M. A., & Goh, P. S. (2019). Recent progresses of ultrafiltration (UF) membranes and processes in water treatment. *Membrane Separation Principles and Applications* (pp. 85–110). Elsevier.

Murshed, S. B., & Kaluarachchi, J. J. (2018). Scarcity of fresh water resources in the Ganges Delta of Bangladesh. *Water Security, 4*, 8–18.

Murtaza, I., Dutt, A., & Ali, A. (2002). Biomolecular Engineering of Escherichia coli organomercurial lyase gene and its expression.

Ndeddy Aka, R. J., & Babalola, O. O. (2016). Effect of bacterial inoculation of strains of Pseudomonas aeruginosa, Alcaligenes feacalis and Bacillus subtilis on germination, growth and heavy metal (Cd, Cr, and Ni) uptake of Brassica juncea. *International Journal of Phytoremediation, 18*(2), 200–209.

Nema, A. K. (2004). Collection and transport of municipal solid waste. *Training program on solid waste management. Delhi: Springer.*

Ng, S. P., Davis, B., Polombo, E. A., & Bhave, M. (2009). A Tn5051-like mer-containing transposon identified in a heavy metal tolerant strain *Achromobacter* sp. AO22. *BMC Research Notes, 7*, 2–38.

Nielsen, K. M., Johnsen, P. J., Bensasson, D., & Daffonchio, D. (2007). Release and persistence of extracellular DNA in the environment. *Environmental Biosafety Research, 6*(1-2), 37–53.

Njuguna, S. M., Makokha, V. A., Yan, X., Gituru, R. W., Wang, Q., & Wang, J. (2019). Health risk assessment by consumption of vegetables irrigated with reclaimed waste water: A case study in Thika (Kenya). *Journal of Environmental Management, 231*, 576–581.

Nmaya, M. M., Agam, M. A., Matias-Peralta, H. M., Yabagi, J. A., & Kimpa, M. I. (2017). Freshwater green microalga for bioremediation of river melaka heavy-metals contamination. *Journal of Science and Technology, 9*(3), 118–123.

Odjegba, V. J., & Fasidi, I. O. (2004). Accumulation of trace elements by Pistia stratiotes: Implications for phytoremediation. *Ecotoxicology, 13*(7), 637–646.

Ogawa, M., Ohtsubo, T., Tsuda, S., & Tsuji, K. (1993). Simplified method to measure suspensibility of water-dispersible powder: Use of microanalytical techniques to reduce wastewater. *Journal of AOAC International, 76*(1), 83–89.

Okoh, A. I., Sibanda, T., & Gusha, S. S. (2010). Inadequately treated wastewater as a source of human enteric viruses in the environment. *International Journal of Environmental Research and Public Health, 7*(6), 2620–2637.

Oligae Guide, (2014). http://www.oilgae.com/blog/2014/01/commonly-used-algae-strains-for-waste-water-treatment.html

Olsen, A., Murrell, K. D., Dalsgaard, A., Johansen, M. V., & Van De, N.Fish-Borne Zoonotic Parasites in Vietnam (FIBOZOPA) project. (2006). Cross-sectional parasitological survey for helminth infections among fish farmers in Nghe An province, Vietnam. *Acta Tropica, 100*(3), 199–204.

Omahmid, O., & Bouboum, K. (2005). Assessment of the health hazards associated with wastewater reuse: Transmission of geohelminthic infections (Marrakech, Morroco). *International Journal of Environmental Health Research, 15*, 127–133.

Ongerth, H. J., & Ongerth, J. E. (1982). Health consequences of wastewater reuse. *Annual Review of Public Health, 3*(1), 419–444.

Orooji, Y., Ghanbari, M., Amiri, O., & Salavati-Niasari, M. (2020). Facile fabrication of silver iodide/graphitic carbon nitride nanocomposites by notable photo-catalytic performance through sunlight and antimicrobial activity. *Journal of Hazardous Materials, 389*, 122079.

Pavione, D. M. S., Bastos, R. K. X., & Bevilacqua, P. D. (2013). Quantitative microbial risk assessment applied to irrigation of salad crops with waste stabilization pond effluents. *Water Science and Technology, 67*(6), 1208–1215.

Pham-Duc, P., Nguyen-Viet, H., Hattendorf, J., Zinsstag, J., Phung-Dac, C., Zurbrügg, C., & Odermatt, P. (2013). Ascaris lumbricoides and Trichuris trichiura infections associated with wastewater and human excreta use in agriculture in Vietnam. *Parasitology International, 62*(2), 172–180.

Pilon, M., Owen, J. D., Garifullina, G. F., Kurihara, T., Mihara, H., Esaki, N., & Pilon-Smits, E. A. (2003). Enhanced selenium tolerance and accumulation in transgenic Arabidopsis expressing a mouse selenocysteine lyase. *Plant Physiology, 131*(3), 1250–1257.

Poerio, T., Piacentini, E., & Mazzei, R. (2019). Membrane processes for microplastic removal. *Molecules, 24*(22), 4148.

Raouf, M. E. A., Maysour, N. E., & Farag, R. K. (2019). Wastewater treatment methodologies, review article. *International Journal of Environment and Agricultural Science, 3*(1), 18.

Reeves, R. D., & Brooks, R. R. (1983). Hyperaccumulation of lead and zinc by two metallophytes from mining areas of Central Europe. *Environmental Pollution Series A, Ecological and Biological, 31*(4), 277–285.

Rosser, S. J., French, C. E., Basran, A., Murray, J. A., Nicklin, S., & Bruce, N. C. (2001). Phytodetoxification of TNT by transgenic plants expressing a bacterial nitroreductase. *Nature Biotechnology, 19*(12), 1168–1172.

Rugh, C. L., Wilde, H. D., Stack, N. M., Thompson, D. M., Summers, A. O., & Meagher, R. B. (1996). Mercuric ion reduction and resistance in transgenic Arabidopsis thaliana plants expressing a modified bacterial merA gene. *Proceedings of the National Academy of Sciences, 93*(8), 3182–3187.

Russ, R., Rau, J., & Stolz, A. (2000). The function of cytoplasmic flavin reductases in the reduction of azo dyes by bacteria. *Applied and Environmental Microbiology, 66*(4), 1429–1434.

Sağ, Y. (2001). Biosorption of heavy metals by fungal biomass and modeling of fungal biosorption: A review. *Separation and Purification Methods, 30*(1), 1–48.

Santhanam, N. (2009). Oilgae guide to algae-based wastewater treatment. *Tamilnadu: Home of Algal Energy*. https://repository.uobabylon.edu.iq/2010_2011/4_6558_416.pdf.

Saraswat, S., & Rai, J. P. N. (2009). Phytoextraction potential of six plant species grown in multimetal contaminated soil. *Chemistry and Ecology, 25*(1), 1–11.

Sarma, H. (2011). Metal hyperaccumulation in plants: A review focusing on phytoremediation technology. *Journal of Environmental Science and Technology, 4*(2), 118–138.

Satyanarayana, U. (2010). *Biotechnology.*https://www.thebiomics.com/books/biotechnology/biotechnology-u-satyanarayana.html

Schomaker, M. (1997). *Development of environmental indicators in UNEP. FAO Land and Water Bulletin (FAO).*

Sciencing.com, Define chemical pollution. https://sciencing.com/define-chemical-pollution-6027793.html

Sekhar, K. C., Kamala, C. T., Chary, N. S., Balaram, V., & Garcia, G. (2005). Potential of Hemidesmus indicus for phytoextraction of lead from industrially contaminated soils. *Chemosphere, 58*(4), 507–514.

Selvam, A., & Wong, J. W. C. (2008). Phytochelatin systhesis and cadmium uptake of Brassica napus. *Environmental technology, 29*(7), 765–773.

Selvam, A., & Wong, J. W. C. (2009). Cadmium uptake potential of Brassica napus cocropped with Brassica parachinensis and Zea mays. *Journal of Hazardous Materials, 167*(1-3), 170–178.

Sharholy, M., Ahmad, K., Mahmood, G., & Trivedi, R. C. (2008). Municipal solid waste management in Indian cities–A review. *Waste Management, 28*(2), 459–467.

Sharma, N. C., Gardea-Torresdey, J. L., Parsons, J., & Sahi, S. V. (2004). Chemical speciation and cellular deposition of lead in Sesbania drummondii. *Environmental Toxicology and Chemistry: An International Journal, 23*(9), 2068–2073.

Sharma, S., & Shah, K. W. (2005). Generation and disposal of solid waste in Hoshangabad. In *Book of Proceedings of the Second International Congress of Chemistry and Environment* (pp. 749–751).

Shekdar, A. V., Krishnaswamy, K. N., Tikekar, V. G., & Bhide, A. D. (1992). Indian urban solid waste management systems—jaded systems in need of resource augmentation. *Waste Management, 12*(4), 379–387.

Shen, Y., Taikan, O., Shinjiro, K., Naota, H., Nobuyuki, U., & Masashi, K. (2014). Projection of future world water resources under SRES scenarios: An integrated assessment. *Hydrological Sciences Journal, 59*(10), 1775 1793. doi:10.1080/02626667.2013.862338.

Solano-Serena, F., Marchal, R., Casarégola, S., Vasnier, C., Lebeault, J. M., & Vandecasteele, J. P. (2000). A Mycobacterium strain with extended capacities for degradation of gasoline hydrocarbons. *Applied and Environmental Microbiology, 66*(6), 2392–2399.

Song, W. Y., Sohn, E. J., Martinoia, E., Lee, Y. J., Yang, Y. Y., Jasinski, M., et al. (2003). Engineering tolerance and accumulation of lead and cadmium in transgenic plants. *Nature Biotechnology, 21*(8), 914–919.

Sors, T. G., Ellis, D. R., Na, G. N., Lahner, B., Lee, S., Leustek, T., & Salt, D. E. (2005). Analysis of sulfur and selenium assimilation in Astragalus plants with varying capacities to accumulate selenium. *The Plant Journal, 42*(6), 785–797.

Srinivas, T. (2008). *Environmental biotechnology*. New Age International.

Sriprang, R., Hayashi, M., Ono, H., Takagi, M., Hirata, K., & Murooka, Y. (2003). Enhanced accumulation of Cd2+ by a Mesorhizobium sp. transformed with a gene from Arabidopsis thaliana coding for phytochelatin synthase. *Applied and Environmental Microbiology, 69*(3), 1791–1796.

Stanbury, W. T. (2003). *Accountability to Citizens in the Westminster Model of Government: More Myth Than Reality*. Fraser Institute.

Sun, Q., Ye, Z. H., Wang, X. R., & Wong, M. H. (2005). Increase of glutathione in mine population of Sedum alfredii: A Zn hyperaccumulator and Pb accumulator. *Phytochemistry, 66*(21), 2549–2556.

Sun, R., Jin, C., & Zhou, Q. (2010). Characteristics of cadmium accumulation and tolerance in Rorippa globosa (Turcz.) Thell., a species with some characteristics of cadmium hyperaccumulation. *Plant Growth Regulation, 61*(1), 67–74.

SWRCB. (2011). *Order No. R3-2011-0222: Waste discharge requirements NPDES general permit for discharges of highly treated groundwater to surface waters*. NPDES NO. CAG993002, California State Water Quality Control Board.

Takeuchi, H., & Tanaka, H. (2020). Water reuse and recycling in Japan—History, current situation, and future perspectives—. *Water Cycle, 1*, 1–12.

Tango, M. S., & Gagnon, G. A. (2003). Impact of ozonation on water quality in marine recirculation systems. *Aquacultural Engineering, 29*(3-4), 125–137.

Tanji, K. K., & Kielen, N. C. (2002). *Agricultural drainage water management in arid and semi-arid areas*. Roma (Italia): FAO.

Tembata, K., & Takeuchi, K. (2018). Collective decision making under drought: An empirical study of water resource management in Japan. *Water Resources and Economics, 22*, 19–31.

Teodosiu, C., Gilca, A. F., Barjoveanu, G., & Fiore, S. (2018). Emerging pollutants removal through advanced drinking water treatment: A review on processes and environmental performances assessment. *Journal of Cleaner Production, 197*, 1210–1221.

Thakur, M. S., Kennedy, M. J., & Karanth, N. G. (1991). An environmental assessment of biotechnological processes. *Advances in Applied Microbiology, 36*, 67–86.

Tian, J. L., Zhu, H. T., Yang, Y. A., & He, Y. K. (2004). Organic mercury tolerance, absorption and transformation in Spartina plants. *Zhi wu Sheng li yu fen zi Sheng wu xue xue bao= Journal of Plant Physiology and Molecular Biology, 30*(5), 577–582.

Trang, D. T., Molbak, K., Cam, P. D., & Dalsgaard, A. (2007b). Incidence of and risk factors for skin ailments among farmers working with wastewater-fed agriculture in Hanoi, Vietnam. *Transactions of the Royal Society of Tropical Medicine and Hygiene, 101*, 502–510. doi:10.1016/j.trstmh.2006.10.005.

Trang, D. T., van der Hoek, W., Cam, P. D., Vinh, K. T., Van Hoa, N., & Dalsgaard, A. (2006). Low risk for helminth infection in wastewater-fed rice cultivation in Vietnam. *Journal of Water and Health, 4*(3), 321–331.

Trang, D. T., van der Hoek, W., Tuan, N. D., Cam, P. D., Viet, V. H., Luu, D. D., et al. (2007a). Skin disease among farmers using wastewater in rice cultivation in Nam Dinh, Vietnam. *Tropical Medicine & International Health, 12*, 51–58. doi:10.1111/j.1365-3156.2007.01941.x.

Truong, P. (1999). *Vetiver grass technology for mine rehabilitation*. Bangkok: Office of the Royal Development Projects Board.

Tsagarakis, K. P., Tsoumanis, P., Chartzoulakis, K., & Angelakis, A. N. (2001). Water resources status including wastewater treatment and reuse in Greece: Related problems and prospectives. *Water International, 26*(2), 252–258.

Tserendorj, A., Anceno, A. J., Houpt, E. R., Icenhour, C. R., Sethabutr, O., Mason, C. S., et al. (2011). Molecular techniques in ecohealth research toolkit: Facilitating estimation of aggregate gastroenteritis burden in an irrigated periurban landscape. *EcoHealth, 8*(3), 349–364.

Valls, M., Atrian, S., de Lorenzo, V., & Fernández, L. A. (2000). Engineering a mouse metallothionein on the cell surface of Ralstonia eutropha CH34 for immobilization of heavy metals in soil. *Nature Biotechnology, 18*(6), 661–665.

Van Huysen, T., Abdel-Ghany, S., Hale, K. L., LeDuc, D., Terry, N., & Pilon-Smits, E. A. (2003). Overexpression of cystathionine-gamma- synthase enhances selenium volatilization in Brassica juncea. *Planta, 218*, 71–78.

Verbyla, M. E., Cairns, M. R., Gonzalez, P. A., Whiteford, L. M., & Mihelcic, J. R. (2015). Emerging challenges and synergies for pathogen control and resource recovery in natural wastewater treatment systems. *WIREs Water, 2*, 701–714. doi:10.1002/wat2.1101.

Verbyla, M. E., Symonds, E. M., Kafle, R. C., Cairns, M. R., Iriarte, M., Mercado, G. A., et al. (2016). Managing microbial risks from indirect wastewater reuse for irrigation in urbanizing watersheds. *Environmental Science & Technology, 50*, 6803–6813. doi:10.1021/acs.est.5b05398.

Verle, P., Kongs, A., De, N. V., Thieu, N. Q., Depraetere, K., Kim, H. T., et al. (2003). Prevalence of intestinal parasitic infections in northern Vietnam. *Tropical Medicine & International Health, 8*(10), 961–964.

Volesky, B. (1994). Advances in biosorption of metals: Selection of biomass types. *FEMS Microbiology Reviews, 14*(4), 291–302.

Wang, J., & Chen, H. (2020). Catalytic ozonation for water and wastewater treatment: Recent advances and perspective. *Science of the Total Environment, 704*, 135249.

Wang, J., Liu, G. F., Lu, H., Jin, R. F., Zhou, J. T., & Lei, T. M. (2012). Biodegradation of Acid Orange 7 and its auto-oxidative decolorization product in membrane-aerated biofilm reactor. *International Biodeterioration & Biodegradation, 67*, 73–77.

Wangeline, A. L., Burkhead, J. L., Hale, K. L., Lindblom, S. D., Terry, N., Pilon, M., et al. (2004). Overexpression of ATP sulfurylase in Indian mustard: Effects on tolerance and accumulation of twelve metals. *Journal of Environmental Quality, 33*(1), 54–60.

Wasewar, K. L., Singh, S., & Kansal, S. K. (2020). Process intensification of treatment of inorganic water pollutants. *Inorganic Pollutants in Water* (pp. 245–271). Elsevier.

Wei, S. H., Zhou, Q. X., Wang, X., Cao, W., Ren, L. P., & Song, Y. F. (2004). Potential of weed species applied to remediation of soils contaminated with heavy metals. *Journal of Environmental Sciences, 16*(5), 868–873.

World Health Organization. (1996). *The world health report: 1996: Fighting disease, fostering development*. World Health Organization.

World Health Organization. (2006). *WHO guidelines for the safe use of wasterwater excreta and greywater*: 1. World Health Organization.

Spina, F., Anastasi, A. E., Prigione, V. P., Tigini, V., & Varese, G. (2012). Biological treatment of industrial wastewaters: A fungal approach.

World Resources Institute (WRI), (2020). Aqueduct country rankings: https://www.wri.org/applications/aqueduct/country-rankings/ (Accessed September 3, 2020).

Xiao, R., Wang, S., Li, R., Wang, J. J., & Zhang, Z. (2017). Soil heavy metal contamination and health risks associated with artisanal gold mining in Tongguan, Shaanxi, China. *Ecotoxicology and Environmental Safety, 141*, 17–24.

Xu, D., Lee, L. Y., Lim, F. Y., Lyu, Z., Zhu, H., Ong, S. L., et al. (2020a). Water treatment residual: A critical review of its applications on pollutant removal from stormwater runoff and future perspectives. *Journal of Environmental Management, 259*, 109649.

Xu, X., Yang, Y., Wang, G., Zhang, S., Cheng, Z., Li, T., et al. (2020b). Removal of heavy metals from industrial sludge with new plant–based washing agents. *Chemosphere, 246*, 125816.

Yajima, A., & Kurokura, H. (2008). Microbial risk assessment of livestock-integrated aquaculture and fish handling in Vietnam. *Fisheries Science, 74*(5), 1062–1068.

Zahmatkesh, M., Spanjers, H., & van Lier, J. B. (2018). A novel approach for application of white rot fungi in wastewater treatment under non-sterile conditions: Immobilization of fungi on sorghum. *Environmental Technology, 39*(16), 2030–2040.

Zhao, X. W., Zhou, M. H., Li, Q. B., Lu, Y. H., He, N., Sun, D. H., & Deng, X. (2005). Simultaneous mercury bioaccumulation and cell propagation by genetically engineered Escherichia coli. *Process Biochemistry, 40*(5), 1611–1616.

Zhou, X., & Wang, G. (2010). Nutrient concentration variations during Oenanthe javanica growth and decay in the ecological floating bed system. *Journal of Environmental Sciences, 22*(11), 1710–1717.

Zhu, L., Li, Z., & Ketola, T. (2011). Biomass accumulations and nutrient uptake of plants cultivated on artificial floating beds in China's rural area. *Ecological Engineering, 37*(10), 1460–1466.

Zhu, Y. L., Pilon-Smits, E. A., Tarun, A. S., Weber, S. U., Jouanin, L., & Terry, N. (1999). Cadmium tolerance and accumulation in Indian mustard is enhanced by overexpressing γ-glutamylcysteine synthetase. *Plant Physiology, 121*(4), 1169–1177.

Chapter 9

Genetically engineered microorganism to degrade waste and produce biofuels and other useful products

Suchitra Kumari Panigrahy[a], Dharm Pal[b] and Awanish Kumar[c]
[a] *Department of Biotechnology, Guru GhasidasVishwavidyalaya (A Central University), Bilaspur, Chhattisgarh, India,* [b] *Associate Professor, Department of Chemical Engineering, National Institute of Technology Raipur, Raipur, Chhattisgarh, India,* [c] *Associate Professor, Department of Biotechnology, National Institute of Technology Raipur, Raipur, Chhattisgarh, India*

9.1 Introduction

Urbanization, modernization, industrial development, luxurious lifestyle lead to overuse and misuse of natural resources. The depletion of natural resources causes pollution, which becomes a matter of concern worldwide. Pollution at various levels such as untreated sewage water, industrial waste, solid waste, and portable water contaminated with nondegradable/xenobiotic compounds adversely affects many lives. The major contributors of pollution are oil spills, fertilizers, garbage, sewage disposals, and toxic chemicals (Sanghvi, Thanki, Pandey, & Singh, 2020). Release of toxic materials affects entire ecosystem, including deterioration of the quality of groundwater, soil properties, human and animal health (Zhao & Kaluarachchi, 2002). So it is necessary to degrade these pollutants for a healthy environment. Various methods such as filtration, oxidation and reduction, evaporation, solidifications, incineration, reverse osmosis, landfill deposition, electrochemical treatment, physiochemical treatment, lagooning treatment and biological methods using microorganisms and their novel enzyme systems have been widely used to remove pollutants (Bilal et al., 2019; Kanadasan & Razak, 2015; Liu, Bilal, Duan, & Iqbal, 2019; Shi, Tal, Hankins, & Gitis, 2014). The expense, final product and its complexity, chemically reactive nature, exposure to the contact persons, toxic/expensive reagent necessities, and generation of large quantities of secondary environmental

Antimicrobial Resistance in Wastewater and Human Health.
DOI: https://doi.org/10.1016/B978-0-323-96124-0.00012-X

pollutants largely restrict their applications (Dasgupta, Sikder, Chakraborty, Curcio, & Drioli, 2015; Vidali, 2001).

The bioremediation method has become one of the key approaches for the degradation of toxic pollutants using microbes due to its low investment, good efficiency, complete mineralization, and no secondary pollution (Bharagava, Chowdhary, & Saxena, 2017; Garcia-Garcia, Sanchez-Thomas, & Moreno-Sanchez, 2016; Rasheed, Bilal, Nabeel, Adeel, & Iqbal, 2019; Pillay, 1992). Enzyme complexity, versatile nature, and diverse metabolic pathways make the microorganisms as model for degrade pollutants using this as energy sources for their metabolism (Dash, Mangwani, & Das, 2014). Microorganisms form the major biomass and taxonomically include bacteria, fungi, virus, algae, protozoa, and nematodes (Annika, Arvind, & Sayali, 2012). The most commonly used microorganisms for the bioremediation process are Arthrobacter, Acromobacter, Alcaligens, Cinetobacter, Corneybacterium, Flavobacterium, Micrococcus, Mycobacterium, Nocardia, Pseudomonas, Rhodococcus, Sphingomonas, and Vibrio species (Gupta, Shrivastava, & Jain, 2001; Jayashree, Nithya, Rajesh, & Krishnaraju, 2012; Kim et al., 2007).

Natural and wild microbial strains degrade pollutants slowly and less efficiently so genetically engineered (GE) microbial strains are introduced into the polluted sites for efficient and speedy degradation (Yadav, Chowdhary, Kaithwas, & Bharagava, 2017). Nowadays genetic engineering is widely adopted to enhance the natural capacity of microorganisms for the remediation process with advanced scientific discoveries (Pant et al., 2020). For the successful in situ bioremediation using GMOs, the knowledge of biotechnology and ecology along with the field engineering techniques and biochemical processes is very much useful (Liu et al., 2019). It is a very intricate job to choose a correct microbial strain in terms of its potential, rapid growth, and nutrition responses and then, engineering it for effective waste degradation (Pant et al., 2020). So a specific bacterial strain selected and engineered should hold genes for metal homeostasis, biodegradative enzymes, metal uptake, synthesis of metal chelators, genes for survival in biotic and abiotic stress conditions, etc. (Kamthan, Chaudhuri, Kamthan, & Datta, 2016).

From the wastewater treatment, waste-activated sludge is a major by-product and becomes challenge for water management authorities (Zhang et al., 2019). Because of its high organic matter and nutrient content, it can logically replace fossil fuels with biomass (organic waste or energy crops) as a source of both energy and materials (e.g., plastics). Organic waste materials are interesting renewable resources that can be converted into different value-added products, such as bioethanol or biochemicals, bioplastics, biopesticides, etc. using microorganisms (Pagliano, Ventorino, Panico, & Pepe, 2017). Conversion of waste-activated sludge into value-added products is important to protect the ecosystem and end-users.

This chapter deals with the effective treatment of the waste using GE microbes for scale-up degradation. It simultaneously addresses the utilization of

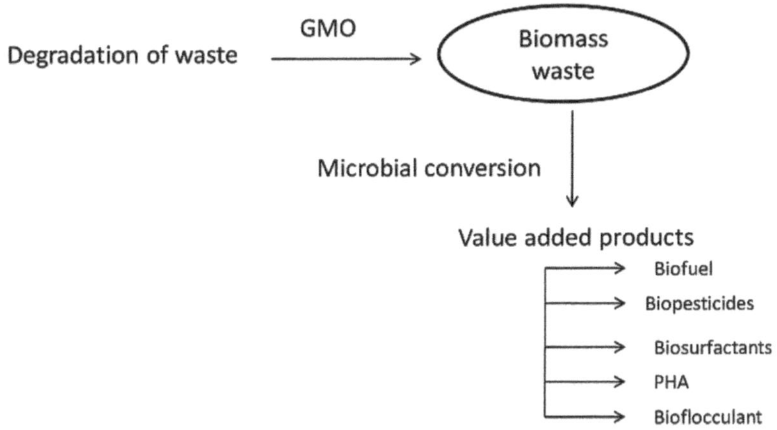

FIGURE 9.1 Conversion of waste into value-added products.

waste residues into conversion of biofuels and other useful products for a healthy and pollution-free environment (Fig. 9.1).

9.2 Development of genetically modified organisms (GMO)

Microorganisms generally deficits specific catabolic pathways for the degradation of xenobiotic compounds (Urgun-Demirtas, Stark, & Pagilla, 2006). So to overcome this, microorganisms engineered genetically to produce strains capable of large-scale bioremediation (Furukawa, 2000). Recombinant DNA technology uses a vector (phage, plasmid, or virus) into which the gene of interest has been incorporated and allowed to be expressed in an appropriate host of choice (Gupta, Sengupta, Prakash, & Tripathy, 2016). The various techniques used for that are discussed below and outlined in Fig. 9.2.

9.2.1 By using molecular tools

9.2.1.1 Molecular cloning

Here the DNA fragment having the specific function being copied and expressed in a vector called plasmid, able to replicate independently. The engineered plasmids carrying desired DNA fragments are reinserted into host bacteria and allowed for replication. The expression of the generated protein (recombinant protein) can be stimulated or suppressed by certain environmental factors (Kumar, Muthukumaran, Sharmila, & Gurunathan, 2018).

9.2.1.2 Electroporation

Here the input of high voltage electric pulses induces temporary permeation in the plasma membranes which facilitates uptake of the DNA. After removal of the pulses, resealing of pores occurs and gene remains inside the cell.

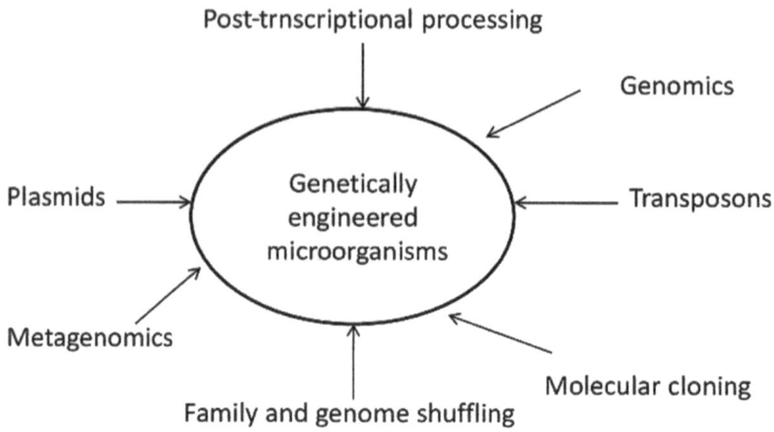

FIGURE 9.2 Important methods for construction of GMO.

9.2.1.3 Protoplast transformation

It mainly involves two steps:

a. Treatment of DNA under hypotonic conditions.
b. PEG treatment to complete the process.

9.2.1.4 Biolistic transformation

The gene of the interest is coated in a chemically inert tungsten or gold beads and introduced into the host bacterial cells with the help of the helium gas acceleration and fired through a stopping screen. The bead-associated DNA molecule leaves foreign DNA inside by passing through the bacterial cells.

9.2.2 By using recombinant DNA technology

9.2.2.1 Plasmids

The specific gene sets responsible for pollutant degradation can be found at chromosomal and extrachromosomal level. The extrachromosomal genetic material is called plasmids and used for cloning as well as an expression system. Mostly the plasmids are linked with a specific substrate metabolism found in nature (Sanghvi et al., 2020).

9.2.2.2 Transposons

Transposons are unique DNA sequences that move to random locations in a chromosome. The modifications caused by it are nonspecific but stable and also make more clones as compared to plasmid (Ramos et al., 1994).

9.2.2.3 Post-transcriptional processing

The genes linked with the parts of the enzyme are expressed and controlled by using cistronic mRNA called post-transcriptional modification

(Carrier & Keasling, 1997a). Here the foreign DNA is present in the 5′ untranslated region of the gene of interest. To make the system stable, hairpin structure was formed in m-RNA at 5′ end by introducing AG in the secondary structure. The m-RNA stability and increased protein level shown by the structure having large AG formation.

9.2.2.4 *Family and genome shuffling*

In the case of family shuffling, the shuffling of DNA occurs to the groups of related genes and genes including directed evolution (Crameri, Raillard, Bermudez, & Stemmer, 1998). Whereas the recombination of chromosomes in several bacteria to improve activity of the whole organism is called genome shuffling. It generates mutated strain having better phenotypic characteristics (Sanghvi et al., 2020).

9.2.2.5 *Genomics*

The analysis of the total genetic information in a microbial cell is done by the genomic technique. Genomic data is reliable as it also helps in the identification of specific sets of genes present in the operon which may be useful in the bioremediation.

9.2.2.6 *Metagenomics*

The complete bacterial communities found at the contaminated site could be identified by using this approach. This technique involves screening, assembling, sequencing, fluorescence-activated cell sorting, high throughput screening technology under a single niche. This tool is used to get rid of the contaminants present in the environment (Roling, 2015). Further development of DNA microarrays (He et al., 2007), mRNA expression profile of microbes (Jennings et al., 2009) and proteomics analysis (Kim et al., 2004) led to the discovery of engineered microbe strains present in the contaminated sites.

9.2.3 Techniques to identify GMOs

9.2.3.1 *PCR-based technique*

Both dead and live cells quantified by PCR-based amplification of nucleic acid whereas metabolically active cells can be detected by using RT-PCR.

9.2.3.2 *Fluorescent-based DNA hybridization technique*

Fluorescent labeled specific DNA probes of a specific strain can be used to detect the engineered microbes. The metabolic state of the cells at specific time can be detected easily as the ribosomal RNA increases along with the growth rate (Boye, Ahl, & Molin, 1995).

9.2.3.3 *Bioluminescence mediated technique*

Bioluminscent or production of colored product is used to detect specific phenotype characteristics of GMO.

9.2.3.4 *DNA microarray technique*

It uses both DNA and r-RNA as probes to identify and specify the nonrecombinant cells as well as GMO (Kumar et al., 2018).

9.3 Waste degradation by genetically modified microbes

Genetically modified microbes degrade a wide variety of compounds that are discussed below and listed in Table 9.1.

9.3.1 Heavy metal degradation

A transposon identified from the chromosome of *Bacillus megaterium* MB1, single polymerase chain reaction primer has the ability to encode broad spectrum Hg resistance (Huang, Chien, & Lin, 2010). Di-rhamnolipid biosurfactant produced by *Pseudomonas aeruginosa* strain BS 2 removed Cd and Pb effectively from artificially contaminated soil. It also increased the microbial population of the contaminated soil (Juwarkar, Nair, Dubey, Singh, & Devotta, 2007). *Bacillus subtilis* 168 GE to express arsenite-S-adenosylmethionine methyltransferase (CmarsM) gene which cause methylation and volatilization of As. This modified strain significantly enhanced As methylation and volatilization in As contaminated organic manure (Huang, Chen, Shen, Rosen, & Zhao, 2015). In another study, arsM gene from *Rhodopseudomonas polustris* when expressed in *Bacillus idriensis* increased methylation of As by 10-fold (Liu, Zhang, Chen, & Sun, 2011). The gene pcPCS 1 (Phytochelatin synthase PCS) from the bean pear (Pyrus calleryana Dene.) was overexpressed in *E. coli* and this engineered microbe had shown increased tolerance to Cd, Cu, Na, Hg (Li, Cong, Lin, & Chang, 2015). Simultaneously AtPCS gene from *Arabidopsis thaliana* expressed in *E. coli* have shown increased degradation of Cd and As by 20- and 50-fold, respectively (Sauge-Merle et al., 2003). The mer A gene from *E. coli* BL 308 was cloned and expressed in *Deinococcus radiodurans* efficiently remediate mercury-contaminated site (Brim, McFarlan, & Fredrickson, 2000).Genetically engineered *E. coli* containing MerR protein from *Shigella flexneri* which encodes the Hg binding domain enhanced mercury detoxification (Qin, Song, Brim, Daly, & Summers, 2006). *Mesorhizobium huakuii* and *E. coli* cells expressing the gene encoding PCS from Arabidopsis thaliana have reported increased Cd accumulation (Sriprang et al., 2003). A GE *E. coli* strain expressing nickel transport has shown a fast accumulation of nickel (Deng, Li, Lu, & He, 2005). Four different engineered strains of *Deinococcus radiodurans* have shown remediation of Hg in a mixed radioactive waste (Brim, McFarlan, Fredrickson, Minton, & Zhai, 2000). *E. coli* cells engineered by surface-displayed MerR protein, which showed sixfold increased biosorption of Hg (Bae, Wu, & Kostal, 2003). Engineered

TABLE 9.1 List of modified organisms in degradation of waste.

Modified organism	Application	References
Deinococcus radiodurans	Degrades Hg	Gupta, Chatterjee, Datta, Voronina, & Walther (2017)
Acidithiobacillus ferrooxidans	Degrades Hg	Ouyang et al. (2013)
Mesorhizobium huakuii	Degrades Cd	Porter, Chang, Conow, Dunham, and Friesen (2017)
Pseudomonas K-62	Degrades Hg	Chang et al. (2015)
Rhodopseudomonas palustris	Removal of Hg from wastewater	Ye, Rensing, Rosen, and Zhu (2012)
Pseudomonas fluorescens	Decolorization of different dye	Godlewska, Przystaś, and Sota (2014)
Streptomyces lividans	Enhanced thermostability and catalytic efficiency	Dubé, Shareck, Hurtubise, Daneault, & Beauregard (2008)
Pseudomonas diminuta	Degrades organo phosphorus	Bigley & Raushel (2019)
Pandoraea sp.	Enhanced xenobiotic degradation	Peeters, De Canck, and Cnockaert (2019)
Pseudomonas sp.	Enhanced biodegradation of pentafluoro sulfanyl substituted aminophenol	Saccomanno, Hussain, and O'Connor (2018)
Alcaligenes sp.	Biodegradation of 2,4 dichlorophenoxy acetic acid	Undugoda, Kannangara, and Sirisena (2016)
Pseudomonas aeruginosa	Degrades crude oil	Aybey and Demirkan (2016)
Deinococcus radiodurans R1	Removal of radioactive iodine	Choi et al. (2017)
Deinococcus radiodurans	Removal of radioactive compound	Gogada et al. (2015)
Burkholderia cepacia	Phytoremediation of volatile organic compounds	Barac et al. (2004)
Burkholderia cepacia	Degradation of toluene	Barac et al. (2004)

strain of *Cupriavidus metallidurans* MSR33 containing novel merB, merG and other mer genes from *C. metallidurans* CH34 strain exhibited broad-spectrum Hg resistance (Rojas et al., 2011). Similarly cloning and expression of merA gene from Hg resistance *E. coli* into Hg-sensitive *E. coli* resulted in increased Hg degradation (Zeyaullah, Haque, Nabi, Nand, & Ali, 2010).

9.3.2 Xenobiotic compounds degradation

lcc1 cDNA isolated from the fungus *Trametes trogii* expressed in *Pichia pastoris* to produce fungal laccase. The recombinant laccase has shown enhanced bioremediation of various xenobiotic compounds (Colao, Lupino, Garzillo, Buonocore, & Ruzzi, 2006). GE strains of fungus *Fusarium solani* with improved dehalogenase activity was raised by parasexual hybridization. Recombinants have shown enhanced DDT degradation quality (Mitra, Mukherjee, Kale, & Murthy, 2001).

An engineered strain of *Pseudomonas putida* by overexpression of Triazophos hydrolase encoded by gene tpd can degrade various organophosphorus pesticides and aromatic hydrocarbons (Gu et al., 2006). Prom1 from *Cochliobolus heterostrophas* and trpc terminator from *Aspergillus nidulans* opd gene was transformed and expressed in the fungus *Gliocladium virens*. These recombinants results in the degradation of organophosphate-derived pollutants (Dave, Lauriano, Xu, Wild, & Kenerley, 1994).

9.3.3 Organic compounds degradation

The promoters like pm from engineered TOL plasmid is responsible in the degradation of toluene and naphthalene (de Lorenzo, Fernaˊndez, Herrero, Jakubzik, & Timmis, 1993). Family shuffling in the biphenyl dioxygenase (bphA) gene isolated from *Pseudomonas* and *Bacillus* strain have shown significant increase in the degradation of PCBs, toluene (Kumamaru, Suenaga, Mitsuoka, Watanabe, & Furukawa, 1998). The same phenomenon also applicable to the bphA genes isolated from *Burkholderia* sp. and *Rhodococcus* sp. (Barriault, Plante, & Sylvestre, 2002).

The genes encoding enzymes which metabolizes chlorobenzoic acid (CBA) from *P. aeruginosa* and *Arthobacter globiformis* cloned and expressed in *Comamonas testosterone* VP44 which results in complete mineralization of monochlorobiphenyls (Hrywna, Tsoi, Maltseva, Quensen, & Tiedje, 1999). The genes encoding the 2.4 dinitrotoluene degradation pathway from *Burkbolderia* sp. were engineered into *Pseudomonas fluorescens* and this recombinant strain had superior activity in degrading DNT (Monti, Smania, Fabro, Alvarez, & Argarana, 2005). The recombinant strain was obtained by cloning the *Arthrobacter* sp. FG1 dehalogenase encoding genes in *P.putida* degrade 4-CBA rapidly as compared to the indigenous bacterial strain (Massa, Infantin, & Radice, 2009).

Recombinant *E. coli* cells encapsulating AtzA (Atrazine chlorohydrolase) gene significantly increased degradation of atrazine (Strong, McTavish, Sadowsky, & Wackett, 2000). *Pseudomonas putida* was GE using a plasmid containing naphthalene degrading gene. This engineered bacteria is stable and more effective in naphthalene degradation (Filonov, Akhmetov, & Puntus, 2005).

A plant species poplar was inoculated with the endophyte *Burkholderia cepacia* have shown horizontal gene transfer and increased remediation of toluene (Taghavi, Barac, & Greenberg, 2005). In another experiment pTOM, toluene degradation plasmid of *Burkholderia cepacia* introduced into natural endophyte B. cepacia. These engineered endophytic bacteria have shown improved efficiency in degrading volatile organic contaminant toluene (Barac et al., 2004). *Deinococcus radiodurans* was engineered by cloned expression of tod and xyl genes from Pseudomonas putida, which were able to oxidize toluene (Brim et al., 2006).

9.3.4 Dye degradation

The GE strain *E. coli* JM109 decolorize azo dye with high efficiency (Jin, Yang, Zhang, Wang, & Liu, 2009).

9.4 Conversion of biomass into value-added products

During mass-scale waste management, valuable wastes are often wasted instead of being consumed in a meaningful way due to lack of proper knowledge. These residues can replace the raw material used in various researches and industries and help to reduce production cost as well as the pollution load from the environment. Synthesis of various products from bioconversion of biomass using microbial biotechnology is generally reflected as a greenest technique (Dessie, Luo, & Wang, 2020). So the conversions of waste to various useful products using microorganisms are discussed below and listed in Table 9.2.

9.4.1 Biofuel production

Shortage of fossil combustibles, lead to the use of renewable resources such as solar, wind, tidal, and biomass for the generation of energy (Eswari et al., 2020). In the past few years for producing energy from renewable sources, anaerobic digestion is promoted for treating organic waste (Lema et al 2001; Lettinga et al., 2001). The anaerobic species belonging to the families *Streptococcaceae* and *Enterobacteriaceae* as well as the genera *Bacteroides, Clostridium, Butyrivibrio, Eubacterium, Bifidobacterium*, and *Lactobacillus* are most commonly involved in the anaerobic digestion process (Novaes, 1986).

Species belonging to the families *Clostridiaceae, Streptococcaceae, Sporolactobacillaceae, Lachnospiraceae,* and *Thermoanaerobacteriacea* produce biohydrogen from organic waste (Angenent, Khursheed, Al-dahhan, Wrenn, & Dominguez- Espinoza, 2004). Different rumen bacteria, such as *Clostridia,* methylotrophs, methanogenic archae or facultative anaerobic bacteria (*E. coli, Enterobacter* spp., *Citrobacter* spp.) and aerobic bacteria (*Alcaligenes* spp., *Bacillus* spp.) have been studied to perform dark fermentation. *Clostridium butyricum* and *Clostridium articum* specifically produce butyric acid and propionate as major products, respectively, both of which are interest for hydrogen production (Hawkes, Hussy, Kyazze, Dinsdale, & Hawkes, 2007). *Clostridium*

TABLE 9.2 List of value added products produced by microorganisms from waste.

Type of by-product	Microbial strain	References
Biohydrogen	*Clostridiaceae, Streptococcaceae, Sporolactobacillaceae, Lachnospiraceae,* and *Thermoanaerobacteriacea*	Angenent et al. (2004)
Bioplastic	*Ralstonia eutropha, Burkholderia sacchari, Bacillus megaterium, Pseudomonas resinovorans* and *Burkholderia cepacia*	Pan, Perrotta, Stipanovic, Nomura, & Nakas (2012), Cesário et al. (2014), Annamalai et al. (2018)
Biopesticide	*Bacillus thuringiensis, Trichoderma* sp.	Brar et al. (2006), Sanchis and Bourguet (2008), Verma et al. (2005)
Bioflocculants	*Achromobacter* sp., *Agrobacterium* sp., *Bacillus cereus, Exiguobacterium acetylicum, Enterobacter* sp., *Acinetobacter* sp., *Haemophilus* sp., *Citrobacter* sp., *Galactomyces* sp., *Klebsiella* sp., *Ochrobactium cicero, Pichia membranifaciens, Rhodococcus erythropolis, Saccharomycete* spp., *Solibacillus silverstris*	Batta et al. (2013), Guo et al. (2013), Wan et al. (2013), Wang et al. (2014)
Biosurfactant	*Acinetobacter, Arthrobacter, Bacillus, Pseudomonas, Rhodococcus,* and *Enterobacter*	Liang et al. (2014)
Organic acid	*Clostridium stercorarium* subsp., *Thermolacticum thermoautotrophica, Methanothermobacter thermoautotrophicus*	Collet et al. (2003)

butyricum produced biohydrogen by hydrolyzing waste biomass at various processing conditions (varied concentrations, temperature, and reaction times) (Pattra, Sangyoka, Boonmee, & Reungsang, 2008).

9.4.2 Bioplastics production

Polyhydroxyalkanoates (PHAs) are alternative to modern plastic due to their biodegradable and environmentally secure nature (Usmani et al., 2021). Recently, waste biomass and residues from various sources such as forests, agricultural land, marine, industries, and municipal solid waste gained huge interest in the viable production of PHA (Al-Battashi et al., 2019). Till now, numerous studies have been reported to produce PHA by utilizing agribiomass waste using *Ralstonia eutropha, Burkholderia sacchari, Bacillus megaterium, Pseudomonas resinovorans,* and *Burkholderia cepacia* (Annamalai, Al-Battashi, Al-Bahry, & Sivakumar, 2018; Cesário, Raposo, & de Almeida, 2014; Pan, Perrotta, Stipanovic, Nomura, & Nakas, 2012). Different bacteria (e.g., *Alcaligenes* spp., *Azotobacter* spp., methylotrophs, *Pseudomonas* spp., *Bacillus* spp. and recombinant *Escherichia coli*) have been used in PHA production using different low-cost substrates, such as organic waste and by products (Pagliano et al., 2017). The cost of PHA production substantially lowered by recombinant *Baccilus subtilis* by utilizing malt waste in the medium as a carbon source better than glucose (Law et al., 2003). *R. eutrophus* produces PHB by using industrial fruit and vegetable waste (Ganzeveld, Van Hagen, Van Agteren, de Koning, & Schoot Uiterkamp, 1999).The excess activated sludge from a wastewater treatment plant fed with industrial waste streams as a substrate for PHB accumulation by using *C. necator* species (Kumar, Mudliar, Reddy, & Chakrabarti, 2004). Wastewater from food processing and starch-rich grain-based alcohol industries was also used as a substrate for PHB production using this strain (Khardenavis et al., 2007).

9.4.3 Biopesticide production

During wastewater fermentation biopesticide is another valuable product from the microorganisms. It is highly target-specific, leave no toxic residues with lesser harmful impact on the environment than chemical pesticides (Liu et al., 2011). A commonly known biopesticide is derived from *Bacillus thuringiensis* (Bt) and consists of crystal delta endotoxins and other pesticidal substances (Brar et al., 2006; Sanchis & Bourguet, 2008). The necessary nutritional elements for sustainable growth, sporulation and crystal formation by Bt is provided by wastewater.

Trichoderma sp. also produced bioherbicides/biopesticides by using wastewater as a raw material which has a broader spectrum activity in comparison to Bt (Verma, Brar, Tyagi, Valéro, & Surampalli, 2005).

9.4.4 Bioflocculant production

Bioflocculant, an extracellular biopolymer including proteins, glycoproteins, polysaccharides, lipids, and glycolipids is excreted by microorganisms (Salehizadeh & Shojaosadati, 2003). Bioflocculants production from organic wastewater directly has been performed in several studies successfully (Li, Zhong, Lei, Chen, & Bai, 2008; Wang et al., 2007; Zhang, Lin, Xia, Wang, & Yang, 2007). Various strains of bioflocculants producing microorganisms have successfully been isolated from wastewater, including *Achromobacter* sp., *Agrobacterium* sp., *Bacillus cereus, Exiguobacterium acetylicum, Enterobacter* sp., *Acinetobacter* sp., *Haemophilus* sp., *Citrobacter* sp., *Galactomyces* sp., *Klebsiella* sp., *Ochrobactium cicero, Pichia membranifaciens, Rhodococcus erythropolis, Saccharomycete* spp., *Solibacillus silverstris*, etc. (Batta, Subudhi, Lal, & Devi, 2013; Guo, Yang, & Zeng, 2013; Wan, Zhao, Guo, Alam, & Bai, 2013; Wang, Lee, Ma, Wang, & Ren, 2014).

9.4.5 Biosurfactant production

Biosurfactants are amphipathic compounds excreted by different bacterial genera (e.g., *Acinetobacter, Arthrobacter, Bacillus, Pseudomonas, Rhodococcus,* and *Enterobacter*) (Liang et al., 2014) with low toxicity, surface, and interfacial tension but higher efficiency at extreme conditions (Nitschke & Pastore, 2006). The biosurfactant family includes a variety of compounds like glycolipids, lipopeptides, fatty acids, polysaccharide-protein complexes, peptides, phospholipids, and neutral lipids (Yin et al., 2009). The type and quantity of the biosurfactant production varies significantly with the microorganism species and the wastewater composition (Das, Mukherjee, & Sen, 2009). Some microorganisms, such as *P. aeruginosa, R. eutropha, A. beijerinckii, A. chroococcum,* and *P. mendocina,* are able to concurrently produce PHAs and biosurfactants using the same type of organic substrate (Pagliano et al., 2017).

9.4.6 Organic acids and chemicals production

The first and key step of the bioconversion process is the conversion of biomass waste to energy and biochemicals, done by pretreatments of biomass to help in conversion of sugars (hydrolysate). These hydrolysates (sugars) are converted further into biofuels such as biohydrogen, methane, ethanol, methanol, butanol, and other biochemicals or organic acids such as butyric, lactic, acetic, and propionic acids (Liu & Wu, 2016). Various microbial strains such as *Clostridium stercorarium* subsp., *Thermolacticum thermoautotrophica, Methanothermobacter thermoautotrophicus* produced acetic acid from dairy waste products (Collet et al., 2003).

9.5 Future prospects and conclusion

The Use of GE bacteria for the degradation of waste has been widely studied all over the world, which reported the high prospective of genetically modified bacteria for the bioremediation of pollutants as compared to the conventional methods. Various engineering strains are constructed by introducing the genes involved in pollutant degradation into the recipients. This enhances the adaptability and efficiency of bacteria to degrade heavy metals and other environmental pollutants. So to fulfill the demand for waste remediation more attention should be given for designing efficient GE bacteria.

However, the field applications of GE bacteria are limited due to various ecological and environmental factors. After entering to the natural environment, GE bacteria with high degradation capacity will have a long-term impact on the atmosphere and human beings. Due to the detrimental effects of engineered strains on human survival and the stability of ecological environment, its broad use in the natural environment is still controversial.

So further researches should be focused on increasing the efficiency and optimization strategies of GE bacteria. Detailed studies of the field and various comparative life cycle assessments also needed to minimize the field risks for improved microbial benefits.

Extensive studies needed to be carried out for increasing the survival rate of the microbes when released in the polluted sites. More work is still required to determine their side effects at the remediation site and nearby ecological niches by studying the metabolic engineering of modified microorganisms. More credence needs to be given to know the combinatorial effect of plants and microbes and make it economical. Furthermore, the omics data and transcriptomics at mRNA level should be studied to know and check the expression of genes during the degradation of compounds.

Recovering valuable resources and value-added bioproducts provides an alternative treatment pathway for increasing and problematic urban waste. Recovery of metabolic products (bioplastics, biopesticides, bioflocculants, and biosurfactants) remains in its early stage and further research on optimization of operational parameters and selection of nontoxic strains are needed to strengthen this area. To scale up the biorefinery approach using waste, further research must be carried out at larger scales (pilot- and ultimately full-scale) to optimize each biorefinery process. This will help to understood techno-economic performance of these processes.

References

Al-Battashi, H. S., Annamalai, N., Sivakumar, N., Al-Bahry, S., Tripathi, B. N., Nguyen, Q. D., et al. (2019). Lignocellulosic biomass (LCB): A potential alternative biorefinery feedstock for polyhydroxyalkanoates production. *Reviews in Environmental Science and Biotechnology, 18*, 183–205.

Angenent, L. T., Khursheed, K., Al-dahhan, M., Wrenn, B. A., & Dominguez- Espinoza, R. (2004). Production of bioenergy and biochemicals from industrial and agricultural wastewater. *Trends in Biotechnology, 22*, 477–485.

Annamalai, N., Al-Battashi, H., Al-Bahry, S., & Sivakumar, N. (2018). Biorefinery production of poly-3-hydroxybutyrate using waste office paper hydrolysate as feedstock for microbial fermentation. *Journal of Biotechnology, 265*, 25–30.

Annika, A. D., Arvind, R. G., & Sayali, R. N. (2012). Decolorization of textile dyes and biological stains by bacterial strains isolated from industrial effluents. *Advances in Applied Science Research, 3*(5), 2660–2671.

Aybey, A., & Demirkan, E. (2016). Inhibition of quorum sensing-controlled virulence factors in Pseudomonas aeruginosa by human serum paraoxonase. *Journal of Medical Microbiology, 65*, 105–113.

Bae, W., Wu, C., & Kostal, J. (2003). Enhanced mercury biosorption by bacterial cells with surface-displayed MerR. *Applied and Environmental Microbiology, 69*, 3176–3180.

Barac, T., Taghavi, S., Borremans, B., Provoost, A., Oeyen, L., Colpaert, J. V., et al. (2004). Engineered endophytic bacteria improve phytoremediation of water-soluble, volatile, organic pollutants. *Nature Biotechnology, 22*(5), 583–588.

Barriault, D., Plante, M. M., & Sylvestre, M. (2002). Family shuffling of a targeted bphA region to engineer biphenyl dioxygenase. *Journal of Bacteriology, 184*(14), 3794–3800.

Batta, N., Subudhi, S., Lal, B., & Devi, A. (2013). Isolation of a lead tolerant novel bacterial species, Achromobacter sp. TL-3: Assessment of bioflocculant activity. *Indian Journal of Experimental Biology, 51*, 1004–1011.

Bharagava, R. N., Chowdhary, P., & Saxena, G. (2017). Bioremediation an eco-sustainable green technology, its applications and limitations. In R. N. Bharagava (Ed.), Environmental pollutants and their bioremediation approaches (pp. 1–22). Boca Raton: CRC Press, Taylor & Francis Group.

Bigley, A. N., & Raushel, F. M. (2019). The evolution of phosphotriesterase for decontamination and detoxification of organophosphorus chemical warfare agents. *Chemico-Biological Interactions, 308*, 80–88.

Bilal, M., & Iqbal, H. M. (2019). Microbial-derived biosensors for monitoring environmental contaminants: Recent advances and future outlook. *Process Safety and Environment Protection, 124*, 8–17.

Boye, M., Ahl, T., & Molin, S. (1995). Application of strain-specific rRNA oligonucleotide probe targeting *Pseudomonas fluorescens* Ag1 in a mesocosm study of bacterial release into the environment. *Applied and Environmental Microbiology, 61*, 1384–1390.

Brar, S. K., Verma, M., Barnabé, S., Tyagi, R. D., Valéro, J. R., & Surampalli, R. Y. (2006). Efficient centrifugal recovery of Bacillus thuringiensis biopesticides from fermented wastewater and wastewater sludge. *Water Research, 40*, 1310–1320.

Brim, H., McFarlan, S. C., Fredrickson, J. K., Minton, K. W., Zhai, M., Wackett, L. P., & Daly, M. J. (2000). Engineering Deinococcus radiodurans for metal remediation in radioactive mixed waste environments. *Nature Biotechnology, 18*, 85–90.

Brim, H., Osborne, J. P., Kostandarithes, H. M., Fredrickson, J. K., Wackett, L. P., & Daly, M. J. (2006). Deinococcus radiodurans engineered for complete toluene degradation facilitates Cr (VI) reduction. *Microbiology (Reading, England), 152*, 2469–2477.

Carrier, T. A., & Keasling, J. D. (1997a). Engineering mRNA stability in E. coli by the addition of synthetic hairpins using a 50 cassette system. *Biotechnology and Bioengineering, 55*(3), 577–580.

Cesário, M. T., Raposo, R. S., de Almeida, M. C. M. D., et al. (2014). Production of poly(3-hydroxybutyrate-co-4-hydroxybutyrate) by *Burkholderia sacchari* using wheat straw hydrolysates and gamma-butyrolactone. *International Journal of Biological Macromolecules, 71*, 59–67.

Chang, S., Wei, F., Yang, Y., Wang, A., Jin, Z., Li, J., et al. (2015). Engineering tobacco to remove mercury from polluted soil. *Applied Biochemistry and Biotechnology, 175*, 3813–3827.

Choi, M. H., Jeong, S. W., Shim, H. E., Yun, S. J., Mushtaq, S., Choi, D. S., et al. (2017). Efficient bioremediation of radioactive iodine using biogenic gold nanomaterial-containing radiation-resistant bacterium, Deinococcus radiodurans R1. *Chemical Communications, 53*, 3937–3940.

Colao, M. C., Lupino, S., Garzillo, A. M., Buonocore, V., & Ruzzi, M. (2006). Heterologous expression of lcc1 gene from Trametes trogii in Pichia pastoris and characterization of the recombinant enzyme. *Microbial Cell Factories, 5*, 31.

Collet, C., et al. (2003). Improvement of acetate production from lactose by growing Clostridium thermolacticum in mixed batch culture. *Journal of Applied Microbiology, 95*, 824–831.

Crameri, A., Raillard, S. A., Bermudez, E., & Stemmer, W. P. (1998). DNA shuffling of a family of genes from diverse species accelerates directed evolution. *Nature, 391*(6664), 288.

Das, P., Mukherjee, S., & Sen, R. (2009). Substrate dependent production of extracellular biosurfactant by a marine bacterium. *Bioresource Technology, 100*, 1015–1019.

Dasgupta, J., Sikder, J., Chakraborty, S., Curcio, S., & Drioli, E. (2015). Remediation of textile effluents by membrane based treatment techniques: A state of the art review. *Journal of Environmental Management, 147*, 55–72.

Dash, H. R., Mangwani, N., & Das, S. (2014). Characterization and potential application in mercury bioremediation of highly mercury-resistant marine bacterium Bacillus thuringiensis PW-05. *Environmental Science and Pollution Research, 21*(4), 2642–2653.

Dave, K. I., Lauriano, C., Xu, B., Wild, J. R., & Kenerley, C. M. (1994). Expression of organophosphate hydrolase in the filamentous fungus *Gliocladium virens*. *Applied Microbiology and Biotechnology, 41*, 352–358.

de Lorenzo, V., Ferna´ndez, S., Herrero, M., Jakubzik, U., & Timmis, K. N. (1993). Engineering of alkyl- and haloaromatic-responsive gene expression with mini-transposons containing regulated promoters of biodegradative pathways of Pseudomonas. *Gene, 130*(1), 41–46.

Deng, X., Li, Q. B., Lu, Y. H., & He, N. (2005). Genetic engineering of Escherichia coli SE5000 and its potential for Ni21 bioremediation. *Process Biochemistry, 40*, 425–430.

Dessie, W., Luo, X., Wang, M., Feng, L., Liao, Y., Wang, Z., et al. (2020). Current advances on waste biomass transformation into value-added products. *Applied Microbiology and Biotechnology, 104*, 4757–4770.

Dubé, E., Shareck, F., Hurtubise, Y., Daneault, C., & Beauregard, M. (2008). Homologous cloning, expression, and characterization of a laccase from Streptomyces coelicolor and enzymatic decolourisation of an indigo dye. *Applied Microbiology and Biotechnology, 79*, 597–603.

Eswari, A. P., Meena, R. A., Kannah, R. Y., Sakthinathan, G., Karthikeyan, O. P., & Banu, J. R. (2020). Chapt22- Bioconversion of marine waste biomass for biofuel and value-added products recovery. In R. Praveen Kumar, Edgard Gnansounou, Jegannathan Kenthorai Raman, & Gurunathan Baskar (Eds.), *Refining Biomass Residues for Sustainable Energy and Bioproducts* (pp. 481–507) Eds. Academic Press.

Filonov, A. E., Akhmetov, L. I., Puntus, I. F., et al. (2005). The construction and monitoring of genetically tagged, plasmid-containing, naphthalene-degrading strains in soil. *Microbiology, 74*, 526–532.

Furukawa, K. (2000). Biochemical and genetic bases of microbial degradation of polychlorinated biphenyls (PCBs). *Journal of General and Applied Microbiology, 46*, 283–296.

Ganzeveld, K. J., Van Hagen, A., Van Agteren, M. H., de Koning, W., & Schoot Uiterkamp, A. J. M. (1999). Upgrading of organic waste: production of the copolymer poly-3-hydroxybutyrate-co-valerate by *Ralstonia eutropha* with organic waste as sole carbon source. *Journal of Cleaner Production, 7*, 413–419.

Garcia-Garcia, J. D., Sanchez-Thomas, R., & Moreno-Sanchez, R. (2016). Bio-recovery of nonessential heavy metals by intra-and extracellular mechanisms in free-living microorganisms. *Biotechnology Advances, 34*(5), 859–873.

Godlewska, E. Z., Przystaś, W., & Sota, E. G. (2014). Decolourisation of different dyes by two pseudomonas strains under various growth conditions. *Water, Air, & Soil Pollution, 225*, 1846.

Gogada, R., Singh, S. S., Lunavat, S. K., Pamarthi, M. M., Rodrigue, A., Vadivelu, B., et al. (2015). Engineered Deinococcus radiodurans R1 with NiCoT genes for bioremoval of trace cobalt from spent decontamination solutions of nuclear power reactors. *Applied Microbiology and Biotechnology, 99*, 9203–9213.

Gu, L., He, J., Huang, X., Jia, K., & Li, S. (2006). Construction of a versatile degrading bacteria Pseudomonas putida KT2440-DOP and its degrading characteristics. *Acta Microbiologica Sinica, 46*, 763–766.

Guo, J., Yang, C., & Zeng, G. (2013). Treatment of swine wastewater using chemically modified zeolite and bioflocculant from activated sludge. *Bioresource Technology, 143*, 289–297.

Gupta, V., Sengupta, M., Prakash, J., & Tripathy, B. C. (2016). Basic and applied aspects of biotechnology. Springer 2016.

Gupta, V. K., Shrivastava, A. K., & Jain, N. (2001). Biosorption of chromium (VI) from aqueous solutions by green algae Spirigyra species. *Water Research, 35*(17), 4079–4085.

Gupta, D.K., Chatterjee, S., Datta, S., Voronina, A.V., Walther, C. (2017). Radionuclides: Accumulation and Transport in Plants. *Reviews of Environmental Contamination and Toxicology 241*, 139–160.

Hawkes, F. R., Hussy, I., Kyazze, G., Dinsdale, R., & Hawkes, D. L. (2007). Continuous dark fermentative hydrogen production by mesophilic microflora: Principles and progress. *International Journal of Hydrogen Energy, 32*, 172–184.

He, Z., Gentry, T. J., Schadt, C. W., Wu, L., Liebich, J., Chong, S. C., … Zhou, J. (2007 May). GeoChip: a comprehensive microarray for investigating biogeochemical, ecological and environmental processes. *ISME J, 1*(1), 67–77.

Hrywna, Y., Tsoi, T. V., Maltseva, O. V., Quensen, J. F., & Tiedje, J. M. (1999). Construction and characterization of two recombinant bacteria that grow on *ortho*-and *para*–substituted chlorobiphenyls. *Applied and Environmental Microbiology, 65*, 2163–2169.

Huang, C. C., Chien, M. F., & Lin, K. H. (2010). Bacterial mercury resistance of TnMERI1 and its' application in bioremediation. *Interdisciplinary Studies on Environmental Chemistry, 3*(11), 21–29.

Huang, K., Chen, C., Shen, Q., Rosen, B. P., & Zhao, F. J. (2015). Genetically engineering Bacillus subtilis with a heat resistant arsenite methyltransferase for bioremediation of arsenic-contaminated organic waste. *Applied and Environmental Microbiology, 81*, 6718–6724.

Jayashree, R., Nithya, S. E., Rajesh, P. P., & Krishnaraju, M. (2012). Biodegradation capability of bacterial species isolated from oil contaminated soil. *Journal of Academia and Industrial Research, 1*(3), 127–135.

Jennings, L. K., Chartrand, M. M., Lacrampe-Couloume, G., Lollar, B. S., Spain, J. C., & Gossett, J. M. (2009). Proteomic and transcriptomic analyses reveal genes upregulated by cisdichloroethene in Polaromonas sp. strain JS666. *Applied and Environmental Microbiology, 75*(11), 3733–3744.

Jin, R., Yang, H., Zhang, A., Wang, J., & Liu, G. (2009). Bioaugmentation on decolorization of C.I. direct blue 71 using genetically engineered strain Escherichia coli JM109 (pGEX-AZR). *Journal of Hazardous Materials, 163*, 1123–1128.

Juwarkar, A. A., Nair, A., Dubey, K. V., Singh, S. K., & Devotta, S. (2007). Biosurfactant technology for remediation of cadmium and lead contaminated soils. *Chemosphere, 68*, 1996–2000.

Kamthan, A., Chaudhuri, A., Kamthan, M., & Datta, A. (2016). Genetically modified (GM) crops: Milestones and new advances in crop improvement. *Theoretical Applied Genetics, 129*, 1639–1655.

Kanadasan, J., & Razak, H. A. (2015). Engineering and sustainability performance of self-compacting palm oil mill incinerated waste concrete. *Journal of Cleaner Production, 89*, 78–86.

Khardenavis, A. A., Kumar, M. S., Mudliar, S. N., & Chakrabarti, T. (2007). Biotechnological conversion of agro-industrial wastewaters into biodegradable plastic, poly β-hydroxybutyrate. *Bioresource Technology, 98*(18), 3579–3584.

Kim, S. J., Jones, R. C., Cha, C. J., Kweon, O., Edmondson, R. D., & Cerniglia, C. E. (2004). Identification of proteins induced by polycyclic aromatic hydrocarbon in Mycobacterium vanbaalenii PYR-1 using two-dimensional polyacrylamide gel electrophoresis and de novo sequencing methods. *Proteomics, 4*(12), 3899–3908.

Kim, S. U., Cheong, Y. H., Seo, D. C., Hu, J. S., Heo, J. S., & Cho, J. S. (2007). Characterization of heavy metal tolerance and biosorption capacity of bacterium strains CPB4 (Bacillus Sp.). *Water Science and Technology, 55*(1), 105–111.

Kumamaru, T., Suenaga, H., Mitsuoka, M., Watanabe, T., & Furukawa, K. (1998). Enhanced degradation of polychlorinated biphenyls by directed evolution of biphenyl dioxygenase. *Nature Biotechnology, 16*(7), 663.

Kumar, N. M., Muthukumaran, C., Sharmila, G., & Gurunathan, B. (2018). Genetically modified organisms and its impact on the enhancement of bioremediation. In S. Varjani, A. Agarwal, E. Gnansounou, & B. Gurunathan (Eds.), Bioremediation: Applications for environmental protection and management. Energy, Environment, and Sustainability. Singapore: Springer.

Kumar, S. R., Mudliar, S. N., Reddy, K. M. K., & Chakrabarti, T. (2004). Production of biodegradable plastics from activated sludge generated from a food processing industrial wastewater treatment plant. *Bioresource Technology, 95*, 327–330.

Law, K. H., Chenga, Y. C., Leungb, Y. C., Lob, W. H., Chuac, H., & Yua, H. F. (2003). Construction of recombinant *Bacillus subtilis* strains for polyhydroxyalkanoates synthesis. *Biochemical Engineering Journal, 16*, 203–208.

Lema, J. M., & Omil, F. (2001). Anaerobic treatment: A key technology for a sustainable management of wastes in Europe. *Water Science and Technology, 44*, 33–140.

Lettinga, G. (2001). Digestion and degradation, air for life. *Water Science and Technology, 44*, 57–176.

Li, H., Cong, Y., Lin, J., & Chang, Y. (2015). Enhanced tolerance and accumulation of heavy metal ions by engineered Escherichia coli expressing Pyrus calleryana phytochelatin synthase. *Journal of Basic Microbiology, 55*, 398–405.

Li, Z., Zhong, S., Lei, H. Y., Chen, R. W., & Bai, T. (2008). Production and application of a bioflocculant by culture of Bacillus licheniformis X14 using starch wastewater as carbon source. *Journal of Biotechnology, 136*, 313.

Liang, T. W., Wu, C. C., Cheng, W. T., Chen, Y. C., Wang, C. L., Wang, I. L., et al. (2014). Exopolysaccharides and antimicrobial biosurfactants produced by *Paenibacillus macerans* TKU029. *Applied Biochemistry and Biotechnology, 172*, 933–950.

Liu, C. M., & Wu, S. Y. (2016). From biomass waste to biofuels and biomaterial building blocks. *Renewable Energy, 96*, 1056–1062.

Liu, L., Bilal, M., Duan, X., & Iqba, H. M. N. (2019). Mitigation of environmental pollution by genetically engineered bacteria - Current challenges and future perspectives. *Science of the Total Environment, 667*, 444–454.

Liu, S., Zhang, F., Chen, J., & Sun, G. (2011). Arsenic removal from contaminated soil via biovolatilization by genetically engineered bacteria under laboratory conditions. *Journal of Environmental Sciences, 23*, 1544–1550.

Massa, V., Infantin, O. A., Radice, F., Orlandi, V., Tavecchio, F., Giudici, R., et al. (2009). Efficiency of natural and engineered bacterial strains in the degradation of 4-chlorobenzoic acid in soil slurry. *International Biodeterioration & Biodegradation, 63*, 112–115.

Mitra, J., Mukherjee, P. K., Kale, S. P., & Murthy, N. B. K. (2001). Bioremediation of DDT in soil by genetically improved strains of soil fungus *Fusarium solani*. *Biodegradation, 12*, 235–245.

Monti, M. R., Smania, A. M., Fabro, G., Alvarez, M. E., & Argarana, C. E. (2005). Engineering *Pseudomonas fluorescens* for biodegradation of 2,4-dinitrotoluene. *Applied and Environmental Microbiology, 71*, 8864–8872.

Nitschke, M., & Pastore, G. M. (2006). Production and properties of a surfactant obtained from Bacillus subtilis grown on cassava wastewater. *Bioresource Technology, 97*, 336–341.

Novaes, R. F. V. (1986). Microbiology of anaerobic digestion. *Water Science and Technology, 18*, 1–14.

Ouyang, J., Guo, W., Li, B., Gu, L., Zhang, H., & Chen, X. (2013). Proteomic analysis of differential protein expression in Acidithiobacillus ferrooxidans cultivated in high potassium concentration. *Microbiological Research, 168*, 455–460.

Pagliano, G., Ventorino, V., Panico, A., & Pepe, O. (2017). Integrated systems for biopolymers and bioenergy production from organic waste and by–products: A review of microbial processes. *Biotechnology for Biofuels, 10*, 113.

Pan, W., Perrotta, J. A., Stipanovic, A. J., Nomura, C. T., & Nakas, J. P. (2012). Production of polyhydroxyalkanoates by *Burkholderia cepacia* ATCC 17759 using a detoxified sugar maple hemicellulosic hydrolysate. *Journal of Industrial Microbiology & Biotechnology, 39*, 459–469.

Pant, G., Garlapati, D., Agrawal, U., Prasuna, R. G., Mathimani, T., & Pugazhendhi, A. (2020). Biological approaches practised using genetically engineered microbes for a sustainable environment: A review. *Journal of Hazardous Materials, 405*, 124631.

Pattra, S., Sangyoka, S., Boonmee, M., & Reungsang, A. (2008). Bio-hydrogen production from the fermentation of sugarcane bagasse hydrolysate by Clostridium butyricum. *International Journal of Hydrogen Energy, 33*(19), 5256–5265.

Peeters, C., De Canck, E., Cnockaert, M., Brandt, E., Snauwaert, C., Verheyde, B., et al. (2019). Comparative genomics of Pandoraea, a genus enriched in xenobiotic biodegradation and metabolism. *Frontiers in Microbiology, 10*, 2556.

Pillay, T.V.R. (1992). Aquaculture and the Environment. Fishing News Books, 1992.

Porter, S. S., Chang, P. L., Conow, C. A., Dunham, J. P., & Friesen, M. L. (2017). Association mapping reveals novel serpentine adaptation gene clusters in a population of symbiotic Mesorhizobium. *The ISME Journal, 11*, 248.

Qin, J., Song, L., Brim, H., Daly, M. J., & Summers, A. O. (2006). Hg(II) sequestration and protection by the MerR metal-binding domain (MBD). *Microbiology (Reading, England), 152*, 709–719.

Ramos, J. L., Díáz, E., Dowling, D., de Lorenzo, V., Molin, S., O'Gara, F., et al. (1994). The behavior of bacteria designed for biodegradation. *Biotechnology, 12*(12), 1349.

Rasheed, T., Bilal, M., Nabeel, F., Adeel, M., & Iqbal, H. M. (2019). Environmentally-related contaminants of high concern: Potential sources and analytical modalities for detection, quantification, and treatment. *Environment International, 122*, 52–66.

Rojas, L. A., Yanez, C., Gonzalez, M., Lobos, S., Smalla, K., & Seeger, M. (2011). Characterization of the metabolically modified heavy metal-resistant Cupriavidus metallidurans strain MSR33 generated for mercury bioremediation. *Plos One, 6*, 1–10.

Roling, W. F. (2015). Maths on microbes: Adding microbial ecophysiology to metagenomics. *Microbial Biotechnology, 8*(1), 21.

Saccomanno, M., Hussain, S., O'Connor, N. K., Beier, P., Somlyay, M., Konrat, R., et al. (2018). Biodegradation of pentafluorosulfanyl substituted aminophenol in Pseudomonas spp. *Biodegradation, 29*(3), 259–270.

Salehizadeh, H., & Shojaosadati, S. A. (2003). Removal of metal ions from aqueous solution by polysaccharide produced from Bacillus firmus. *Water Research, 37*, 4231–4235.

Sanchis, V., & Bourguet, D. (2008). Bacillus thuringiensis: Applications in agriculture and insect resistance management: A review. *Agronomy for Sustainable Development, 28*, 11–20.

Sanghvi, G., Thanki, A., Pandey, S., & Singh, N. K. (2020). Engineered bacteria for bioremediation. In V. C. Pandey, & V. Singh (Eds.), *Bioremediation of Pollutants* (pp. 359–374). Elsevier.

Sauge-Merle, S., Cuine, S., Carrier, P., Lecomte-Pradines, C., Luu, D. T., & Peltier, G. (2003). Enhanced toxic metal accumulation in engineered bacterial cells expressing Arabidopsis thaliana phytochelatin synthase. *Applied and Environmental Microbiology, 69*(1), 490–494.

Shi, X., Tal, G., Hankins, N. P., & Gitis, V. (2014). Fouling and cleaning of ultrafiltration membranes: A review. *Journal of Water Process Engineering, 1*, 121–138.

Sriprang, R., Hayashi, M., Ono, H., Takagai, M., Hirata, K., & Murooka, Y. (2003). Enhanced accumulation of Cd^{2+} by a Mesorhizobium sp. transformed with a gene from Arabidopsis thaliana coding for phytochelatin synthase. *Applied and Environmental Microbiology, 69*, 1791–1796.

Strong, L. C., McTavish, H., Sadowsky, M. J., & Wackett, L. P. (2000). Field-scale remediation of atrazine-contaminated soil using recombinant *Escherichia coli* expressing atrazine chlorohydrolase. *Environmental Microbiology, 2*, 91–98.

Taghavi, S., Barac, T., Greenberg, B., et al. (2005). Horizontal gene transfer to endogenous endophytic bacteria from poplar improves phytoremediation of toluene. *Applied and Environmental Microbiology, 71*, 8500–8505.

Undugoda, L. J. S., Kannangara, S., & Sirisena, D. M. (2016). Genetic basis of naphthalene and phenanthrene degradation by phyllosphere bacterial strains *Alcaligenes faecalis* and *Alcaligenes* sp.11SO. *Journal of Bioremediation & Biodegradation, 7*(2), 1–5.

Urgun-Demirtas, M., Stark, B., & Pagilla, K. (2006). Use of genetically engineered microorganisms (GEMs) for the bioremediation of contaminants. *Critical Reviews in Biotechnology, 26*, 145–164.

Usmani, Z., Sharma, M., Awasthi, A. K., Sivakumar, N., Lukk, T., Pecoraro, L., et al. (2021). Bioprocessing of waste biomass for sustainable product development and minimizing environmental impact. *Bioresource Technology, 322*, 124548.

Verma, M., Brar, S. K., Tyagi, R., Valéro, J., & Surampalli, R. (2005). Wastewater sludge as a potential raw material for antagonistic fungus (Trichoderma sp.): Role of pre-treatment and solids concentration. *Water Research, 39*, 3587–3596.

Vidali, M. (2001). Bioremediation: An overview. *Journal of Applied Chemistry, 73*(7), 1163–1172.

Wan, C., Zhao, X. Q., Guo, S. L., Alam, M. A., & Bai, F. W. (2013). Bioflocculant production from Solibacillus silvestris W01 and its application in cost-effective harvest of marine microalga Nannochloropsis oceanica by flocculation. *Bioresource Technology, 135*, 207–212.

Wang, L., Lee, D. J., Ma, F., Wang, A., & Ren, N. (2014). Bioflocculants from isolated strain or mixed culture: Role of phosphate salts and Ca2+ ions. *Journal of the Taiwan Institute of Chemical Engineers, 45*, 527–532.

Wang, S. G., Gong, W. X., Liu, X. W., Tian, L., Yue, Q. Y., & Gao, B. Y. (2007). Production of a novel bioflocculant by culture of Klebsiella mobilis using dairy wastewater to decrease the costs associated with the production of bioflocculants. *Biochemical Engineering Journal, 36*, 81–86.

Yadav, A., Chowdhary, P., Kaithwas, G., & Bharagava, R. (2017). Toxic metals in environment, threats on ecosystem and bioremediation approaches. *Handbook of Metalmicrobe Interactions and Bioremediation* p. 813. Boca Raton, FL: CRC Press, Taylor & Francis Group.

Ye, J., Rensing, C., Rosen, B. P., & Zhu, Y. G. (2012). Arsenic biomethylation by photosynthetic organisms. *Trends in Plant Science, 17*, 155–162.

Yin, H., Qiang, J., Jia, Y., Ye, J. S., Peng, H., Qin, H. M., et al. (2009). Characteristics of biosurfactant produced by Pseudomonas aeruginosa S6 isolated from oil-containing wastewater. *Process Biochemistry, 44*, 302–308.

Zeyaullah, M., Haque, S., Nabi, G., Nand, K. N., & Ali, A. (2010). Molecular cloning and expression of bacterial mercuric reductase gene. *African Journal of Biotechnology, 9*, 3714–3718.

Zhang, A., Venkatesh, V. G., Liu, Y., Wan, M., Qu, T., & Huisingh, D. (2019). Barriers to smart waste management for a circular economy in China. *Journal of Cleaner Production, 240*, 118198.

Zhang, Z. Q., Lin, B., Xia, S. Q., Wang, X. J., & Yang, A. M. (2007). Production and application of a bioflocculant by multiple-microorganism consortia using brewery wastewater as carbon source. *Journal of Environmental Sciences, 19*, 660–666.

Zhao, Q., & Kaluarachchi, J. J. (2002). Risk assessment at hazardous waste-contaminated sites with variability of population characteristics. *Environment International, 28*(1-2), 41–53.

Chapter 10

Human health hazards due to antimicrobial resistance spread

Shom Prakash Kushwaha[a], Syed Misbahul Hasan[a], Arun Kumar[a], Muhammad Arif[a] and Munendra Mohan Varshney[b]
[a] *Faculty of Pharmacy, Integral University, Dasauli, Kursi Road, Lucknow, Uttar Pradesh, India,*
[b] *Raj Kumar Goel Institute of Technology (Pharmacy), 5 Km Stone, Delhi-Meerut Road, Ghaziabad, Uttar Pradesh, India*

10.1 Introduction

A golden age of antibiotic discovery spanned from the 1950s through the 1970s, when new classes of antibiotics were found to cure illnesses like tuberculosis (TB) and syphilis that had previously been inoperable. Since then, new classes of antibiotics have been hard to come by, which is concerning given the bacterial robustness that has been shown throughout time, as well as the widespread misuse of antibiotics in medicine (Aminov & Mackie, 2007).

10.1.1 Antimicrobial resistance

Antimicrobial resistance (AMR) has emerged as a major public health issue of the 21st century, posing a threat to the effective prevention and treatment of an ever-increasing number of infections caused by bacteria, viruses, fungi, and parasites that are no longer treatable with common antibiotics. This AMR develops over time, mainly as a result of genetic mutations. Since the late 1960s, just a few new classes of antibiotics have been discovered and research has essentially come to a halt. *Acinetobacter baumannii, Pseudomonas aeruginosa* and certain enterobacterial species (such *Escherichia coli* and *Klebsiella* species) have progressively restricted therapy choices for Gram-negative bacteria (Aminov, 2009). People, animals, food, plants, and the environment all include antimicrobial-resistant microbes. They may transmit from person to person or from animal to animal, including via the consumption of animalderived foods. Antibiotic resistance in bacteria makes the issue of AMR even more serious. Bacteria that cause common or serious diseases have become more resistant to new antibiotics during the last few decades. In light of this, immediate action is required to avoid a global healthcare catastrophe from escalating. There is a significant burden of

Antimicrobial Resistance in Wastewater and Human Health.
DOI: https://doi.org/10.1016/B978-0-323-96124-0.00007-6

infectious illness in India, and a new analysis shows that antimicrobial medicines are being used inappropriately and irrationally against these infections, leading to a rise in AMR development. Thus, antibiotic resistance must be monitored and managed in human medicine as well as veterinary, agri-food production, and aquaculture (Normark & Normark, 2002).

10.2 Spread of antimicrobial resistance

Antibiotic-resistant bacteria may spread by a variety of channels and are influenced by a variety of variables depending on the situation. When it comes to the transmission of resistant bacteria in healthcare institutions, agriculture and the community: inadequate sanitation, hygiene, and infection control are all significant factors.

When animals are given antibiotics, it is possible that resistance may develop. Poorly cooked food, closeness, and poor hygiene are all sources of resistant bacteria spreading to humans and other animals. Resistant bacteria may spread to humans and animals through polluted water or excrement. A person in the hospital or the community may be administered antibiotics and antibiotic resistance may develop. Resistant bacteria may transmit to other individuals via poor hygiene and closeness. Bacteria that are resistant to antibiotics have been found in people and other animals that have come into contact with them. When bacteria adapt to resist an antibiotic, resistance develops. The more antibiotics we use, the greater the risk that bacteria may evolve resistance to them (Carvalho, Forestier, & Mathias, 2019).

10.2.1 From face to face

Bacteria are widely dispersed, and as a result, humans are constantly exposed to them. Our bacterial components differs from person to person; certain kinds of bacteria are found in all populations, while others are unique to individuals. People may get bacteria from each other if they come into touch. Indirect transmission, such as when someone coughs, is another possibility. Bacteria may be spread to another person who comes into contact with a contaminated surface (such as a door knob). It is critical to practice good hand hygiene in order to prevent the transmission of diseases and the development of antibiotic resistant bacteria. Bacteria are nonetheless a regular element of our environment to which we are constantly exposed, even with appropriate cleanliness standards (Prestinaci, Pezzotti, & Pantosti, 2015).

10.2.2 From animal to human

When it comes to bacteria, it may go both ways: from animals to people, as well as humans to animals. Diseases in animals become increasingly difficult to treat when germs develop resistant to first-line antibiotics, just as they are in people. Keeping pets or raising livestock for food means that many people spend a lot of time with animals. There's also a chance you'll run across some

FIGURE 10.1 How antibiotics resistance is developed and spreads.

wildlife. An illness called zoonosis may be transmitted from animal to human by ticks or mosquitoes, for example. Farmers and their families have been shown to possess the same strains of resistant bacteria as their livestock in some cases. Veterinarians who treat animals are also at risk of spreading germs that are resistant to antibiotics because of exposure to livestock. The bacterium might then go on the move and infect other people (Chopra, Hesse, O'Neill, & van der Goot, 2002).

10.2.3 From food and water

Bacteria may be found in and on the bodies of all animals. Antibiotics are often used in animal farms to treat and prevent illnesses, as well as to promote growth. Antibiotic-resistant bacteria may then invade the farm's animals and spread across the herd. Bacteria may be transmitted to meat during slaughter or processing and then end up in the final product. Animal dung may infect crops, as can crops that come into touch with it. Infections including diarrhea caused by some bacteria like *Salmonella, Campylobacter, and Enterohaemorrhagic. Escherichia coli* may be caused by eating food that has been contaminated with germs. Genes encoding resistance may be conveyed to the normal gut flora of the consumer without producing an infection by resistant bacterial strains or resistance genes (Fig. 10.1). The germs that have developed a resistance to antibiotics may go on to infect other persons. Microorganisms like bacteria may move from person to person via the use of water for irrigation, washing dishes, or other sanitary functions. Many water sources, such as drinking wells, rivers, and wastewater treatment plant effluents, include resistant microorganisms. A number of bacterial infections, including typhoid fever and cholera, may spread via polluted water. Resistant bacteria may get into water in a variety of ways, one of which is by the discharge of untreated waste from animals and people (Memish, Venkatesh, & Shibl, 2003).

10.2.4 From healthcare facilities

As a consequence of the large number of ill patients in close proximity and the widespread use of antibiotics, healthcare facilities are breeding grounds for antibiotic-resistant bacteria. The transmission of resistant germs may be facilitated by poor hygiene habits among physicians, nurses, and other healthcare providers, as well as patients and visitors. Unclean equipment, dirty facilities, and inadequate sanitation are all sources of danger in the workplace. Overcrowding in the wards and a lack of isolation rooms exacerbate the spread of infection (Wright, 2010).

10.2.5 From travels and trade

The microorganisms which they bring back from their travels, spread all across the planet. Every day, millions of individuals travel by plane, and if someone has a resistant bacteria with them, they will carry it with them. International visitors are more likely than domestic ones to pick up resistant germs when they go to places where resistant bacteria are common. This has been shown in several researches. Hospitalized patients have a significantly greater risk because of the extra risk factors they are exposed to. Patients moved from another hospital with a high frequency of resistance have been the source of many hospital epidemics. The list of items imported and exported to and from nations throughout the globe may be vast, including meat, fruits, vegetables, seeds, grain, and animals. Any of these may be a vehicle for the propagation of bacteria.

Due to the rising use of antimicrobials in agriculture, as well as the practice of discharging raw sewage into receiving waterways, the number of antibiotic-resistant bacteria in aquatic settings has grown significantly. In addition to serving as a repository for clinical resistance genes, the environment may also serve to facilitate the dissemination and development of those genes and the vectors that carry them (Young, 1993).

10.3 Concern for microbial resistance

10.3.1 Global perspectives

Expanding antibiotic AMR internationally in poor nations has challenged the scientific and pharmaceutical industries as a result of increasing use throughout the globe of antibiotics. A wide range of public health organizations in almost every field consider AMR to be a severe worldwide problem. Antimicrobial-resistant microorganisms do not stay in the same place they initially appeared. International medical travel, international food production, the export and import of food and animals across nations, as well as travel, all make it easier for antimicrobial-resistant organisms to spread around the globe. Antibiotic-resistant bacteria infect about 2.8 million individuals each year in the United States, killing at least 35,000 people as a consequence, according to a recent study. According to WHO (Levy, 1995), AMR might be at blame for 350 million deaths worldwide by 2050. Some examined strains of *E. coli* bacteria have shown resistance rates ranging from 5% to 80%, depending on the drug used. Globally, AMR is seen as a serious health threat. In response to the issue, there are requests for international treaties on AMR, as well as public pleas for global collective action. Antibiotic resistance is not yet fully understood on a global scale; however, it is more prevalent in nations with less developed healthcare systems (Amábile-Cuevas, 2010).

Resistance to antibiotics is spreading alarmingly, and it may do so at a faster rate in underdeveloped nations. There are other factors that contribute to this fast development and dissemination, including antibiotic resistance in bacteria, which we are still learning about, as well as regulatory and financial concerns related to antibiotic misuse. Moreover, bacteria are evolving to be better equipped to deal with environmental stress, which includes current and future antibiotics, separate from the more difficult to cure side of resistance. Antibiotic self-prescription and poor sanitary conditions, even in hospitals, are common in developing countries, posing a threat from specific multiresistant pathogens that are not common in developed countries and against which no new antibiotics are being investigated. In developing countries, we face these peculiarities. Resistance movements now a days may readily traverse boundaries due to the effects of globalization, as well as the local ramifications of these idiosyncrasies. If we are to get off to a good start in the post-antibiotic age, we need a global plan that is created and implemented (Walsh, 2010). Southern Europe risks being particularly affected. Italy, Greece, and Portugal are forecast to top the list of OECD countries with the highest mortality rates from AMR while the United States, Italy, and France would have the highest absolute death rates (Table 10.1), with almost 30,000 AMR deaths a year forecast in the US alone by 2050 (Stemming the Superbug Tide: Just A Few Dollars More, OECD Health Policy Studies, 2018).

Human medicine and agricultural food have both made AMR a top priority concern in their daily routine systems, and many governments see it as a serious rising danger to global public health and food security. Global efforts have been underway in recent years to raise awareness about this problem and gain political will for action. While the emergence of AMR is a normal process, excessive and improper use of these drugs has the potential to exacerbate the situation very quickly. There are medicinal and nontherapeutic uses for antimicrobials, which are a critical component of our food and agricultural production systems.

TABLE 10.1 Predicted deaths due to antimicrobial resistance by 2050.

Country	AMR mortality rate per lakh persons
Italy	18.17
United States	8.98
France	8.61
Germany	2.64
Iceland	0.28

Antimicrobial usage in the food and agricultural industry is predicted to increase significantly in certain regions of the globe due to rising demand for food, especially animal derived food, to fulfill the needs of a growing global population (Martinez et al., 2009).

10.3.2 Asian perspectives

The Asia-Pacific region is most sensitive to AMR's dangers. There are several obstacles preventing AMR control in the region's lower middle-income economies and high-income nations from progressing at the same rate. There are forecasted to be 27 megacities in the area by 2030, and these highly populated cities might act as massive reservoirs for the development of drug resistant diseases. Lower middle-income economies with unplanned urbanization, create settings where sanitation is poor, wastewater management is inadequate, and where air pollution causes respiratory diseases that are often mistreated with antibiotics, are especially at risk. Weak regulatory enforcement encourages the sale of almost all antimicrobials "over the counter" across Asia. Many cases of unlicensed health practitioners prescribing antibiotics for self-medication and treating infections by prescribing antibiotics are well known. Antimicrobials offered in Asia are of uncertain quality as well. Over 78% of all counterfeit medications are made in Asia, and 44% of all counterfeit drugs are utilized in Asia as well. The worldwide worth of these pharmaceuticals, which is estimated at $75 billion per year, shows the scale of the counterfeit drug problem. Falsified medications sometimes contain inadequate amounts of antimicrobials, providing microorganisms an edge in their interactions with the antimicrobial.

10.3.3 Indian perspectives

India was dubbed "the AMR capital of the world" because of its high levels of antibiotic resistance (Chaudhry & Tomar, 2017). While the discovery of novel, multidrug-resistant organisms challenges contemporary diagnostic and therapeutic approaches, India is still striving to combat the old pathogens that are responsible for TB, malaria, and cholera, all of which have very

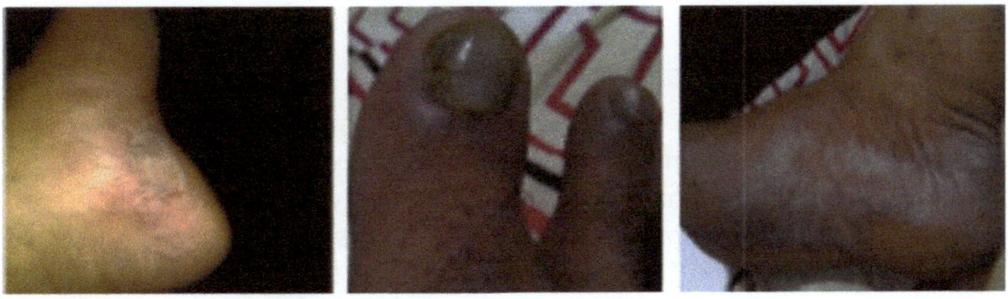

FIGURE 10.2 Incurable *Candida auris.*

high drug resistance. Many factors, such as indigence, illiteracy, overcrowding and famine have contributed to the current state of affairs. Illness perception, self-prescription of antibiotics, infection prevention inadequacy, and infection control guidelines have been seen in the general population endorsement of multidrug-resistant pathogens in hospital premises settings (Swaminathan et al., 2017). However, easy availability to over the counter medications adds to AMR (Bate et al., 2009) as well. When it comes to overall consumption of antibiotics for human purposes, India tops the list. Antibiotics were consumed in India in 2010 at a rate of roughly 10.7 units per person. However, despite the fact that these numbers are worrisome, they are not altogether unexpected in India, where antibiotics are routinely manipulated (Holloway, Kotwani, Batmanabane, Puri, & Tisocki, 2017). In spite of the fact that medical practitioners (Goswami, Gandhi, Patel, & Dikshit, 2013) are woefully uninformed about the sensible use of antibiotics, especially fixed drug combinations, it is generally known that India is the leading country in the production and use of substandard and counterfeit antimicrobial agents. As per a scoping report on AMR in India (Https://Cddep.Org/Wp-Content/Uploads/2017/11/Scoping-Report-on-Antimicrobial-Resistance-in-India.Pdf, n.d.), more than 70% of Gram-negative bacterial strains of *Escherichia coli* were found to be resistant to fluoroquinolone antibiotics, third-generation cephalosporin antimicrobials, drug combination of piperacillin-tazobactum and carbapenemases among Gram-negative bacteria under the Indian government's auspices.

10.4 Health hazards due to bacteria and fungi

10.4.1 Urgent threats

- Acinetobacter
- *Candida auris*
- *Clostridioides difficile*
- Enterobacterales
- *Neisseria gonorrhoeae*

10.4.2 Serious threats

10.4.2.1 Enterobacterales

Escherichia coli (*E. coli*) and *Klebsiella pneumoniae* are two examples of bacteria in the Enterobacterales group. Extended spectrum beta lactamases are produced by certain Enterobacterales (ESBLs). Many popular antibiotics, such as penicillins and cephalosporins, are rendered useless by ESBL enzymes, which break them down and kill them. Because of this, the number of antibiotics available to treat infections caused by Enterobacterales that produce ESBLs is decreasing. Even routine illnesses caused by ESBL-producing bacteria, such as urinary tract infections, may require more involved treatment options. Patients with these illnesses may need hospitalization and intravenous (IV) carbapenem medications instead of taking oral antibiotics at home. Only a few medicines, such as carbapenems, can treat bacteria that produce the enzyme ESBL, although enzymes that destroy these antibiotics are becoming more common.

10.4.2.2 Enterococci

Important nosocomial infections and a rising clinical issue are multidrug-resistant (MDR) enterococci. These microorganisms have employed a wide range of genetic techniques to gain resistance to almost all antimicrobials now used in clinical practise. Due to their capacity to attract antibiotic resistance determinants, MDR enterococci exhibit a diverse array of antibiotic resistance mechanisms, such as the alteration of drug targets, the inactivation of therapeutic agents, the overexpression of efflux pumps, and a complex adaptive response to the cell envelope that aids in survival in the human host and the nosocomial environment. Under pressure from antibiotics, MDR enterococci, which are highly adapted to live in the gastrointestinal system, can overtake other bacteria and become the dominant flora, putting the extremely sick and immunocompromised patient at risk for invasive infections.

10.4.2.3 Pseudomonas aeruginosa

Bacteria (germs) known as enterococci are often found in the human gut and female genital system, as well as in the soil and water. Its resistance is at an alarming level Table 10.2 (The Center for Disease, Dynamics Economics & Policy. ResistanceMap: Antibiotic Resistance., 2021). These microorganisms have the potential to cause an infection. *Pseudomonas aeruginosa* is the most common form of Pseudomonas to cause infections in the blood, lungs (pneumonia) or elsewhere in the body after surgery.

10.4.2.4 Nontyphoidal salmonella

About 1.35 million people become sick from salmonella infections each year, which results in 26,500 hospitalizations and 420 fatalities. The majority of

TABLE 10.2 Antibiotic resistance in the *Escherichia coli* isolates in India.

Antibiotics	Percentage of resistant isolates
Carbapenems	18
Cephalosporins (3rd generation)	77
Fluoroquinolones	84
Piperacillin-tazobactam	28
Aminopenicillins	92
Amoxicillin-clavulanate	60
Aminoglycosides	17

these diseases may be traced back to certain foods. Salmonella usually causes Diarrhea, fever, and stomach pains in those who become sick. When an infection occurs, symptoms may appear anywhere from 6 hours to 6 days later and persist anywhere from 4 to 7 days. The majority of patients heal on their own, without the need of antibiotics. The only time antibiotics are prescribed is when a patient has a life-threatening disease or is at high risk of developing one. Depending on the severity of their condition, some patients may need inpatient care.

10.4.2.5 *Salmonella serotype typhi*

Salmonella Typhi bacteria produce typhoid fever, a potentially fatal disease. Infection with *Salmonella paratyphi* bacteria causes paratyphoid fever, which may be fatal. Typhoid and paratyphoid fever afflict between 11 and 21 million individuals annually across the world. Typhoid fever and paratyphoid fever are most prevalent in areas with hazardous water, food, and sanitation. Contamination of food and water by human waste and direct touch spreads these illnesses. *Salmonella typhi* and *Salmonella paratyphi* may be transmitted by sick individuals and those who have recovered but are still passing the germs in their faeces. A person may get typhoid fever or paratyphoid fever by eating or drinking anything that has been touched by someone who is excreting *Salmonella typhi* in their stool, or who has not fully cleaned their hands after using the bathroom and / or the toilet. It is possible to get infected with Salmonella by drinking water that has been polluted with sewage. Water used to rinse raw food is polluted with sewage, which contains *Salmonella typhi* or *Salmonella paratyphi*. The germs that cause typhoid fever and paratyphoid fever may grow and spread into your bloodstream if you eat food or drink that has been infected with *Salmonella typhi* or *Salmonella paratyphi*.

10.4.2.6 *Shigella*

A disease known as shigellosis is caused by Shigella bacteria. Diarrhea (often bloody), fever, and cramps are the most common symptoms in those infected with Shigella. They typically appear one to two days after infection and persist for 7 days in the majority of cases. Antibiotics are not necessary for the majority of patients to recover. Antibiotics should, however, be administered to individuals who are very sick or have underlying diseases that compromise their immune systems. Because of this, antibiotics may help patients feel well faster and prevent the disease from spreading to others.

10.4.2.7 Staphylococcus aureus

Methicillin-resistant *Staphylococcus aureus* (MRSA) is the most common cause of skin infections in the community and may progress to sepsis. When it comes to healthcare facilities MRSA may create serious issues in healthcare facilities like hospitals and nursing homes, such as infections of the blood, lungs, or surgical sites.

10.4.2.8 Streptococcus pneumoniae

Pneumococcal infections were responsible for 60,000 new instances of invasive illness per year up to the year 2000. Pneumococcal bacteria were responsible for up to 40% of these illnesses, and these germs were resistant to at least one antibiotic. These figures have plummeted as a result of vaccination against pneumococcal disease in children. In 2008, the term "non-susceptibility" to penicillin was redefined. Invasive pneumococcal illness affected about 31,400 people in 2018. More than a third of pneumococcal germs tested so far have been shown to be resistant to at least one antibiotic. Streptococcus pneumoniae is found in different proportions throughout the country.

10.4.2.9 Tuberculosis

Bacteria that cause TB are airborne and may transmit from person to person. Lung cancer is the most common site of tuberculosis, although it may also spread to the brain, kidneys, or spine. Tuberculosis is usually treatable and curable, but if it isn't treated properly, individuals with tuberculosis may die. A rare but serious complication of tuberculosis treatment is drug-resistant strains. As a result, the medication is no longer effective against tuberculosis germs. Tuberculosis spreads in the same manner as drug-susceptible tuberculosis does: via direct contact with infected people. A person may get tuberculosis from someone else by breathing in someone else's contaminated air. When someone with tuberculosis illness of the lungs or throat coughs, sneezes, talks, or sings, the tuberculosis germs are released into the air. Breathing in these germs may cause infection in those who are in close proximity.

10.4.2.10 Campylobacter

Campylobacter is the source of campylobacteriosis and causes diarrhea, fever, and cramping. They often appear 2–5 days after infection and persist for approximately a week before disappearing completely. A few individuals have side effects including temporary paralysis or arthritis. Campylobacter may sometimes travel to the bloodstream and produce a life-threatening illness in individuals with compromised immune systems such as those with AIDS or chemotherapy. Campylobacter may be found in animal's intestines, livers, and other organs and it can be spread into the water wastes after slaughter. When a cow has a Campylobacter infection in her udder or if the milk is contaminated with dung, it may contaminate the product. Contamination of fruits and vegetables may occur when they come into touch with soil or water that contains cow, bird, or other animal excrement (poop). Lakes and streams may get contaminated by animal excrement.

10.4.2.11 Candida

Yeast infection, or Candidiasis, is caused by Candida (a kind of fungus). *Candida albicans* is the most prevalent species that causes infection in humans (Fig. 10.2) It is common for candida to reside on the skin, as well as within the body in areas like the mouth, throat, intestines, and the vagina. As it spreads and gets deeper into the internal organs like kidney and brain. Yeast fungus Candida is becoming more resistant to antifungal medications due to this issue. The antifungal medication Fluconazole is only effective against around 7% of the Candida blood samples. Rarely are Fluconazole and Echinocandin medicines effective in treating patients with Candida infections. Amphotericin B is the most common therapy, although it may be harmful to people who are already extremely ill. Patients with drug-resistant Candida blood stream infections (commonly known as candidemia) may have a worse chance of survival than those with candidemia that is treatable with antifungal medications. Fluconazole resistance is found in almost all United States. *C. auris* samples and Amphotericin B resistance are found in about a third of those samples.

10.4.3 Concerning threat

10.4.3.1 Streptococcus

Antibiotic-resistant illnesses cause an estimated 2.8 million hospitalizations and 35,000 deaths per year in the United States, according to distinct estimates. More than $4.6 billion in annual healthcare costs have increased and price estimates to treat illnesses caused by six multidrug resistant pathogens. Antibiotics are well known for their function in killing pathogenic microorganisms. However, bacteria have the ability to alter or modify it, resulting in antibiotics not working as intended. As a result, people develop an immunity to antibiotics. One or more antibiotic medicines are resistant to many bacteria, including certain *Streptococcus pneumoniae* (pneumococcus). Pneumococcal infections were responsible for thousands of deaths due to invasive illness in the year 2000. Only 40% of patients who were infected with pneumococcal germs were able to be treated with an antibiotic (Https://Www.Cdc.Gov/Abcs/Reports-Findings/Surv-Reports.Html, n.d.). To minimize infections, initiatives such as pneumococcal conjugate vaccinations for children and variations in penicillin resistance in 2008 were intended to be helpful. Invasive pneumococcal disease was shown to be active in over 30K people in 2018. More over 30% of infected patients showed pneumococcal bacterial resistance to at least one antibiotic, according to records that are readily available. Drug resistant *Streptococcus pneumoniae* (DRSP) infections are more expensive to treat than those caused by nonresistant pneumococcus bacterial species because of their higher frequency of drug resistance. New antibiotic drug research is needed due to the requirement of many costly antibiotics, and monitoring is required to follow the resistance route. People who work in child care centers are at increased risk for DRSP infection because of recurrence of illness owing to treatment failures, as well as educational needs for patients, doctors, and microbiologists. A resistant infection is more common in pneumococcal infection patients who have recently taken antibiotics than in healthy persons. As a result of the broad misuse of antibiotic drugs and other factors, it is difficult to prevent pneumococcal drug resistance. The emergence of drug-resistant bacteria is also due to not using the 23 valent polysaccharide vaccination (PPSV23) recommended for people at high risk for pneumococcal disease. Some clinical labs have not adopted standard methodologies (NCCLS recommendations) for diagnosing and characterizing DRSP, and there is no universal pneumococcal vaccine available (Kim, McGee, Tomczyk, & Beall, 2016).

10.4.4 Watch list

10.4.4.1 Aspergillus fumigatus

It is important to note that the advent of azole-resistant *A. fumigatus* affects aspergillar infections treatment. This is a critical problem, given the limited number of highly effective agents available. A high infection rate is linked to resistant forms of bacteria and leads to a treatment failure when compared to illnesses caused by bacteria that are more susceptible (Howard et al., 2009; Thors, Bierings, Melchers, Verweij, & Wolfs, 2011). The clinical findings have been confirmed by numerous aspergillosis experimental models (Denning et al., 1997; Mavridou, Bruggemann, Melchers, Verweij, & Mouton, 2010). There was a substantially greater incidence of invasive aspergillosis due to resistant strain (88%) (Vander-Linden et al., 2011) as compared to invasive aspergillosis as a result of a susceptible strain (Baddley et al., 2009).

10.4.4.2 Mycoplasma genitalium

Antibiotic therapy's effectiveness has declined since it was originally shown to be present in *M. genitalium* in 2008 (Jensen, Bradshaw, Tabrizi, Fairley, & Hamasuna, 2008), with treatment success rates as low as 40% in certain investigations

(Workowski & Bolan, 2015). According to recent meta-analyses, Moxifloxacin's mediated cure rates have fallen from 100% in research done before 2010 to 89% in studies published after 2010 (Li, Le, Li, Cao, & Su, 2017). Each of the five single nucleotide azithromycin resistance mutations at locations 2058 and 2059 in the 23S rRNA gene's macrolide resistance-determining region is linked to the drug's failure due to macrolide resistance (Chaudhry & Tomar, 2017). Quinolone resistance has been linked to many mutations in the topoisomerase gene parC's quinone resistance-determining regions (QRDR), although the role of individual parC mutations in treatment failure remains unknown. Nucleotide mutations in the QRDR of the DNA gyrase gene gyrA may also have a role in clinical treatment failure; however, the exact nature of these changes is yet uncertain (Bebear et al., 1999; Tagg, Jeoffreys, Couldwell, Donald, & Gilbert, 2013).

10.4.4.3 Bordetella pertussis

The use of antibiotics and prophylaxis is critical in the fight against the spread of pertussis in the United States. As a result, erythromycin is the drug of choice for patients with confirmed or suspected pertussis, as well as close contacts of such patients (Hoppe, Halm, Hagedorn, & Kraminer-Hagedorn, 1988). A case of *B. pertussis* that was resistant to Erythromycin was discovered in Yuma, Arizona in 1994 (Lewis et al., 1995). To date, the United States has seen four more instances of *B. pertussis* that was resistant to erythromycin (Hill, Baker, & Tenover, 2000). *B. pertussis* strains were recently shown to be resistant to erythromycin in different ways, according to a study of 1030 isolates (Wilson, Cassiday, Popovic, & Sanden, 2002). Identification of effective antibiotic treatment and measures to reduce the spread of resistance depends on understanding the mechanism of resistance. Because molecular approaches such as PCR are increasingly being used exclusively for the detection of *B. pertussis*, it is necessary to develop molecular assays to identify resistance even in the absence of an isolate.

10.5 Conclusion

Antibiotic-resistant microorganisms, resistance genes, and antibiotic residues are released into the environment when treated sewage is discharged. On the other hand, it is not apparent if a rise in the number of antibiotic resistance genes found in sewage and the habitats that it affects is the consequence of selection pressure exerted on-site by residual antibiotics, or merely of fecal contamination with resistant bacteria. Resistant microorganisms are found in water supplies because of human and animal waste. Genes from these bacteria have been found in a variety of water-borne microorganisms, including those with resistance genes. Instead, many antibiotics derived from industrial sources are found in waterways, where they have the potential to disrupt microbial communities. Reducing the number of resistant bacteria in wastewaters necessitates the use of risk assessment techniques based on improved technologies for antibiotic detection and antibiotic-resistance microbial source monitoring.

References

Amábile-Cuevas, C. F. (2010). Global perspectives of antibiotic resistance, *Antimicrobial Resistance in Developing Countries* (Vol. 9780387893709, pp. 3–13). New York, NY: Springer. https://doi.org/10.1007/978-0-387-89370-9_1.

Aminov, R. I. (2009). The role of antibiotics and antibiotic resistance in nature. *Environmental Microbiology, 11*(12), 2970–2988. https://doi.org/10.1111/j.1462-2920.2009.01972.x.

Aminov, R. I., & Mackie, R. I. (2007). Evolution and ecology of antibiotic resistance genes. *FEMS Microbiology Letters, 271*(2), 147–161.

Baddley, J. W., Marr, K. A., Andes, D. R., Walsh, T. J., Kauffman, C. A., Kontoyiannis, D. P., et al. (2009). Patterns of susceptibility of Aspergillus isolates recovered from patients enrolled in the Transplant-Associated Infection Surveillance Network. *Journal of Clinical Microbiology, 47*(10), 3271–3275. https://doi.org/10.1128/JCM.00854-09.

Bate, R., Tren, R., Mooney, L., Hess, K., Mitra, B., Debroy, B., et al. (2009). Pilot study of essential drug quality in two major cities in India. *Plos One, 4*(6), e6003. https://doi.org/10.1371/journal.pone.0006003.

Bebear, C. M., Renaudin, J., Charron, A., Renaudin, H., de Barbeyrac, B., Schaeverbeke, T., et al. (1999). Mutations in the gyrA, parC, and parE genes associated with fluoroquinolone resistance in clinical isolates of Mycoplasma hominis. *Antimicrobial Agents and Chemotherapy, 43*(4), 954–956.

Carvalho, G., Forestier, C., & Mathias, J.-D. (2019). Antibiotic resilience: A necessary concept to complement antibiotic resistance? *Proceedings Biological Sciences, 286*(1916), 20192408. https://doi.org/10.1098/rspb.2019.2408.

Chaudhry, D., & Tomar, P. (2017). Antimicrobial resistance: The next BIG pandemic. *International Journal of Community Medicine and Public Health, 4*(8). doi:10.18203/2394-6040.Ijcmph20173306.

Chopra, I., Hesse, L., O'Neill, A., & van der Goot, H. (2002). Discovery and development of new anti-bacterial drugs, *Trends in Drug Research III* (32, pp. 213–225). Amsterdam, The Netherlands: Elsevier. https://doi.org/10.1016/S0165-7208(02)80022-8.

Denning, D. W., Venkateswarlu, K., Oakley, K. L., Anderson, M. J., Manning, N. J., Stevens, D. A., et al. (1997). Itraconazole resistance in Aspergillus fumigatus. *Antimicrobial Agents and Chemotherapy, 41*(6), 1364–1368.

Goswami, N., Gandhi, A., Patel, P., & Dikshit, R. (2013). An evaluation of knowledge, attitude and practices about prescribing fixed dose combinations among resident doctors. *Perspectives in Clinical Research, 4*(2), 130–135. https://doi.org/10.4103/2229-3485.111797.

Hill, B. C., Baker, C. N., & Tenover, F. C. (2000). A simplified method for testing Bordetella pertussis for resistance to erythromycin and other antimicrobial agents. *Journal of Clinical Microbiology, 38*(3), 1151–1155.

Holloway, K. A., Kotwani, A., Batmanabane, G., Puri, M., & Tisocki, K. (2017). Antibiotic use in South East Asia and policies to promote appropriate use: Reports from country situational analyses. *BMJ (Clinical Research Ed.), 358*, j2291. https://doi.org/10.1136/bmj.j2291.

Hoppe, J. E., Halm, U., Hagedorn, H. J., & Kraminer-Hagedorn, A. (1988). Comparison of erythromycin ethylsuccinate and co-trimoxazole for treatment of pertussis. *Infection, 17*(4), 227–231.

Howard, S. J., Cerar, D., Anderson, M. J., Albarrag, A., Fisher, M. C., Pasqualotto, A. C., et al. (2009). Frequency and evolution of Azole resistance in Aspergillus fumigatus associated with treatment failure. *Emerging Infectious Diseases, 15*(7), 1068–1076. https://doi.org/10.3201/eid1507.090043.

https://cddep.org/wp-content/uploads/2017/11/scoping-report-on-antimicrobial-resistance-in-india.pdf. n.d.

https://www.cdc.gov/abcs/reports-findings/surv-reports.html. (n.d.).

Jensen, J. S., Bradshaw, C. S., Tabrizi, S. N., Fairley, C. K., & Hamasuna, R. (2008). Azithromycin treatment failure in Mycoplasma genitalium-positive patients with nongonococcal urethritis is associated with induced macrolide resistance. *Clinical Infectious Diseases: An Official Publication of the Infectious Diseases Society of America, 47*(12), 1546–1553. https://doi.org/10.1086/593188.

Kim, L., McGee, L., Tomczyk, S., & Beall, B. (2016). Biological and epidemiological features of antibiotic-resistant streptococcus pneumoniae in pre- and post-conjugate vaccine eras: A United States perspective. *Clinical Microbiology Reviews, 29*(3), 525–552. https://doi.org/10.1128/CMR.00058-15.

Levy, S. B. (1995). *Antimicrobial Resistance: A Global Perspective* (pp. 1–13). Boston, MA.: Springer Science and Business Media LLC. https://doi.org/10.1007/978-1-4757-9203-4_1.

Lewis, K., Saubolle, M. A., Tenover, F. C., Rudinsky, M. F., Barbour, S. D., & Cherry, J. D. (1995). Pertussis caused by an erythromycin-resistant strain of Bordetella pertussis. *The Pediatric Infectious Disease Journal, 14*(5), 388–391.

Li, Y., Le, W.-J., Li, S., Cao, Y.-P., & Su, X.-H. (2017). Meta-analysis of the efficacy of moxifloxacin in treating Mycoplasma genitalium infection. *International Journal of STD & AIDS, 28*(11), 1106–1114. https://doi.org/10.1177/0956462416688562.

Martinez, J. L., Fajardo, A., Garmendia, L., Hernandez, A., Linares, J. F., Martínez-Solano, L., et al. (2009). A global view of antibiotic resistance. *FEMS Microbiology Reviews, 33*(1), 44–65. https://doi.org/10.1111/j.1574-6976.2008.00142.x.

Mavridou, E., Bruggemann, R. J. M., Melchers, W. J. G., Verweij, P. E., & Mouton, J. W. (2010). Impact of cyp51A mutations on the pharmacokinetic and pharmacodynamic properties of voriconazole in a murine model of disseminated aspergillosis. *Antimicrobial Agents and Chemotherapy, 54*(11), 4758–4764. https://doi.org/10.1128/AAC.00606-10.

Memish, Z. A., Venkatesh, S., & Shibl, A. M. (2003). Impact of travel on international spread of antimicrobial resistance. *International Journal of Antimicrobial Agents, 21*(2), 135–142.

Normark, B. H., & Normark, S. (2002). Evolution and spread of antibiotic resistance. *Journal of Internal Medicine, 252*(2), 91–106. https://doi.org/10.1046/j.1365-2796.2002.01026.x.

Prestinaci, F., Pezzotti, P., & Pantosti, A. (2015). Antimicrobial resistance: A global multifaceted phenomenon. *Pathogens and Global Health, 109*(7), 309–318. https://doi.org/10.1179/2047773215Y.0000000030.

Stemming the Superbug Tide: Just A Few Dollars More, OECD Health Policy Studies. (2018). OECD. 10.1787/9789264307599-en.

Swaminathan, S., Prasad, J., Dhariwal, A. C., Guleria, R., Misra, M. C., Malhotra, R., et al. (2017). Strengthening infection prevention and control and systematic surveillance of healthcare associated infections in India. *BMJ (Clinical Research Ed.), 358*, j3768. https://doi.org/10.1136/bmj.j3768.

Tagg, K. A., Jeoffreys, N. J., Couldwell, D. L., Donald, J. A., & Gilbert, G. L. (2013). Fluoroquinolone and macrolide resistance-associated mutations in Mycoplasma genitalium. *Journal of Clinical Microbiology, 51*(7), 2245–2249. https://doi.org/10.1128/JCM.00495-13.

The Center for Disease (2021). Dynamics Economics & Policy. ResistanceMap: Antibiotic resistance. https://resistancemap.cddep.org/AntibioticResistance.php.

Thors, V. S., Bierings, M. B., Melchers, W. J. G., Verweij, P. E., & Wolfs, T. F. W. (2011). Pulmonary aspergillosis caused by a pan-azole-resistant Aspergillus fumigatus in a 10-year-old boy. *The Pediatric Infectious Disease Journal, 30*(3), 268–270. https://doi.org/10.1097/INF.0b013e3182037879.

Vander-Linden, J. W. M., Snelders, E., Kampinga, G. A., Rijnders, B. J. A., Mattsson, E., Debets-Ossenkopp, Y. J., et al. (2011). Clinical implications of azole resistance in Aspergillus fumigatus, The Netherlands, 2007-2009. *Emerging Infectious Diseases, 17*(10), 1846–1854. https://doi.org/10.3201/eid1710.110226.

Walsh, T. R. (2010). Emerging carbapenemases: A global perspective. *Antimicrobial Resistance: Focus on Stewardship Strategies to Optimise the Activity of Currently Available Agents, 36*, S8–S14. https://doi.org/10.1016/S0924-8579(10)70004-2.

Wilson, K. E., Cassiday, P. K., Popovic, T., & Sanden, G. N. (2002). Bordetella pertussis isolates with a heterogeneous phenotype for erythromycin resistance. *Journal of Clinical Microbiology, 40*(8), 2942–2944.

Workowski, K. A., & Bolan, G. A. (2015). Sexually transmitted diseases treatment guidelines, 2015. *MMWR. Recommendations and Reports: Morbidity and Mortality Weekly Report. Recommendations and Reports, 64*(RR-03), 1–137.

Wright, G. D. (2010). Antibiotic resistance in the environment: A link to the clinic? *Current Opinion in Microbiology, 13*(5), 589–594. https://doi.org/10.1016/j.mib.2010.08.005.

Young, H. K. (1993). Antimicrobial resistance spread in aquatic environments. *The Journal of Antimicrobial Chemotherapy, 31*(5), 627–635.

Chapter 11

Acquired knowledge and identified gaps in resistance and human health risk

Kumud Nigam and Somali Sanyal
Amity Institute of Biotechnology, Amity University, Lucknow, Uttar Pradesh, India

11.1 Introduction

Drugs have magnificently helped the human civilization to flourish. Many ethnic populations were wiped out by diseases, endemic, pandemic as there was non-availability of medicine or least alternate. In predrug ages, the average human life was less than 20 years and high birth rate was necessary for existence of races. Drug can be defined as "Any substance (other than food) that is used to prevent, diagnose, treat or relieve symptoms of a disease or abnormal condition. Drugs can also affect how the brain and the rest of the bodywork and cause changes in mood, thoughts, behavior, or feelings. Some types of drugs, such as opioids, may be abused or lead to addiction." Moses "learned in all the wisdom of the Egyptians," codified his sanitary rules and regulations in the form of religious rites and ceremonies and thus secured their observance among the faithful even down to the present time (Sir & Lankester, 1905).

From the stone age to the modern era, there is the number of examples of diseases (from malaria plug to flu and corona) which derailed the growth of humanity. This can be ruled off by the availability of drugs and its effectiveness. The Nobel period of human race started in 1928 with the invention of true antibiotic Penicillin by **Alexander Fleming**. In 1090, **Enrilich** discovered that a chemical Arsphenamine was effective in the treatment of syphilis and he termed his discovery as "chemotherapy" (The use of a chemical to treated disease). Microbiologist **Selman Waksman** who first discovered Streptomycin uses the term antibiotic in 1943. During his lifetime, he had been discover over 20 antibiotics.

According to **Charles Darwin's** every organism try to survive, reproduce, and possess heritable traits that enable them to better adopt to their environment. Similarly, microbes also evolved with time to adopt better to the microenvironment and become resistant to these useful drugs.

Antimicrobial Resistance in Wastewater and Human Health.
DOI: https://doi.org/10.1016/B978-0-323-96124-0.00006-4

Antimicrobial resistance (AMR) occurs when microorganism shows no response to medicine this makes the infections more intensive which may lead to terminal illness and ultimate death. The successful use of any therapeutic agent that is curing that disease tend to develop tolerance after a period of time. This makes most of our health agencies to consider this resistance to therapy as a serious problem. Today a drug that is performing excellent may not be useful in future by the development of resistant strains. This problem is so serious that experts are worried of returning to the preantibiotic era (WHO & Resistance, 2020).

A database in 2010 enlisted the existence of more than 20,000 potential resistance gene (R gene) in nearly 400 different types bacterial genome sequences (Davies & Davies, 2010). Keeping this in mind, WHO published a catalog of 12 families of bacteria that has antibiotic resistance and are greatest threat to the human health and enlisted them "Priority Pathogen." This list is divided into three categories: critical, high, and medium priority (WHO, 2020).

Worldwide, there is no system to collect the data and particular use of different antibiotics in health care and animal products. Only 42 countries worldwide systematically collect the data on antibiotic resistance (Animal Production | Antimicrobial Resistance Food and Agriculture Organization of the United Nations, 2019). It makes the gaps in knowledge about antibiotic resistance which eventually contribute more to AMR (Animal Production |Antimicrobial Resistance Food and Agriculture Organization of the United Nations, 2019). Today's world needs a surveillance system which collect the data from all over and gives an early alert for changes in antimicrobial exposure. This will also help us to standardize the use of antimicrobial agents to address AMR. Many questioners-based studies in developed and developing countries established that most of the people have limited knowledge about correctly using antibiotic and its hazard (WHO Library Cataloguing-in-Publication Data Global Action Plan on Antimicrobial Resistance, 2015; Aslam et al., 2018; Carter, Sun, & Jump, 2016).

11.2 Unjustified use of antibiotics

Drug resistance is increasing worldwide but in the third-world country it is more common due to unjustified use of antibiotics. It is more in the third-world country due to the following reasons:

1. Nonavailability of qualified doctors in comparison to population.
2. Free availability of medicine without prescription.
3. Government policies.

The WHO recommend standard ratio of 1:1000 doctor to population ratio but in India in 2020 doctors population ratio is 1:1456 which are way behind WHO criteria. Among this 52% of the doctors are practicing in just five states which makes the problem more severe (Business Line. 52 Percent of India's Allopathic Doctors Are Practicing in Just 5 States. Radheshyam Jadhav, 2020). The doctor should be the first to response for the appropriate and careful use as they are trained and taught to do so. If the patient actually has bacterial infection it is important to prescribe right antibiotic, in right doses and in last for significant duration in which the microbes lower its existence using (Using Medication: Using Antibiotics Correctly and Avoiding Resistance, 2008). If any discrepancy in selection of right antibiotic, right dose for appropriate duration occur bacteria can develop resistance to antibiotic. So the need of the moment is to establish a justified doctor–patient ratio. In the developed country like India if they want to achieve the goal to fulfill D:P ratio by 2030 we need 2.07 million more doctors (Potnuru, 2017).

Government policies for health sector also contribute to antibiotic resistance in the developing countries like India. In European nation, government expenditure on heath sector is more than global average that is 6% but India have only 1.3% (Potnuru, 2017). If we spare the expenditure part then also government relaxed policies on distribution of drugs is of major concern. Government should imply the policy "no medication without prescription" and if not abide a hefty penalty should be charged. India has highest rates of antibiotic use in the world which are significantly increasing by the time as it increased 100% from 2000 to 2015 (Klein, Boeckel, Martinez, & Laxminarayan, 2018).

Free availability and self-medication are the major causes for antibiotic resistance. A combination of nonavailability of doctors and rare of availability of pharmacist has contributed to the use of over-the-counter antibiotics in the country (Shehab et al., 2016). The problem is on global level as 62% of pharmacies around the world provide antibiotics without prescription (Morgan, Okeke, Laxminarayan, Perencevich, & Weisenberg, 2011). Most common conditions where, of self-treatment with antibiotics is practiced are sore throat, running nose and cough condition that would get better without any antibiotic treatment. It has observed that approx. 2 million people in America every year get infected with antibiotic-resistant pathogen, and 23,000 of them dies. Despite being illegal systemic antibiotics are available over store without a prescription.

11.3 Global picture of antibacterial resistance

11.3.1 Drug resistance in bacteria with special reference to mycobacterium tuberculosis

Like other living organisms, microbes continuously fight and acquired strong metabolic activity to survive during million years of evolution. There incredible adaptability makes them to acquire a variety of biochemical mechanism which gives them chances for their survival. Their flexible metabolic power helped them to adopt the changes in their environment (Bérdy, 2012). In turn, this

TABLE 11.1 Leading antimicrobial-resistant disease.

S.no.	Disease	Causal organism	Drug resistance
1	Tuberculosis	*M. Tuberculosis*	Isoniazid and Rifampin
2	Clostridium difficile infections (CDIS)	*Clostridium difficile*	Aminoglyosides, Tetracyclin, Erythromycin, Clindamycin, Penicilin
3	Vancomycin resistant enterococci infection	*Enterococci*	Vancomycin
4	Methicillin-resistant Staphylococcus aureus (MRSA) infection	*Staphylococcus aureus*	Erythromycin clindamycin, aminoglycosides, β-lactams, etc.
5	Gonorrhea	*Gonorrhea*	Cefixime, Ceftriaxone, Azithromycin, Aminoglycosides, Tetracyclin
6	Carbapenem-resistant Enterobacteriaceae	*Enterobacteria*	Carbapenem

No permission required.

incredible power of adaptability makes them resistance to antibiotic which affects the mankind. A list of leading antimicrobial resistant diseases are detailed in Table 11.1. Most of the antibiotics are originated from soil-dwelling bacteria. Penicillin was the first antibiotic discovered accidentally when Penicillin fungus contaminated a culture plate. After the discovery of Penicillin few scientist Rene Dubos and Selman Waksman, started the search for antibacterial agent from soil microorganism (bacteria and fungi). Current antibiotics affect almost every process in the bacterial cell based on their structure and mode of action. They can be divided in seven major groups. These include β-lactams (inhibit cell wall synthesis), aminoglycosides (protein synthesis), macrolides (protein synthesis), tetracyclines (protein synthesis), daptomycin (cell membrane function), platensimycin (fatty acid biosynthesis), and glycopeptides (cell wall synthesis) (Peterson & Kaur, 2018).

Tuberculosis is caused by grampositive acid fast eubacteria. Mycobacterium tuberculosis strains which are antibiotic resistance are the concern for global tuberculosis epidemic. Mycobacterium tuberculosis is intrinsically resistant to many antibiotic due to number of mechanism such as the presence of thick, waxy, hydrophobic cell envelop, and the presence of drug degrading and modifying enzymes Fig. 11.1 (Gygli, Borrell, Trauner, & Gagneux, 2017). Mycobacterium tuberculosis strain are classified as multidrug resistant (MDR). Success rate of tuberculosis treatment are alarmingly low as most of the cases (54%) are MDR and some (28%) are even XDR (extensive drug resistant).

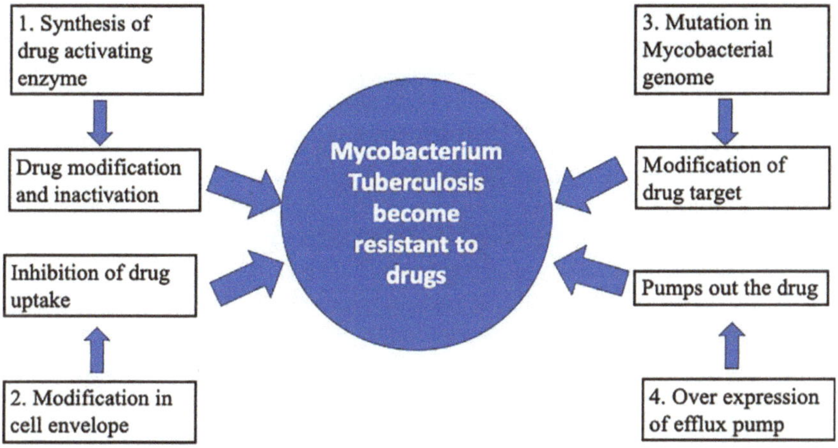

FIGURE 11.1 Drug-resistant mechanism in Mycobacterium tuberculosis. Drug resistant in Mycobacterium tuberculosis can be caused by several mechanisms including drug modification and inactivation, inhibition of drug uptake, modification of drug target, and overexpression of efflux pump.

Mycobacterium tuberculosis is able to acquire resistance to the antibiotic to which they were previously susceptible. This "acquired antibiotic resistance" occurs through either mutation and horizontal gene transformation. Among this horizontal gene transfer which involve transfer of gene between two organisms of same species is responsible for the transfer of antibacterial resistant gene among bacteria and fuel the process of evolution. In horizontal gene transfer, newly acquired DNA is incorporated into the genome of the recipient through either recombination or insertion. Three principal mechanisms can mediate horizontal gene transfer-

- Transformation (uptake of free DNA).
- Conjugation (plasmid-mediated transfer).
- Transduction (phage-mediated transfer).

11.3.2 Drug resistance in virus

One of the major cause of death worldwide are viral infection. Few major viral infection include HIV, Hepatitis B&C, Influenza in which only HIV causes million casualties in the year of 2020 worldwide. While others like flu causes 20,000 deaths in united states only in 2019–2020 period (Reed et al., 2015). Antiviral drug are now available for number of virus as HIV (Arts & Hazuda, 2012), Hepatitis B&C (Rajbhandari & Chung, 2016; Webster, Klenerman, & Dusheiko, 2015), and Influenza A&B viruses (Ison, 2013), herpes simplex viruses (Birkmann & Zimmermann, 2016), cytomegalovirus (Campos, Ribeiro, Boutolleau, & Sousa, 2016), epstein barr virus (AlDabbagh et al., 2017), varicella zoster virus, etc. (Kim et al., 2014). Antiviral drug uses two different approaches either they directly target the viruses or they target the host cells. Antiviral drugs that directly targets viruses include the inhibition of virus attachment, inhibition of virus entry, uncoating inhibition, polymerase inhibition, protease inhibition, inhibition of nucleoside, and nucleotide reverse transcriptase and the inhibition of integrase (Kausar et al., 2021). Viral structure are very simple having only DNA and RNA as the genetic material. Antiviral treatment that is initially successful at reducing viral loads may eventually be rendered ineffective in individual patient or in entire population by the emergence of drug resistance strains (Campos et al., 2016; Siliciano & Siliciano, 2013; Tana & Ghany, 2013; Pawlotsky, 2011; Li, Chan, & Lee, 2015; Piret & Boivin, 2014; Piret & Boivin, 2016). When a viral strains mutated that allows successful replication of the virus despite of the presence of antiviral drug and subsequently this type of strain out number entire viral population which ultimately cause drug resistance for entire viral population. These mutations interfere the drug molecule ability to inhibit the viral target by altering the shape and chemical properties of viral protein either by preventing drug binding or virus ability to enter the host cell in the presence of drug. Bacteria are single celled microorganism and are competent to reproduce itself without using host cell in comparison to virus which are smaller than bacteria and need living host to multiply and survive. This is the basic reason that dealing viruses is more tedious then bacteria. To establish an infection and to create new virions a virus should follow many stages including uncoating of the viral particle, transcription and translation of the viral genome, copying the viral genome, assembly of the cell protein (Herz, Bonhoeffer, Anderson, May, & Nowak, 1996) and cell cycle dependent events. In drug-resistant virus when antiviral drug is administered and the drug is present at high drug level, virus stop to replicate and when the drug level decay and reach a lower level in the host body virus complete the target phase of its life cycle. Antiviral drug has more chance to get infective as a virus can mutate itself frequently and without mutation also can escape by synchronization (Fridman, Goldberg, Ronin, Shoresh, & Balaban, 2014; Levin-Reisman et al., 2017).

11.3.3 Drug resistance in malaria parasites

Malaria is a very common and serious disease in Asia and Africa. It is caused by protozoan parasite *Plasmodium* species which uses mosquito as a carrier to infect humans. Infected person shows symptoms like high fever, chills and flu like illness which sometime be fatal. Globally approx. 359 million cases are reported every year and nearly 1.5–2.0 million death occurs annually (Menard, Djalle, Yapou, Manirakiza, & Talarmin, 2006). India launches National Malaria Control program in 1953 with which the incident of malaria sharply declined but after 1965 it again increased due to drug-resistant

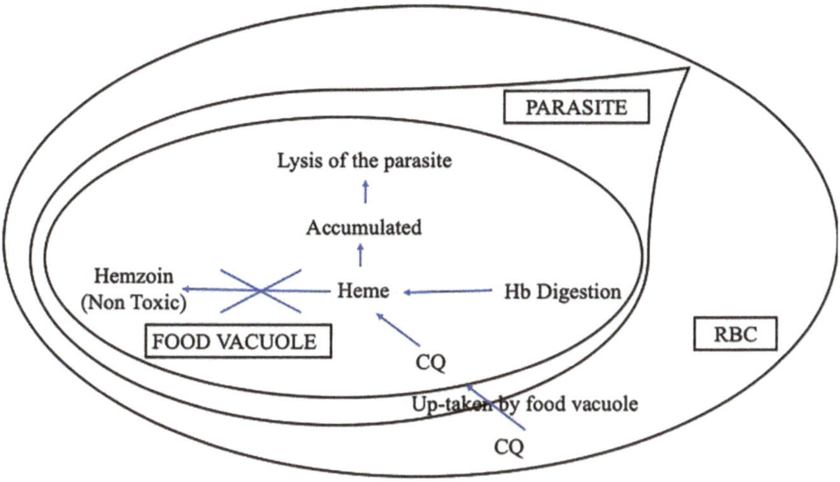

FIGURE 11.2 Mechanism of action of Chloroquine (CQ). Food vacuole is a lysosome like organelle in which the breakdown of haemoglobin and detoxification of heme occurs. Due to low Ph at the food vacuole CQ is taken up by the parasitic food vacuole by protonation and ion trapping mechanisms. In food vacuole, CQ gets accumulated and prevent the formation of Hemzoin. As a results Heme gets accumulated and cause the lysis of parasite within the RBC.

malaria. In India first time chloroquine resistance was reported in 1973 for Plasmodium falciparum. Uncontrolled urbanization without proper sewerage system increases malaria by manyfold. Orissa, Assam, West Bengal, and other Northern states show higher transmission of resistant *P. falciparum* (Bousema et al., 2004; Khatoon, Baliraine, Bonizzoni, Malik, & Yan, 2009). Similar chloroquine resistance in *P. falciparum* and *P. vivax* were reported from Bombay, Indo-Nepal border, Western Mysore (Sharma, 1996; Ghosh, Yadav, & Sharma, 1992; Dua, Kar, & Sharma, 1996). *P. falciparum* strain developed multiple mutations in the transmembrane protein (Chinappi, 2010) which allow them to reduce the accumulation of drug in their digestive vacuole which normally kills the pathogen by binding to subunit of oxidized heme. Normally parasite polymerizes heme subunits released by the digestion of hemoglobin and results into the formation of harmless clumps. Chloroquine prevents the formation of clump Fig. 11.2. The free heme then lyses membrane and cause death of the parasite. Chloroquine resistances develop due to mutation in PfCRT protein of *P. falciparum* which prevents the acculation of chloroquine on the digestive vacuole of the parasite. Researchers develop synthetic alternatives for chloroquine but parasite sooner developed resistance to that (Roper et al., 2004).

11.3.4 Drug resistance in fungi

Any organism which can produce spores and feeds on organic matter is fungus, which can be singled celled or very complex multicellular. Fungi are almost present everywhere from soil, plants, and animals and in the water, air also. Most of the fungal species are found as a normal flora in different anatomical sites include skin, lung, urinary, gastrointestinal track which do not causes any harm to the host. The level of antifungal resistance is not as severe as observed with some bacteria against different antibiotics. Antifungal resistance can occur when drugs are given in low doses and for short treatment course (Lortholary et al., 2011).

Nearly all ecosystem have fungal and bacterial families which play role in the functioning of numerous ecosystem (Deveau et al., 2018) they show interdependency and contribute combinedly to both health and disease. Bacteria and fungi can interact by direct cell–cell contact, chemical interaction, environmental modification, and by use of metabolic by-products. Several gram negative pathogens are capable of inhibiting filament formation and capable of killing Candida albicans (Peleg, Hogan, & Mylonakis, 2010). Any disturbance in equilibrium of bacteria and fungus may lead to diseases an overuse of antibiotic may harm to bacteria of normal flora which controls fungal elements may lead to fungal disease. Treatment choice for antifungal therapy are cause of concern because of very few known antifungal drug classes and by the emergence of antifungal drug resistance. Epidemiological studies find the fact that previously there was predominance of fungal species which were susceptible to all classes of antifungal but widespread use of antifungals made them unsusceptible(Arastehfar et al., 2020). Antifungal therapeutic failure can be due to inappropriate drug load, drug administration for shorter period, various hosts underlying conditions such as abdominal and liver abscesses (Arastehfar et al., 2020; Zhao et al., 2017), and acquisition of resistance by mutations in the drug targets and biofilm formation (Berman & Krysan, 2020; Cowen, Sanglard, Howard, Rogers, & Perlin, 2015).

A detail of drug-resistant viral, fungal, and protozoan diseases are enlisted in Table 11.2.

11.4 Impact of drug resistance on human health

Today human health largely depends on drugs because of growing population uncontrolled urbanization, junk food, unavailability of clean water, air pollution, and soil contaminated with pollutants such as heavy metals and reliability on machines which makes human lazy. The goal of medicine is the relief of pain and suffering promotion of health prevention of disease for stalling of death and promoting of peaceful death. Drugs are the tool that we use to ease our life but it is unimaginable when these tools became ineffective. Today drug resistance is a major cause of concern which increases ability to treat infections and illness in livings. This can lead to the following problems:

- increased human illness, suffering, and death,
- increased cost and length of treatments, and
- increased side effects from the use of multiple and more powerful medications.

TABLE 11.2 Drug resistant in viral, fungal, and protozoan disease.

S. N.		Disease	Causal organism	Drug resistant
1	Viral	Herpes infection	*Herpes simplex* virus	Acyclovir
2		Vericella infection	*Varicella zoster* virus	Acyclovir
3		Cytomegalovirus infection	*Cytomegalovirus*	Ganciclovir
4		Hepatitis B	*Hepatitis B* virus	Lamivndine, Adfovir, Telbivudine
5	Fungal	Aspergillosis	*Aspergilus fumigatus*	Azole antifungals
6		Candidiasis	*Candida auris*	Fluconazole
		Candidamia	*Candida glabrata*	Candida, echinocandin
7	Protozoan	Malaria	*Plsomidum falciparum*	Sulfadoxine, mefloquine, or quinine
				Chloroquine
			Plasmodium vivax	
8		Leishmaniasis	*Leishmania donovani*	Sodium stibogluconate
9		Filariasis	*Wucheria bancrofti*	Ivermactin, Albendazole

No permission required.

Death is the ultimate loss due to drug resistance as an penicillin-resistant *Pneumococci* causes fourfold increased risk of suppurative complication (Metlay et al., 2000), *N.gonorrhoea* strains which are resistant to most antibiotics may cause infertility with other complications and infection with some other viruses such as HIV (Unemo & Shafer, 2014). In dealing with resistant variant of viruses we have to maintain the drug load and intensity which comes with other costs as it causes drug toxicity which ultimately down grade the human health and causes other health-related issues.

11.5 Strategies for its control

1. Strictly prohibit antibiotics dispensing without doctor consultation and prescription.

Strategy to control the emergence and spread of drug resistance strain of viruses we should have work on various levels.

Government level

1. Increase doctor-to-patient availability ratio in all the ethnic group region.
2. Collecting epidemiological data of viral strain in the community level.
3. Conduct public awareness program.

Healthcare professional level

1. Proper test for diagnosing the viral and bacterial strain which are causing disease.
2. Prescribe antibiotic according to the diagnosed strains.
3. Drug doses and proper time interval should be recommended to maintain drug load.

Besides all these parameters we should develop new diagnostic techniques to diagnose resistance strains of microbes and developing new drugs.

11.6 Future prospect

For making a good future to provide healthy life to our community, we have to work jointly on a government level, community level, and health workers level. Antibiotics use has been reduced without significant impact on patient health, the new trend in some countries which are suggested by international data (Bronzwaer et al., 2002). We should work to promote acquired immunity by developing and administrating vaccines against viral infections that can lead to reduction in antibiotic use which was formally intended to treat with antibiotics (Neuzil, Mellen, Wright, Mitchel, & Griffin, 2000). Developing new vaccines against malaria and typhoid has potency to reduce antimicrobial use and resistant relevant to these pathogens. We should focus on making vaccines for resistant pathogens and vaccination drive for the disease prevailed in the specific community or area which can effectively reduce antimicrobial use and promote human health.

11.7 Conclusion

Drug resistance became widely recognized global threats in last two decades. Knowledge of drug resistance, development of new vaccines and medicines, public awareness program, government support (policies and incentive), development of new diagnostic tool, industrial support in medical research are the need of the hour that will help the mankind to fight back and deal with drug resistance.

References

AlDabbagh, M. A., Gitman, M. R., Kumar, D., Humar, A., Rotstein, C., & Husain, S. (2017). The role of antiviral prophylaxis for the prevention of epstein–barr virus–associated posttransplant lymphoproliferative disease in solid organ transplant recipients: A systematic review. *American Journal of Transplantation, 17*(3), 770–781. https://doi.org/10.1111/ajt.14020.

Animal Production | Antimicrobial Resistance Food and Agriculture Organization of the United Nations. (2019). Animal Production | Antimicrobial Resistance Food and Agriculture Organization of the United Nations.

Arastehfar, A., Daneshnia, F., Salehi, M., Yaşar, M., Hoşbul, T., Ilkit, M., et al. (2020). Low level of antifungal resistance of Candida glabrata blood isolates in Turkey: Fluconazole minimum inhibitory concentration and FKS mutations can predict therapeutic failure. *Mycoses, 63*(9), 911–920. https://doi.org/10.1111/myc.13104.

Arastehfar, A., Gabaldón, T., Garcia-Rubio, R., Jenks, J. D., Hoenigl, M., Salzer, H. J. F., et al. (2020). Drug-resistant fungi: An emerging challenge threatening our limited antifungal armamentarium. *Antibiotics, 9*(12), 877. https://doi.org/10.3390/antibiotics9120877.

Arts, E. J., & Hazuda, D. J. (2012). HIV-1 antiretroviral drug therapy. *Cold Spring Harbor Perspectives in Medicine, 2*(4). doi:10.1101/cshperspect.a007161.

Aslam, B., Wang, W., Arshad, M. I., Khurshid, M., Muzammil, S., Rasool, M. H., et al. (2018). Antibiotic resistance: A rundown of a global crisis. *Infection and Drug Resistance, 11*, 1645–1658. https://doi.org/10.2147/IDR.S173867.

Bérdy, J. (2012). Thoughts and facts about antibiotics: Where we are now and where we are heading. *The Journal of Antibiotics, 65*(8), 385–395. https://doi.org/10.1038/ja.2012.27.

Berman, J., & Krysan, D. J. (2020). Drug resistance and tolerance in fungi. *Nature Reviews Microbiology, 18*(6), 319–331. https://doi.org/10.1038/s41579-019-0322-2.

Birkmann, A., & Zimmermann, H. (2016). HSV antivirals—Current and future treatment options. *Current Opinion Virology, 18*, 9–13.

Bousema, J. T., Gouagna, L. C., Drakeley, C. J., Meutstege, A. M., Okech, B. A., Akim, I. N. J., et al. (2004). Plasmodium falciparum gametocyte carriage in asymptomatic children in western Kenya. *Malaria Journal, 3*. doi:10.1186/1475-2875-3-18.

Bronzwaer, S. L. A. M., Cars, O., Buchholz, U., Mölstad, S., Goettsch, W., Veldhuijzen, I. K., et al. (2002). The relationship between antimicrobial use and antimicrobial resistance in Europe. *Emerging Infectious Diseases, 8*(3), 278–282. https://doi.org/10.3201/eid0803.010192.

Business Line. 52 percent of India's allopathic doctors are practicing in just 5 states. Radheshyam Jadhav (2020).

Campos, A. B., Ribeiro, J., Boutolleau, D., & Sousa, H. (2016). Human cytomegalovirus antiviral drug resistance in hematopoietic stem cell transplantation: Current state of the art. *Reviews in Medical Virology, 26*(3), 161–182. https://doi.org/10.1002/rmv.1873.

Carter, R. R., Sun, J., & Jump, R. L. P. (2016). A survey and analysis of the American public's perceptions and knowledge about antibiotic resistance. *Open Forum Infectious Diseases, 3*(3). doi:10.1093/ofid/ofw112.

Chinappi, M. (2010). On the mechanisms of chloroquine resistance in Plasmodium falciparum. *Plos One, 5*(11), e14064.3.

Cowen, L. E., Sanglard, D., Howard, S. J., Rogers, P. D., & Perlin, D. S. (2015). Mechanisms of antifungal drug resistance. *Cold Spring Harbor Perspectives in Medicine, 5*(7). doi:10.1101/cshperspect.a019752.

Davies, J., & Davies, D. (2010). Origins and evolution of antibiotic resistance. *Microbiology and Molecular Biology Reviews, 74*(3), 417–433. https://doi.org/10.1128/MMBR.00016-10.

Deveau, A., Bonito, G., Uehling, J., Paoletti, M., Becker, M., Bindschedler, S., et al. (2018). Bacterial–fungal interactions: Ecology, mechanisms and challenges. *FEMS Microbiology Reviews, 42*(3), 335–352. https://doi.org/10.1093/femsre/fuy008.

Dua, V. K., Kar, P. K., & Sharma, V. P. (1996). Chloroquine resistant Plasmodium vivax malaria in India. *Tropical Medicine and International Health, 1*(6), 816–819. https://doi.org/10.1111/j.1365-3156.1996.tb00116.x.

Fridman, O., Goldberg, A., Ronin, I., Shoresh, N., & Balaban, N. Q. (2014). Optimization of lag time underlies antibiotic tolerance in evolved bacterial populations. *Nature, 513*(7518), 418–421. https://doi.org/10.1038/nature13469.

Ghosh, S. K., Yadav, R. S., & Sharma, V. P. (1992). Sensitivity status of Plasmodium falciparum to chloroquine, amodiaquine, quinine, mefloquine and sulfadoxine/pyrimethamine in a tribal population of district Sundargarh, Orissa. *Indian Journal of Malariology, 29*(4), 211–218.

Gygli, S. M., Borrell, S., Trauner, A., & Gagneux, S. (2017). Antimicrobial resistance in Mycobacterium tuberculosis: Mechanistic and evolutionary perspectives. *FEMS Microbiology Reviews, 41*(3), 354–373. https://doi.org/10.1093/femsre/fux011.

Herz, A. V., Bonhoeffer, S., Anderson, R. M., May, R. M., & Nowak, M. A. (1996). Viral dynamics in vivo: Limitations on estimates of intracellular delay and virus decay. *Proceedings of the National Academy of Sciences, 93*(14), 7247–7251. https://doi.org/10.1073/pnas.93.14.7247.

Ison, M. G. (2013). Clinical use of approved influenza antivirals: Therapy and prophylaxis. *Influenza and Other Respiratory Viruses, 7*(1), 7–13. https://doi.org/10.1111/irv.12046.

Kausar, S., Said Khan, F., Ishaq Mujeeb Ur Rehman, M., Akram, M., Riaz, M., Rasool, G., et al. (2021). A review: Mechanism of action of antiviral drugs. *International Journal of Immunopathology and Pharmacology, 35*, 205873842110026. https://doi.org/10.1177/20587384211002621.

Khatoon, L., Baliraine, F. N., Bonizzoni, M., Malik, S. A., & Yan, G. (2009). Short report: Prevalence of antimalarial drug resistance mutations in Plasmodium vivax and P. falciparum from a malaria-endemic area of Pakistan. *American Journal of Tropical Medicine and Hygiene, 81*(3), 525–528. https://doi.org/10.4269/ajtmh.2009.81.525.

Kim, S. R., Khan, F., & Tyring, S. K. (2014). Varicella zoster: An update on current treatment options and future perspectives. *Expert Opinion on Pharmacotherapy, 15*(1), 61–71. https://doi.org/10.1517/14656566.2014.860443.

Klein, E. Y., Boeckel, T. P. V., Martinez, E. M., Laxminarayan, R., et al. (2018). Global increase and geographic convergence in antibiotic consumption between 2000 and 2015. *PNAS, 115*(15), E3463–E3470.

Levin-Reisman, I., Ronin, I., Gefen, O., Braniss, I., Shoresh, N., & Balaban, N. Q. (2017). Antibiotic tolerance facilitates the evolution of resistance. *Science, 355*(6327), 826–830. https://doi.org/10.1126/science.aaj2191.

Li, T. C. M., Chan, M. C. W., & Lee, N. (2015). Clinical implications of antiviral resistance in influenza. *Viruses, 7*(9), 4929–4944. https://doi.org/10.3390/v7092850.

Lortholary, O., Desnos-Ollivier, M., Sitbon, K., Fontanet, A., Bretagne, S., Dromer, F., et al. (2011). Recent exposure to caspofungin or fluconazole influences the epidemiology of candidemia: A prospective multicenter study involving 2,441 patients. *Antimicrobial Agents and Chemotherapy, 55*(2), 532–538. https://doi.org/10.1128/AAC.01128-10.

Menard, D., Djalle, D., Yapou, F., Manirakiza, A., & Talarmin, A. (2006). Frequency distribution of antimalarial drug-resistant alleles among isolates of Plasmodium falciparum in Bangui, Central African Republic. *American Journal of Tropical Medicine and Hygiene, 74*(2), 205–210. https://doi.org/10.4269/ajtmh.2006.74.205.

Metlay, J. P., Hofmann, J., Cetron, M. S., Fine, M. J., Farley, M. M., Whitney, C., et al. (2000). Impact of penicillin susceptibility on medical outcomes for adult patients with bacteremic pneumococcal pneumonia. *Clinical Infectious Diseases, 30*(3), 520–528. https://doi.org/10.1086/313716.

Morgan, D. J., Okeke, I. N., Laxminarayan, R., Perencevich, E. N., & Weisenberg, S. (2011). Non-prescription antimicrobial use worldwide: A systematic review. *The Lancet Infectious Diseases, 11*(9), 692–701. https://doi.org/10.1016/S1473-3099(11)70054-8.

Neuzil, K. M., Mellen, B. G., Wright, P. F., Mitchel, E. F., & Griffin, M. R. (2000). The effect of influenza on hospitalizations, outpatient visits, and courses of antibiotics in children. *New England Journal of Medicine, 342*(4), 225–231. https://doi.org/10.1056/NEJM200001273420401.

Pawlotsky, J. M. (2011). Treatment failure and resistance with direct-acting antiviral drugs against hepatitis C virus. *Hepatology, 53*(5), 1742–1751. https://doi.org/10.1002/hep.24262.

Peleg, A. Y., Hogan, D. A., & Mylonakis, E. (2010). Medically important bacterial–fungal interactions. *Nature Reviews Microbiology, 8*(5), 340–349. https://doi.org/10.1038/nrmicro2313.

Peterson, E., & Kaur, P. (2018). Antibiotic resistance mechanisms in bacteria: relationships between resistance determinants of antibiotic producers, environmental bacteria, and clinical pathogens. *Frontiers in Microbiology, 9*. doi:10.3389/fmicb.2018.02928.

Piret, J., & Boivin, G. (2014). Antiviral drug resistance in herpesviruses other than cytomegalovirus. *Reviews in Medical Virology, 24*(3), 186–218. https://doi.org/10.1002/rmv.1787.

Piret, J., & Boivin, G. (2016). Antiviral resistance in herpes simplex virus and varicella-zoster virus infections: Diagnosis and management. *Current Opinion in Infectious Diseases, 29*(6), 654–662. https://doi.org/10.1097/QCO.0000000000000288.

Potnuru, B. (2017). Aggregate availability of doctors in India: 2014-2030. *Indian Journal of Public Health, 61*(3), 182–187. https://doi.org/10.4103/ijph.IJPH_143_16.

Rajbhandari, R., & Chung, R. T. (2016). Treatment of Hepatitis B: A Concise Review. *Clinical and Translational Gastroenterology, 7*(9), e190. https://doi.org/10.1038/ctg.2016.46.

Reed, C., Chaves, S. S., Daily Kirley, P., Emerson, R., Aragon, D., Hancock, E. B., et al. (2015). Estimating influenza disease burden from population-based surveillance data in the United States. *Plos One, 10*(3), e0118369. https://doi.org/10.1371/journal.pone.0118369.

Roper, C., Pearce, R., Nair, S., Sharp, B., Nosten, F., & Anderson, T. (2004). Intercontinental spread of pyrimethamine-resistant malaria. *Science, 305*(5687), 1124. https://doi.org/10.1126/science.1098876.

Sharma, V. P. (1996). Re-emergence of malaria in India. *Indian Journal of Medical Research, 103*, 26–45.

Shehab, N., Lovegrove, M. C., Geller, A. I., Rose, K. O., Weidle, N. J., & Budnitz, D. S. (2016). US emergency department visits for outpatient adverse drug events, 2013-2014. *JAMA - Journal of the American Medical Association, 316*(20), 2115–2125. https://doi.org/10.1001/jama.2016.16201.

Siliciano, J. D., & Siliciano, R. F. (2013). Recent trends in HIV-1 drug resistance. *Current Opinion in Virology, 3*(5), 487–494. https://doi.org/10.1016/j.coviro.2013.08.007.

Sir E. Ray Lankester: Romanes Lecture, "Nature and Man," Oxford University Press, 1905, p. 21.

Tana, M. M., & Ghany, M. G. (2013). Hepatitis B virus treatment: Management of antiviral drug resistance. *Clinical Liver Disease, 2*(1), 24–28. https://doi.org/10.1002/cld.162.

Unemo, M., & Shafer, W. M. (2014). Antimicrobial resistance in Neisseria gonorrhoeae in the 21st Century: Past, evolution, and future. *Clinical Microbiology Reviews, 27*(3), 587–613. https://doi.org/10.1128/CMR.00010-14.

Using Medication: Using Antibiotics Correctly and Avoiding Resistance (2008).

Webster, D. P., Klenerman, P., & Dusheiko, G. M. (2015). Hepatitis C, *The Lancet* (385, pp. 1124–1135). The Lancet: Lancet Publishing Group. https://doi.org/10.1016/S0140-6736(14)62401-6.

WHO (2020). Neisseria gonorrhoeae Antimicrobial Resistance Surveillance: Consolidated Guidance © Pan American Health Organization, WHO.

WHO Library Cataloguing-in-Publication Data Global Action Plan on Antimicrobial Resistance (2015).

WHO. Antimicrobial Resistance. 13 October 2020.

Zhao, Y., Prideaux, B., Nagasaki, Y., Lee, M. H., Chen, P. Y., Blanc, L., et al. (2017). Unraveling drug penetration of echinocandin antifungals at the site of infection in an intra-abdominal abscess model. *Antimicrobial Agents and Chemotherapy, 61*(10). https://doi.org/10.1128/AAC.01009-17.

Chapter 12

Assessment and monitoring of human health risk during wastewater reuse

Sayali Mukherjee and Niketa Chauhan
Amity Institute of Biotechnology, Amity University, Lucknow, Uttar Pradesh, India

12.1 Introduction

Wastewater is water that has been contaminated by residential, industrial, or commercial activities. The volume of polluted water that must be treated is also augmented by organizations and industry. Domestic wastewater is created by operations such as toilet use, cleansing, cooking and serving, and washing. Commercial wastewater may include hazardous substances, necessitating specialist management and removal. Industrial effluents are produced by industrial or commercial producing processes, such as farming, and are more resistant to treatment than domestic waste. The nature of industrial wastewater differs by industry. Wastewater effluents are discharged into a wide range of habitats, including lakes, ponds, streams, rivers, estuaries, and oceans. Storm runoff is also included in wastewater because dangerous compounds wash off roadways, parking lots, and rooftops. Suspended solids are organic substances that do not disintegrate but remain suspended. The major focus of wastewater treatment plants is to remove as many organic materials as possible. Sodium, copper, lead, and zinc are among the inorganic minerals, metals, and compounds found in sewage and wastewater. Industrial and commercial waste, precipitation, and inflow and infiltration through faulty pipes are also potential sources. Microorganisms detected in sewage come from two places: soil and sanitary waste. Microorganisms in sewage come from two places: soil and sanitary waste. A milliliter of sewage generally includes between 100,000 and 1 million bacteria. While most of these species, such as certain bacteria, serve an important part in the waste breakdown and are considered an essential component of organic matter, some are pathogenic, or disease-carrying, and constitute a hazard to public health.

The environment and biological health may be harmed if wastewater is not appropriately managed. Fish and wildlife populations may be negatively

Antimicrobial Resistance in Wastewater and Human Health.
DOI: https://doi.org/10.1016/B978-0-323-96124-0.00003-9

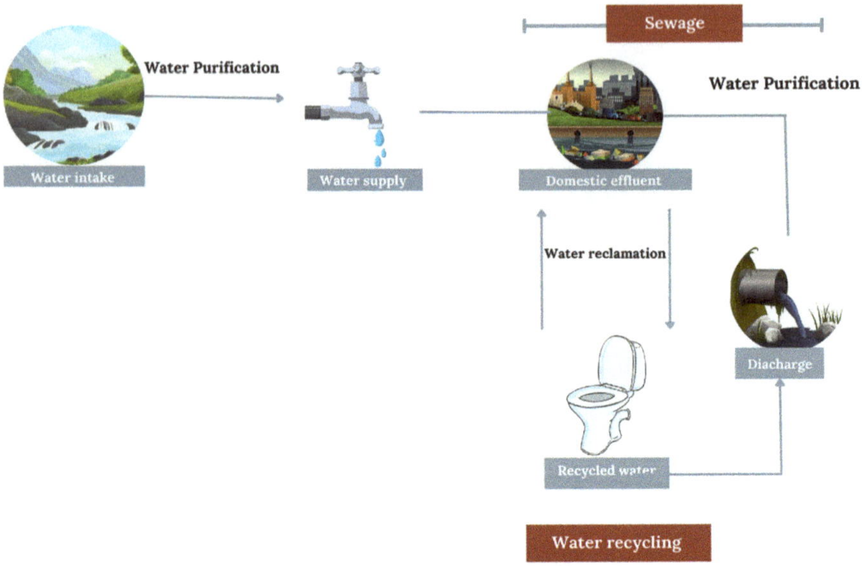

FIGURE 12.1 Water Reclamation and Reuse. Treatment of wastewater from different sources and reusing it for beneficial uses.

affected, oxygen levels may be depleted and drinking water may be contaminated. The dissolved oxygen in a lake can be depleted by decaying organic waste and trash, making fish and other aquatic biotas unable to thrive. Surplus nutrients, such as phosphorus and nitrogen, can induce eutrophication, which can be hazardous to aquatic creatures, stimulate excessive plant growth, diminish available oxygen, disrupt habitat, and lead to the extinction of some species. Bacteria, viruses, and disease-causing diseases can endanger human health and pollute shellfish populations, resulting in bans on human leisure, and drinking water. Heavy metals and additional pollutants, such as some pharmaceutical and personal-care products that enter the environment primarily through wastewater effluents are also a serious threat to biodiversity.

The process of recovering water from a number of sources, purifying it, and reusing it for beneficial uses such as agriculture and irrigation, potable water supply, groundwater replenishment, industrial activities, and environmental restoration is known as water reuse (Fig. 12.1). One facet of an area's water resource management is wastewater reuse. Water recycling, conservation, and new project development are some of the other choices. Economic variables, potential uses for recovered water, the seriousness of waste disposal restrictions, and policymaking, in which the desire to conserve rather than increase current water resources may outweigh economic and public health concerns, all impact whether reuse is appropriate. Domestic wastewater was reused for the first time in the 19th century, when sewage networks were established.

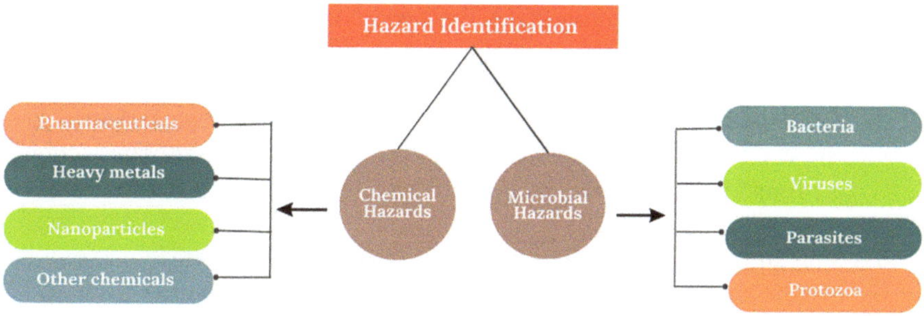

FIGURE 12.2 Potential hazards in wastewater. The different types of chemical and microbial hazards in wastewater.

Many places are already experiencing unprecedented droughts, putting the globe on the verge of a worldwide water disaster. This water scarcity has far-reaching consequences for how everyone on the planet lives, feeds, works. There is a crucial need of technology for purifying water for drinking purposes and the growing occurrences of drought as a result of climate change is a major challenge. Water reuse may be utilized to supplement existing water sources while also improving water security, sustainability, and adaptability. It is a long-term, dependable, and energy-recovery strategy that is a potential way to address global water scarcity. However, because only around 4% of global wastewater is now handled for water reuse, the water reuse industry will require long-term initiatives to grow. In addition, the reclaimed water must fulfill certain standards for human and environmental health safety according to their local water reuse legislation. Wastewater reuse has become an essential part of Integrated Water Resources Management, and it helps to ensure that future generations will have enough water. Wastewater has historically been used for human consumption, either directly or indirectly, due to water shortages and the lack of treatment facilities. After realizing that wastewater might cause huge catastrophic outbreaks of diseases people began to monitor potable water protection and terminal wastewater treatment.

12.2 Hazard identification

Hazard identification is a step in the risk assessment process that identifies potential dangers that need to be investigated further, because once the risks have been recognized, appropriate steps may be made to eliminate them. The major threats in wastewater are chemical hazards and microbial hazards which might pose threat to humans (Fig. 12.2).

12.2.1 Chemical hazards

Pharmaceuticals have been discovered in water, provoking worries about the hazards they may cause to individuals and ecosystems. These pharmaceuticals generally enter waterways through the direct excretion of active pharmaceuticals from patients, followed by inadequate removal of those pharmaceuticals from our wastewater during wastewater treatment. Many commonly used pharmaceuticals were found in wastewater samples, according to the analysis. Antibiotics like azithromycin, ciprofloxacin, and norfloxacin are among them. These chemicals were found in large quantities in the influent wastewater (Botero-Coy et al., 2018).

As a result of their use and disposal, personal care products (PCPs) end up in the aquatic environment. Many investigations have found their prevalence in wastewater, surface water, and groundwater. The presence of these PCPs was observed in all of the aquifers investigated, demonstrating that they are reliable markers of human presence. Their existence might offer early warnings about contaminants that are continually released into surface waters, detect dynamic urban patterns, and recommend urban planning and investment and infrastructure repair methods. Wastewater treatment plants have benzophenone-3, benzophenone-4 phenyl benzimidazole sulfonic acid with values ranging from 27 to 822 ng/L. Triclosan (TCS) and benzophenone-3 removal efficiencies in wastewater treatment plants were found to be 80%–100% and phenylbenzimidazole sulfonic acid and benzophenone-4 removal efficiencies were poor at 0%–40%. The presence of phenylbenzimidazole sulfonic acid and benzophenone-4, which can be up to 560.4 ng/L in wastewater indicates major concern. In surface water, triclosan was found in quantities ranging from 0.2 to 161.0 ng/L (Palmiotto et al., 2018). The presence of microorganisms along with pharmaceuticals and personal care products (PPCPs) are even more dangerous as it reduces the efficiency of water treatment. When *Burkholderia cepacia* (a bacterium found in drinking water) is exposed to these compounds, it changes its behavior and decreases the efficiency of wastewater disinfection (Gomes, Madureira, Simões, & Simões, 2019).

The incidence and distribution of microplastics (MPs) pollution have been investigated in water ecosystems, and wastewater treatment plants. MPs are plastic pieces that, due to their high fragmentation, have a negative impact on the environment. They can take a range of forms, including pieces, fibers, foam, and other materials, and they can come from a variety of places. Microscopic examination, density separation, Raman spectroscopy, and Fourier transform infrared spectroscopy are a few of the methods used to detect MPs. Furthermore, because MPs are consumed by a diverse range of marine creatures, might lead to biomagnification and have consequences on biota and human health (Rezania et al., 2018).

Cytostatic medicines, which are used in chemotherapy, are eliminated unaltered or as metabolites in the urine and faeces. Due to inadequate removal in wastewater treatment systems, cytostatic drug residues, which are often used in chemotherapy, are routinely discharged into surface water. Consumption patterns, excretion proportion, and wastewater treatment efficiency all have a role in their survival in the natural environment. Mycophenolic acid and hydroxycarbamide are some examples of cytostatics that have predicted environmental concentrations more than 10 ng/L (Franquet-Griell, Gómez-Canela, Ventura, & Lacorte, 2015). In cancer therapy, the nitrogen mustard medications cyclophosphamide (CP) and ifosfamide are routinely utilized. They are continuously released into hospital and urban wastewater systems, eventually reaching wastewater treatment plants via excreta causing threats to aquatic and human health (Russo et al., 2018).

The lack of a defined method of action for the disposal of anticancer medications has led to a significant environmental impact. Even in low quantities, they can be exceedingly toxic to living creatures. The growth and development of these plants may be harmed by these substances thus their presence in water indicates them as environmental contaminants, such as in the case of reusing treated wastewater and agricultural biosolids (Lutterbeck, Kern, Machado, & Kümmerer, 2015). The anti-inflammatory medications diclofenac and dexamethasone have a negative impact on the life of aquatic animals, causing testosterone levels to decline and oxidative stress in male fish (Guiloski, Ribas, Pereira, Neves, & Silva de Assis, 2015). Growing chemical and biological micropollutants in wastewater have forced research into new treatment approaches in terms of monitoring, environmental concerns, and remediation (Gavrilescu, Demnerová, Aamand, Agathos, & Fava, 2015).

Pharmaceutical drugs and PCPs, as well as other organic wastewater pollutants, have been seen to undergo transformation and bioaccumulation. Several of the chemicals might be physiologically converted into breakdown products, generating adducts that could be used to monitor exposure biomarkers, while others could influence fathead minnow swimming behavior and interact with the thyroid axis in zebrafish (Mottaleb, Meziani, Matin, Arafat, & Wahab, 2015). Highly toxic metals in water, agricultural soils, and crops, such as zinc (Zn), copper (Cu), lead (Pb), and cadmium (Cd), pose a threat to human health. Heavy metal risk indexes for food crops have raised concerns about public health. Before it can be used for agriculture, the wastewater in the region must be adequately treated (Chaoua, Boussaa, El Gharmali, & Boumezzough, 2019). The bioaccessible metal concentrations in greenhouse and agricultural field fractions of soils treated with industrial and municipal wastewater for growing crops were found to be higher in greenhouse soils than in field soils. Furthermore, as compared to treated municipal wastewater, wastewater irrigated from industrial sources had greater heavy metal contents (Cao et al., 2018). In heavy metal-contaminated soil, the capacity of vegetables to grow, root length, and germination rates also get compromised (Alia et al., 2015). All microbial activities, including glucose fermentation, syntrophic propionate oxidation, and denitrification, were inhibited by elemental copper nanoparticles, and the inhibitory impact was attributable to the release of copper(II) ions associated with nanoparticle breakdown (Gonzalez-Estrella, Puyol, Gallagher, Sierra-Alvarez, & Field, 2015). These nanoparticles could be dangerous to aquatic life and humans as well. Manganese-doped titanium dioxide nanoparticles in unfiltered samples have been reported to kill zebrafish (Ozmen, Güngördü, Erdemoglu, Ozmen, & Asilturk, 2015).

12.2.2 Microbial hazards

Wastewater from different sources contain pathogenic microbial hazards like bacteria, viruses, fungi, and parasites which may cause infections of lung, intestine, digestive tract, etc. The main source of microbes is through fecal contamination in water. The most prominent bacteria being *Escherichia coli*. The presence of microbes in the wastewater reduces the efficiency of the treatment process. Identification of microbe type by next generation sequencing and 16S rRNA have been helpful in understanding the nature of disease threat to human associated with it. Some of the strains often identified are of Clostridium coccoides group, Bacteroides fragilis group, Bifidobacterium, and Prevotella (Matsuki et al., 2002). Bacteriophages have also been reported to be present in wastewater. Several enteric bacteria, viruses, and parasites have also been identified in wastewater. The presence of antibiotic-resistant bacteria in wastewater is a matter of grave concern.

12.3 Monitoring of human health hazards during wastewater reuse from various sources

Measuring and limiting the risk of chemical and biological hazards exposure during wastewater reuse is part of the monitoring of human health risks. Pharmaceuticals and personal care items, heavy metals, nanoparticles, virus, and antibiotic resistant bacteria are all potential risks in wastewater. Monitoring is taking regular measures to identify changes in one's health and can be based on information from a variety of sources. Hazard control requires establishing acceptable levels of exposure and, as a result, health risk, as well as identifying the degrees of control required to maintain exposure below defined limits (Table 12.1) (WHO, 2003).

The reuse of wastewater for irrigation purposes can help in food security in rural areas; however, the presence of toxic chemicals and pathogenic microorganisms poses a major threat to humans. Many of these microbes can stay in the wastewater for a long duration on the crops and may be transmitted to humans. Sewage from households also contains myriads of microorganisms like bacteria, virus, which are not completely eliminated even after wastewater treatment. Consumption of such water may lead to diseases like typhoid, dysentery, diarrhea, etc.

Wastewater reuse has become an essential part of water resource planning in many nations. Some nations, such as Oman, have a national policy of reusing all treated wastewater effluents and have made significant progress in this direction (Al-Ajmy, 2002). However, treated wastewater may include high levels of salts, heavy metals, pathogens, and new contaminants, all of which have unknown

TABLE 12.1 Permissible limits of heavy metals (mg/L) in drinking water and wastewater.

Metals	Permissible limits for drinking water (WHO)	Permissible limits in effluent water (WHO)
Arsenic	0.01	5
Cadmium	0.005	0.003
Chromium	0.1	0.05
Iron	1	2
Lead	0.01	0.05
Mercury	0.006	0.001
Copper	2	0.25
Nickel	0.07	0.02

The acceptable concentrations of heavy metals like arsenic, cadmium, chromium, lead, etc., in drinking water and wastewater. No permission required.

environmental consequences (Mohammad & Mazahreh, 2003). Heavy metals like chromium, cadmium persist in wastewater and can cause health hazards to human including embryonic fatality and cancer. The pollution value is measured by comprehensive pollution index and organic pollution index and it varies in different seasons.

The use of treated wastewater (TWW) on farms can help save fresh water supplies and reduce the need of chemical fertilizers. Treated wastewater has been effectively utilized for irrigation in many regions of the world, and many studies have recognized its advantages (Mujeriego, 1991; Levine & Sanot, 2004). However, the continued use of treated wastewater for irrigation results in the buildup of heavy metals in some plants which through the food chain cause threat to human health (Elfattah, Shehata, & Talab, 2002). Pharmaceuticals and personal care items are essential to society's health, but their presence in a variety of environmental compartments, such as treated wastewaters, has raised worries about potential human and ecological health consequences. When detected in wastewater treatment plant effluents, these products posed the highest threat to human health (Semerjian, Shanableh, Semreen, & Samarai, 2018).

It is generally recognized that viruses expelled by faeces may be detected in wastewater, and that standard sewage secondary treatment may not entirely eradicate them. An environmental monitoring research at a wastewater treatment facility has found adenovirus DNA in all samples taken for monitoring, TTV DNA in 95% (19/20) of raw sewage and 85% (17/20) of exit samples, and HAV DNA in just two samples over 40 samples (5%). The adenovirus results show that it is the best indication for determining the performance of a wastewater depuration facility in removing viruses and monitoring human health. In municipal and agricultural wastewater, microorganisms, notably enteric viruses, can be detected. It is vital to ensure the safety standard for the reclamation, since these have repeatedly failed to reduce virus loads below the danger threshold. As a result, research into harmful viruses (such as HAV) and standard bacterial criteria must be connected to specific viral indicators such somatic coliphages, adenovirus, and TTV for frequent examination of recycled waters (Carducci et al., 2008).

12.4 Risk assessment in wastewater reuse

Risk management entails a series of activities aimed at determining the risk, controlling the risk, and tracking the measures performed. The identification of risk is the initial stage of risk assessment, which leads to the determination of risk factors and the kind of risk and variables that produce it.

Microbial source tracing (MST) techniques have made it possible to detect sewage pollution in recreational waterways, but the risk of increased levels of MST targets such sewage-associated Bacteroides HF183 and other indicators remain unknown. This technique gives a useful framework for water quality managers to assess the hazards of pollution from fresh sewage to human health. MST data may be interpreted in the context of the risk of gastrointestinal (GI) disease induced by pathogen exposure using quantitative microbial risk assessment (QMRA) modeling. To investigate the risks associated with greywater reuse scenarios, QMRA was conducted. The quality of residential greywater was studied from three sources (bathroom, laundry, and kitchen). The results showed that greywater from all three household sources may be safely used for toilet flushing after a simple microfiltration treatment (Shi, Wang, & Jiang, 2018). Quantitative microbiological risk assessment (QMRA) has been adopted by many big municipal water systems. Small water systems are more sensitive to common water system risks and, as a result, have a higher risk of disease breakout (Hamouda, Jin, Xu, & Chen, 2018). The concentrations of PPCPs compounds in wastewater treatment plants effluents, sludge, and their discharge into freshwater systems may be determined by selecting PPCPs, determining their source, estimating inflow, and determining concentration levels (Tarpani & Azapagic, 2018).

12.5 Strategies to minimize risks associated with wastewater reuse

Even after tertiary wastewater treatment, there is still a danger of enteric virus, hazardous contamination, and environmental damage in wastewater reclamation and reuse. The choice to utilize recovered wastewater instead of conventional or other nonconventional water sources for human use is based primarily on public health concerns that must be minimized to an acceptable level, and secondly on environmental issues. Both of these should be evaluated against the economic benefits for a satisfactory solution. The status of utilization of wastewater in different countries has been depicted in Table 12.2. Risk analy-

TABLE 12.2 Status of utilization of wastewater in different nations.

Region	Utilization	% water utilization	References
South Africa	Agriculture	60	Adewumi et al. (2010)
	Household	27	
	Industrial	3	
	Power	4	
	Mining	3	
USA	Thermoelectric plants	41	Kenny et al. (2009)
	Agriculture	37	
	Industry	6	
	Household	14	
India	Industrial	7	CGWB (2011)
	Agriculture	87	
	Energy	2	
	Household	4	
Greece	Industry	3	Frontistis et al. (2011), Tsagarakis et al. (2001)
	Irrigation	83	

The percentage of wastewater which is being used for agriculture, irrigation, industry, households, etc., around the world. No permission required.

sis and multiple criteria decision analysis (MCDA) are particularly important methods for evaluating criteria and standards for wastewater reclamation and reuse (Ganoulis, 2009). Risk management is the most crucial and difficult component of the entire process due to factors influencing human and societal difficulties.

On a small scale, gas-phase pulsed corona discharge oxidation was used to remove a wide range of medicines from raw sewage from a public hospital as well as biologically treated wastewater from a health care institution. With 1 kWh m^3 of raw sewage and 0.5 kWh m^3 of biologically treated wastewater, this nonselective oxidation of identified pharmaceuticals was proven to be effective, with 87% of residual pharmaceuticals being removed and 100% eradication (Ajo et al., 2018).

The use of aluminum (Al) electrodes in ballasted electro-flocculation (BEF) is a new method for removing cadmium and zinc from industrial mining wastewater. By combining microsand and polymer to improve the weight of the flocs and the velocity at which they settle, the BEF technology modifies the electrocoagulation-electroflocculation settling method. The best removal percentage was obtained at a current intensity of 2 A, a flow rate of 20 L/h, a micro-sand dose of 6 g/L, a polyéthylèneimine (PEI) polymer dose of 100 mg, contact times of 30 min, a stirring speed of 50 RPM, a monopolar configuration of the electrodes, and an electrodes number of 10 electrodes, based on the examination of the operation parameters one by one (Brahmi et al., 2018).

The efficiency of microalgae may be investigated via bioremediation. The prevalence of treatment-resistant cytotoxic drugs in aquatic settings has serious consequences for ecosystems and human health. Researchers looked at the efficacy of microalgae in bioremediation of cytotoxic compounds due to the challenges associated with physicochemical procedures in the treatment of effluents containing these chemicals. The anticancer medication Flutamide was biosorbed using *Chlorella vulgaris* biomass, both living and dead. When considering biomass, pH, and adsorption period, the living microalga performed better in drug removal. The technique was optimized using response surface methodology, which showed that at a Flutamide concentration of 50 M, pH 7.4, and a time of 10 min, the live biomass could remove 98.5% of the drug. It shows that using microalgae to clean wastewaters polluted with cytotoxic chemicals might be a viable option (Habibzadeh, Chaibakhsh, & Naeemi, 2018).

Membrane bioreactor system are found to have more efficacy for the removal for antibiotic-resistant bacteria and antibiotic resistant genes and antibiotics compared to conventional activated sludge . The membrane bioreactor system has removal efficacy of > 70%, for antibiotics including Amoxicillin, chloramphenicol, and vancomycin but in the case of conventional activated sludge antibiotics such as trimethoprim and lincomycin are found to be resistant. In secondary effluent of conventional activated sludge antibiotic-resistant bacteria found to be 2–3 orders of magnitude lower than the raw influent whereas for membrane bioreactor system no antibiotic-resistant bacteria were found in the microfiltration fraction (Le, Ng, Tran, Chen, & Gin, 2018).

12.6 Risks associated with wastewater reuse in COVID-19 pandemic

Coronaviruses are a category of viruses that may infect a variety of animals, including humans, and cause moderate to severe respiratory diseases. In 2002 and 2012, two highly pathogenic zoonotic coronaviruses, severe acute respiratory syndrome coronavirus and Middle East respiratory syndrome coronavirus, respectively, emerged in humans and caused fatal respiratory illness, bringing emerging coronaviruses into the 21st century as a new public health concern. At the end of 2019, a new coronavirus known as SARS-CoV-2 emerged in the Chinese city of Wuhan, causing an outbreak of atypical viral pneumonia. Due to its great transmissibility, this novel coronavirus infection, also known as coronavirus disease 2019 (COVID-19), has spread swiftly around the world. It has surpassed SARS and MERS in terms of both the number of sick people and the geographic scope of epidemic regions. The ongoing COVID-19 outbreak has posed a serious threat to global public health. SARS-CoV-2 viral RNA has been found in faeces samples from ill persons, according to research published (He et al., 2020).

According to the findings, COVID-19 can be transmitted via the fecal-oral route. After discovering that some patients infected with the COVID-19 virus had diarrhea rather than a fever in the early stages of sickness, scientists identified viral RNA in patient faeces. Previous studies on the persistence of coronavirus surrogates and SARS in wastewater have indicated that the virus may live in wastewater for hours to days without treatment (Wölfel et al., 2020).

During the COVID-19 epidemic in the Netherlands, SARS-CoV-2 was discovered in sewage. RT-PCR was utilized to test sewage samples from seven cities and the airport against three nucleocapsid protein gene (N1-3) fragments and one envelope protein gene fragment (E). The N1 fragment was identified in five distinct places' sewage. The N1 fragment was discovered in sewage from six different places, while the N3 and E fragments were discovered in five and four different locations, respectively. SARS-CoV-2 was discovered in sewage for the first time. Even when COVID-19 frequency is low, the presence of the virus in sewage implies that sewage surveillance might be a helpful tool for tracking the virus's spread in the population (Zhang, Wang, & Xue, 2020). Infection with SARS-CoV-2 causes viral shedding in the faeces, which is the cause of the current COVID-19 epidemic. As a consequence, the detection of SARS-CoV-2 in wastewater has provided the capacity to monitor the incidence of diseases in the community using wastewater-based epidemiology (WBE). SARS-CoV-2 RNA was extracted from wastewater in an Australian catchment

TABLE 12.3 SARS-CoV-2 detection in wastewater of several countries.

Country	Type of sample in which SARS-CoV2 is detected	References
Brisbane, Australia	Untreated wastewater	Ahmed et al. (2020)
Valencia, Spain	Untreated wastewater	Randazzo et al. (2020)
	Secondary treated	
Yamanashi, Japan	Secondary treated	Haramoto et al. (2020)
Wuhan, China	Effluent of septic tank	Zhang et al. (2020)
Louisiana, USA	Untreated wastewater	Sherchan et al. (2020)
Amersfoort, Tilburg, Utrecht, etc., Netherlands	Untreated wastewater	Medema et al. (2020)
Massachusetts, USA	Untreated wastewater	Wu et al. (2020)
Paris, France	Untreated wastewater	Wurtzer et al. (2020)
Bozeman, USA	Untreated wastewater	Nemudryi et al. (2020)
Istanbul, Turkey	Waste activated sludge	Kocamemi et al. (2020)

The different types of samples in which SARS-CoV-2 is detected and is proposed to be transmitted through it.

and viral RNA copies were quantified using reverse transcriptase quantitative polymerase chain reaction (RT-qPCR), resulting in two positive detections in six days from the same wastewater treatment facility (Ahmed et al., 2020). Based on the estimated viral RNA copy counts discovered in the wastewater, the number of infected people in the watershed was computed. Their findings show that WBE may be used to monitor infectious diseases in communities, such as COVID-19, and that greater methodological and molecular test validation for enveloped viruses in wastewater is needed (Environmental Protection Agency, 2002). The virus has been discovered in many nations' wastewater (Table 12.3). Wastewater containing human excreta can aid the spread of dangerous germs in the environment. As a result, the discharge of untreated wastewater containing such germs is hazardous to human health.

12.7 Conclusion

Due to increased water demand, conventional water supplies have been severely exhausted. As a result, wastewater recovery and reuse are becoming increasingly important. However, the quality of recovered water is a major concern, and it must be managed under strict guidelines. Monitoring of recycled or reclaimed water systems should be a major goal to avoid putting people's health in danger.

References

Adewumi, J. R., Ilemobade, A. A., & Van Zyl, J. E. (2010). Treated wastewater reuse in South Africa: Overview, potential and challenges. *Resources, Conservation & Recycling, 55*, 221–231. https://doi.org/10.1016/j.resconrec.2010.09.012.

Ahmed, W., Angel, N., Edson, J., Bibby, K., Bivins, A., O'Brien, J. W., et al. (2020). First confirmed detection of SARS-CoV-2 in untreated wastewater in Australia: A proof of concept for the wastewater surveillance of COVID-19 in the community. *Science of the Total Environment, 728*. doi:10.1016/j.scitotenv.2020.138764.

Ajo, P., Preis, S., Vornamo, T., Mänttäri, M., Kallioinen, M., & Louhi-Kultanen, M. (2018). Hospital wastewater treatment with pilot-scale pulsed corona discharge for removal of pharmaceutical residues. *Journal of Environmental Chemical Engineering, 6*(2), 1569–1577. https://doi.org/10.1016/j.jece.2018.02.007.

Al-Ajmy (2002). Wastewater in the Sultanate and its effects on environment. In *International Conference on Wastewater Management and its Effect on the Environment in Hot and Arid Countries*.

Alia, N., Sardar, K., Said, M., Salma, K., Sadia, A., Sadaf, S., et al. (2015). Toxicity and bioaccumulation of heavy metals in spinach (Spinacia oleracea) grown in a controlled environment. *International Journal of Environmental Research and Public Health, 12*(7), 7400–7416. https://doi.org/10.3390/ijerph120707400.

Botero-Coy, A. M., Martínez-Pachón, D., Boix, C., Rincón, R. J., Castillo, N., Arias-Marín, L. P., et al. (2018). 'An investigation into the occurrence and removal of pharmaceuticals in Colombian wastewater. *Science of the Total Environment, 642*, 842–853. https://doi.org/10.1016/j.scitotenv.2018.06.088.

Brahmi, K., Bouguerra, W., Harbi, S., Elaloui, E., Loungou, M., & Hamrouni, B. (2018). Treatment of heavy metal polluted industrial wastewater by a new water treatment process: Ballasted electroflocculation. *Journal of Hazardous Materials, 344*, 968–980. https://doi.org/10.1016/j.jhazmat.2017.11.051.

Cao, C., Zhang, Q., Ma, Z. B., Wang, X. M., Chen, H., & Wang, J. J. (2018). Fractionation and mobility risks of heavy metals and metalloids in wastewater-irrigated agricultural soils from greenhouses and fields in Gansu. *China. Geoderma, 328*, 1–9. https://doi.org/10.1016/j.geoderma.2018.05.001.

Carducci, A., Morici, P., Pizzi, F., Battistini, R., Rovini, E., & Verani, M. (2008). Study of the viral removal efficiency in a urban wastewater treatment plant. *Water Science and Technology, 58*(4), 893–897. https://doi.org/10.2166/wst.2008.437.

Chaoua, S., Boussaa, S., El Gharmali, A., & Boumezzough, A. (2019). Impact of irrigation with wastewater on accumulation of heavy metals in soil and crops in the region of Marrakech in Morocco. *Journal of the Saudi Society of Agricultural Sciences*, 429–436. doi:10.1016/j.jssas.2018.02.003.

CGWB (2011). Ground Water Year Book - India 2010-11. Central Ground Water Board, Ministry of Water Resources. Government of India. http://www.cgwb.gov.in/documents/Ground%20Water%20Year%20Book-2010-11.pdf.

Elfattah, A., Shehata, S. M., & Talab, A. S. (2002). Evaluation of irrigation with either raw municipal treated waste or river water on elements up take and yield of lettuce and potato plants. *Egyptian Journal of Soil Science, 42*(4), 705–714.

Environmental Protection Agency, National Recommended Water Quality Criteria: 2002. Office of Water, Office of Science and Technology. (2002). Environmental Protection Agency.

Franquet-Griell, H., Gómez-Canela, C., Ventura, F., & Lacorte, S. (2015). Predicting concentrations of cytostatic drugs in sewage effluents and surface waters of Catalonia (NE Spain). *Environmental Research, 138*, 161–172. https://doi.org/10.1016/j.envres.2015.02.015.

Frontistis, Z., Xekoukoulotakis, N. P., Hapeshi, E., & Venieri, D. (2011). Despo Fatta-Kassinos, Dionissios Mantzavinos, Fast degradation of estrogen hormones in environmental matrices by photo-Fenton oxidation under simulated solar radiation. *Chemical Engineering Journal, 178*, 175–182. ISSN 1385-8947 https://doi.org/10.1016/j.cej.2011.10.041.

Ganoulis, J. (2009). *Risk analysis of water pollution*. John Wiley & Sons.

Gavrilescu, M., Demnerová, K., Aamand, J., Agathos, S., & Fava, F. (2015). Emerging pollutants in the environment: Present and future challenges in biomonitoring, ecological risks and bioremediation. *New Biotechnology, 32*(1), 147–156. https://doi.org/10.1016/j.nbt.2014.01.001.

Gomes, I. B., Madureira, D., Simões, L. C., & Simões, M. (2019). The effects of pharmaceutical and personal care products on the behavior of Burkholderia cepacia isolated from drinking water. *International Biodeterioration and Biodegradation, 141*, 87–93. https://doi.org/10.1016/j.ibiod.2018.03.018.

Gonzalez-Estrella, J., Puyol, D., Gallagher, S., Sierra-Alvarez, R., & Field, J. A. (2015). Elemental copper nanoparticle toxicity to different trophic groups involved in anaerobic and anoxic wastewater treatment processes. *Science of the Total Environment, 512–513*, 308–315. https://doi.org/10.1016/j.scitotenv.2015.01.052.

Guiloski, I. C., Ribas, J. L. C., Pereira, L. d. S., Neves, A. P. P., & Silva de Assis, H. C. (2015). Effects of trophic exposure to dexamethasone and diclofenac in freshwater fish. *Ecotoxicology and Environmental Safety, 114*, 204–211. https://doi.org/10.1016/j.ecoenv.2014.11.020.

Habibzadeh, M., Chaibakhsh, N., & Naeemi, A. S. (2018). Optimized treatment of wastewater containing cytotoxic drugs by living and dead biomass of the freshwater microalga, Chlorella vulgaris. *Ecological Engineering, 111*, 85–93. https://doi.org/10.1016/j.ecoleng.2017.12.001.

Hamouda, M. A., Jin, X., Xu, H., & Chen, F. (2018). Quantitative microbial risk assessment and its applications in small water systems: A review. *Science of the Total Environment, 645*, 993–1002. https://doi.org/10.1016/j.scitotenv.2018.07.228.

Haramoto, E., Malla, B., Thakali, O., & Kitajima, M. (2020). First environmental surveillance for the presence of SARS-CoV-2 RNA in wastewater and river water in Japan. *Sci. Total Environ., 737*, 140405. https://doi.org/10.1016/j.scitotenv.2020.140405.

He, G., Sun, W., Fang, P., Huang, J., Gamber, M., Cai, J., & Wu, J. (2020). The clinical feature of silent infections of novel coronavirus infection (COVID-19) in Wenzhou. *Journal of Medical Virology, 92*(10), 1761–1763. https://doi.org/10.1002/jmv.25861.

Kenny, J. F., Barber, N. L., Hutson, S. S., Linsey, K. S., Lovelace, J. K., & Maupin, M. A. (2009). Estimated use of water in the United States in 2005. *U.S. Geological Survey Circular, 1344*, 52. https://doi.org/10.3133/cir1344.

Kocamemi, B. A., Kurt, H., Sait, A., Sarac, F., Saatci, A. M., & Pakdemirli, B. (2020). SARS-CoV-2 detection in Istanbul wastewater treatment plant sludges. *medRxiv*. 2020.05.12.20099358 https://doi.org/10.1101/2020.05.12.20099358 .

Le, T. H., Ng, C., Tran, N. H., Chen, H., & Gin, K. Y. H. (2018). Removal of antibiotic residues, antibiotic resistant bacteria and antibiotic resistance genes in municipal wastewater by membrane bioreactor systems. *Water Research, 145*, 498–508. https://doi.org/10.1016/j.watres.2018.08.060.

Levine, A, & Sanot, A (2004). Recovering sustainable water from waste water. *Environ. Sci. Technol., 38*(11), 201A–208A.

Lutterbeck, C. A., Kern, D. I., Machado, Ê. L., & Kümmerer, K. (2015). Evaluation of the toxic effects of four anti-cancer drugs in plant bioassays and its potency for screening in the context of waste water reuse for irrigation. *Chemosphere, 135*, 403–410. https://doi.org/10.1016/j.chemosphere.2015.05.019.

Matsuki, T., Watanabe, K., Fujimoto, J., Miyamoto, Y., Takada, T., Matsumoto, K., et al. (2002). Development of 16S rRNA-gene-targeted group-specific primers for the detection and identification of predominant bacteria in human feces. *Applied and Environmental Microbiology, 68*(11), 5445–5451. https://doi.org/10.1128/AEM.68.11.5445-5451.2002.

Medema, G., Heijnen, L., Elsinga, G., Italiaander, R., & Brouwer, A. (2020). Presence of SARS Coronavirus-2 RNA in sewage and correlation with reported COVID-19 prevalence in the early stage of the epidemic in the Netherlands. *Environ. Sci. Technol. Lett., 7*, 511–516. https://doi.org/10.1021/acs.estlett.0c00357.

Mohammad, M. J., & Mazahreh, N. (2003). Changes in soil fertility parameters in response to irrigation of forage crops with secondary treated wastewater. *Communications in Soil Science and Plant Analysis, 34*(9–10), 1281–1294. https://doi.org/10.1081/CSS-120020444.

Mottaleb, M. Abdul, Meziani, Mohammed J., Abdul Matin, M., Musavvir Arafat, M., & Wahab, Mohammad A. (2015). Emerging micro-pollutants pharmaceuticals and personal care products (PPCPs) contamination concerns in aquatic organisms-LC/MS and GC/MS analysis. In: *Emerging micro-pollutants in the environment: Occurrence, fate, and distribution* (pp. 43–74). American Chemical Society.

Mujeriego, R. (1991). Salal Golf course irrigation with reclaimed waste water. *Science and Technology, 24*(9), 161–172.

Nemudryi, A., Nemudraiam, A., Surya, K., Wiegand, T., Buyukyoruk, M., Wilkinson, R., & Wiedenheft, B. (2020). Temporal detection and phylogenetic assessment of SARSCoV-2 in municipal wastewater. *Cell Reports Medicine, 1*, 100098. https://doi.org/10.1016/j.xcrm.2020.100098.

Ozmen, M., Güngördü, A., Erdemoglu, S., Ozmen, N., & Asilturk, M. (2015). Toxicological aspects of photocatalytic degradation of selected xenobiotics with nano-sized Mn-doped TiO2. *Aquatic Toxicology, 165*, 144–153. https://doi.org/10.1016/j.aquatox.2015.05.020.

Palmiotto, M., Castiglioni, S., Zuccato, E., Manenti, A., Riva, F., & Davoli, E. (2018). Personal care products in surface, ground and wastewater of a complex aquifer system, a potential planning tool for contemporary urban settings. *Journal of Environmental Management, 214*, 76–85. https://doi.org/10.1016/j.jenvman.2017.10.069.

Randazzo, W., Truchado, P., Cuevas-Ferrando, E., Simón, P., Allende, A., & Sánchez, G. (2020). SARS-CoV-2 RNA in wastewater anticipated COVID-19 occurrence in a low prevalence area. *Water Res, 181*, 115942. https://doi.org/10.1016/j.watres.2020.115942.

Rezania, S., Park, J., Md Din, M. F., Mat Taib, S., Talaiekhozani, A., Kumar Yadav, K., et al. (2018). Microplastics pollution in different aquatic environments and biota: A review of recent studies. *Marine Pollution Bulletin, 133*, 191–208. https://doi.org/10.1016/j.marpolbul.2018.05.022.

Russo, C., Lavorgna, M., Česen, M., Kosjek, T., Heath, E., & Isidori, M. (2018). Evaluation of acute and chronic ecotoxicity of cyclophosphamide, ifosfamide, their metabolites/transformation products and UV treated samples. *Environmental Pollution, 233*, 356–363. https://doi.org/10.1016/j.envpol.2017.10.066.

Semerjian, L., Shanableh, A., Semreen, M. H., & Samarai, M. (2018). Human health risk assessment of pharmaceuticals in treated wastewater reused for non-potable applications in Sharjah, United Arab Emirates. *Environment International, 121*, 325–331. https://doi.org/10.1016/j.envint.2018.08.048.

Sherchan, S. P., Shahin, S., Ward, L. M., Tandukar, S., Aw, T. G., Schmitz, B., Ahmed, W., & Kitajima, M. (2020). First detection of SARS-CoV-2 RNA in wastewater in North America: a study in Louisiana. *USA. Sci. Total Environ., 743*, 140621. https://doi.org/10.1016/j.scitotenv.2020.140621.

Shi, K.-W., Wang, C.-W., & Jiang, S. C. (2018). Quantitative microbial risk assessment of Greywater on-site reuse. *Science of The Total Environment, 635*, 1507–1519. https://doi.org/10.1016/j.scitotenv.2018.04.197.

Tarpani, R. R. Z., & Azapagic, A. (2018). A methodology for estimating concentrations of pharmaceuticals and personal care products (PPCPs) in wastewater treatment plants and in freshwaters. *Science of the Total Environment, 622–623*, 1417–1430. https://doi.org/10.1016/j.scitotenv.2017.12.059.

Tsagarakis, K. P., Tsoumanis, P., Chartzoulakis, K., & Angelakis, A. N. (2001). Water resources status including wastewater treatment and reuse in Greece. *Water International, 26*, 252–258.

WHO (2003). *Copper in drinking-water. Background document for preparation of WHO Guidelines for drinking-water quality*. Geneva: World Health Organization (WHO/SDE/WSH/03.04/88).

Wölfel, R., Corman, V. M., Guggemos, W., Seilmaier, M., Zange, S., Müller, M. A., et al. (2020). Virological assessment of hospitalized patients with COVID-2019. *Nature, 581*(7809), 465–469. https://doi.org/10.1038/s41586-020-2196-x.

Wu, F., Xiao, A., Zhang, J., Gu, X., Lee, W. L., Kauffman, K., Hanage, W., Matus, M., Ghaeli, N., Endo, N., Duvallet, C., Moniz, K., Erickson, T., Chai, P., Thompson, J., & Alm, E. (2020). SARS-CoV-2 titers in wastewater are higher than expected from clinically confirmed cases. *Applied and Environmental Science, 5*(4), e00614–e00620. https://doi.org/10.1128/mSystems.00614-20.

Wurtzer, S., Marechal, V., Mouchel, J.-M., Maday, Y., Teyssou, R., Richard, E., Almayrac, J. L., & Moulin, L. (2020). Evaluation of lockdown impact on SARS-CoV-2 dynamics through viral genome quantification in Paris wastewaters. *medRxiv*. 2020.04.12.20062679 https://doi.org/10.1101/2020.04.12.20062679 .

Zhang, J. C., Wang, S. B., & Xue, Y. D. (2020). Fecal specimen diagnosis 2019 novel coronavirus–infected pneumonia. *Journal of Medical Virology, 92*(6), 680–682. https://doi.org/10.1002/jmv.25742.

Index

Page numbers followed by "*f*" and "*t*" indicate, figures and tables respectively.

FSC
www.fsc.org
MIX
Papier | Fördert gute Waldnutzung
FSC® C083411

Kolibri 360 GmbH
Ferdinand-Jühlke-Straße 7
99095 Erfurt, Deutschland
produktsicherheit@kolibri360.de